Pomil´io, Ottrino

Airplane

design and construction

Pomil´io, Ottrino

Airplane

design and construction

Inktank publishing, 2018

www.inktank-publishing.com

ISBN/EAN: 9783747749982

AIRPLANE DESIGN AND CONSTRUCTION

BY

OTTORINO POMILIO

CONSULTING AERONAUTICAL ENGINEER FOR THE POMILIO BROTHERS CORPORATION

FIRST EDITION

McGRAW-HILL BOOK COMPANY, INC.
239 WEST 39TH STREET. NEW YORK

LONDON: HILL PUBLISHING CO., LTD.
6 & 8 BOUVERIE ST., E. C.
1919

INTRODUCTION

By far the major part of experimental work in aerodynamics has been conducted in Europe rather than in America, where the feat of flying in a heavier than air machine was first accomplished. This book presents in greater detail than has hitherto been attempted in this country the application of aerodynamic research conducted abroad to practical airplane design.

The airplane industry is now shifting from the design and construction of military types of craft to that of pleasure and commercial types. The publication of this book at this time is, therefore, opportune, and it should go far toward replacing by scientific procedure many of the "cut and try" methods now used. Employment of the data presented should enable designers to save both time and expense. The arrangement, presentation of subject matter, and explanation of the derivation of working formulæ, together with the assumptions upon which they are based, and consequently their limitations, are such that the book lends itself to use as a text in technical schools and colleges.

The dedication of this volume to Wilbur and Orville Wright is at once appropriate and significant; appropriate, in that it is a tangible expression of the keen appreciation of the author for the great work of these two brothers; and significant, in that it is a return, in the form of a rational analysis of many of the problems relating to airplane design and operation, on the part of the product of an older civilization to the product of the new, as a sort of recompense for the daring, courage and inventive genius which made human flight possible.

J. S. MacGregor.

New York, 1919.

CONTENTS

ACKNOWLEDGMENT

The author desires to express his sincere thanks to Mrs. Lester Morton Savell for her valuable assistance in matters pertaining to English and to Mr. Garibaldi Joseph Piccione for his intelligent assistance in drawing the diagrams.

O. P.

AIRPLANE DESIGN
AND
CONSTRUCTION

PART I
STRUCTURE OF THE AIRPLANE

CHAPTER I
THE WINGS

While for birds, and in general for all animals of the air, wings serve to insure both sustentation and propulsion, those of the airplane are used solely to provide the means of sustaining the machine in the air.

The phenomenon of sustentation is easily explained. A body moving through the air produces, because of its motion, a disturbance of the atmosphere which is more or less pronounced and complex in character. In the final analysis, this disturbance is reduced to the formation of zones of positive and negative pressures. The resultant of these pressures may then be classified into its three components:

1. Vertical or sustaining force, called *Lift*,
2. Horizontal component parallel and opposite the line of flight, called *Drag*, and
3. Horizontal component perpendicular to the line of flight, called *Lateral Drift*.

The vertical component may be positive or negative. An example of the negative component is found in the elevator used for the climbing maneuver of an airplane, as will be shown later.

The horizontal component parallel to the line of flight, is always negative; *i.e.*, it tends to retard the motion of the body. "Conservation of energy"[1] is the principle underlying this phenomenon.

The horizontal component perpendicular to the line of flight is called the force of "drift," because it tends to make the body drift from the line of flight. This component, generally not existing in normal flight, is of great importance in the directional maneuvers of airplanes.

For a body having a plane of symmetry and moving through space so that the line of flight is contained in that plane, the force of drift is zero and the only components acting are the lift and the drag.

Observations made of birds' wings and results based upon the experiences of experimenters in aeronautics, have demonstrated the possibility of devising surfaces of such form that by properly moving them through the air they create reactions, of which the vertical component has a far greater magnitude than the horizontal.

Thus, a surface capable of developing high values of lift with small values of drag is called a *wing*.

In actual practice, as will be shown further on in a more detailed study of aerodynamical principles (Chapter 7), the value of the ratio $\frac{\text{Lift}}{\text{Drag}}$ varies from 15 to 23. This means that wings may be built, which, for every 23 lb. of load carried, offer a resistance to motion of but 1 lb. It is natural, then, that designers direct all efforts toward increasing the $\frac{\text{Lift}}{\text{Drag}}$ ratio, which is used to define the efficiency of the wing. Three factors influence such efficiency:

the profile of the wing section,

the ratio of the wing span to its depth or chord (called the Aspect Ratio), and

[1] This principle states that energy can be neither created nor destroyed. If the horizontal component were positive, perpetual motion would ensue, since it would be necessary only to furnish the initial force to set the body in motion. The body would then continue in its path without further application of energy.

the relative position of the wings (in multiplane machines).

The profile of a wing section is its major section at right angles to the span of the wing. Because of the simplicity of modern construction, wings are generally built with

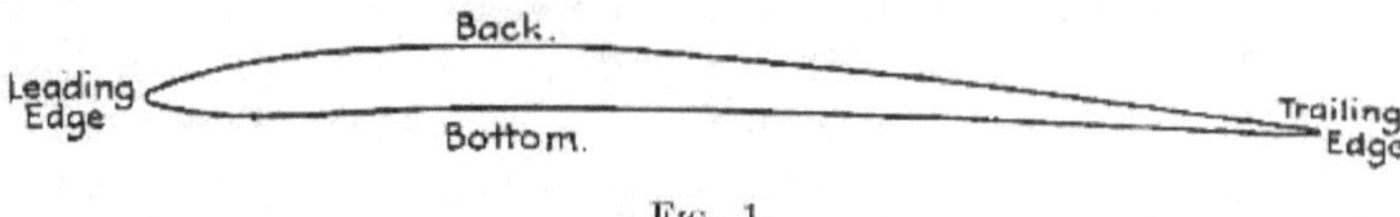

Fig. 1.

a constant section throughout the span. In the early days of aeronautics, however, many types of wings were built with a variable wing section, but the aerodynamical advantages derived from their use were never sufficient to compensate for the complicated construction required.

In the profile of a wing, there are the following distinct elements (Fig. 1): leading edge, back, bottom and trailing edge. The proper use of these elements makes it possible to obtain the highest values of the $\frac{\text{Lift}}{\text{Drag}}$ ratio, as well as to vary the Lift coefficient according to the load to be carried per square foot of wing surface.

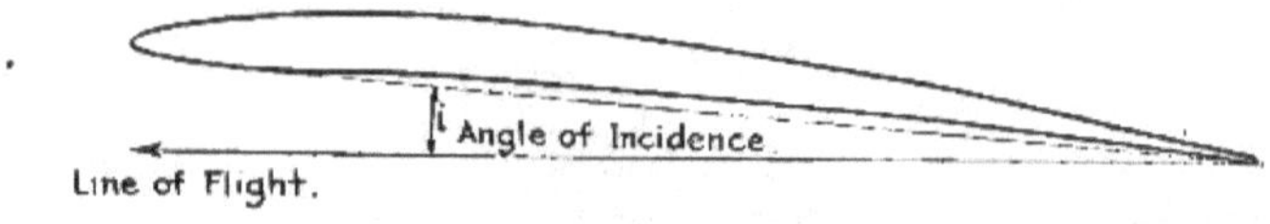

Fig. 2.

The angle between the wing chord and the line of flight, called the angle of incidence of the wing (Fig. 2), may vary between greater or smaller limits. As a result, the distribution and value of the positive and negative pressures will vary, and give different values of Lift, Drag and $\frac{\text{Lift}}{\text{Drag}}$. The laws of variation of these factors are rather complicated and cannot be expressed by means of formulæ. It is possible, however, to express them by means of curves as illustrated in Figs. 3 and 4. These illustrate the laws of variation for the values of the Lift, Drag and $\frac{\text{Lift}}{\text{Drag}}$ coefficients for two types of aerofoils, which, although having the same lengths of chord, differ in other elements.

It is now necessary to introduce a new factor, namely, the speed or velocity of translation of the wing.

All aerodynamical phenomena, when considered with respect to speed, follow the general law that the intensity of the phenomenon increases not in proportion to the speed, but to the *square* of the speed. This is accounted for by the fact that for redoubled speed not only is the velocity of impact of air molecules against the body moving in the air redoubled, but so also is the number of molecules that are struck by the body. Consequently it is seen that the intensity of the phenomenon is quadrupled.

Assuming a wing with an area of A square feet, the following general equations may be written:

$$\left.\begin{aligned} L &= \lambda \times A \times V^2 \\ D &= \delta \times A \times V^2 \end{aligned}\right\} \qquad (1)$$

where

L = total Lift for area A in pounds
D = total Drag for area A in pounds
V = speed of translation in miles per hour (m.p.h.).

In practice it is convenient to refer the coefficients λ and δ to the velocity of 100 m.p.h., whence the equation (1) becomes

$$\left.\begin{aligned} L &= \lambda \times A \times \left(\frac{V}{100}\right)^2 \\ D &= \delta \times A \times \left(\frac{V}{100}\right)^2 \end{aligned}\right\} \qquad (2)$$

If A = 1 sq. ft., and V = 100 m.p.h., then

$$\left.\begin{aligned} L_1 &= \lambda \\ D_1 &= \delta \end{aligned}\right\} \qquad (3)$$

that is, λ is the load in pounds carried by a wing with an area of 1 sq. ft. and moving at a velocity of 100 m.p.h.,

δ λ
1.75 35
1.50 30
1.25 25
1.00 20
0.75 15
0.50 10
0.25 05
0
λ/δ
25
22.5
20
17.5
15
12.5
10
7.5
-3 -2 -1 0 1 2 3 4 5 6 7 8 9
Degrees
λ/δ
δ
λ

Fig. 3.

δ λ
2.25 35
2.00 30
1.75 25
1.50 20
1.25 15
1.00 10
0.75 5
0
λ/δ
25
22.5
20
17.5
15
12.5
10
2.5
-3 -2 -1 0 1 2 3 4 5 6 7 8 9
Degrees.
λ
δ
λ/δ

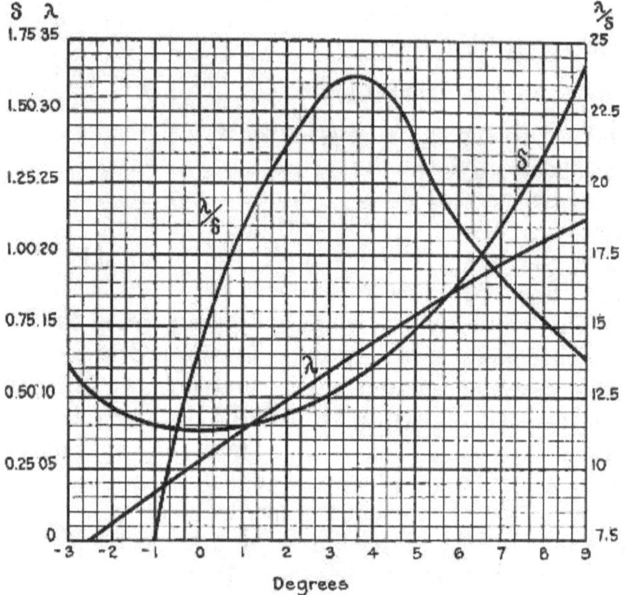

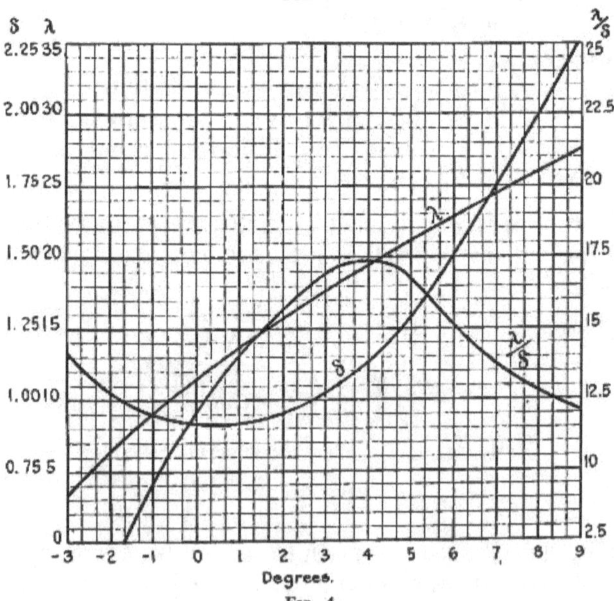

Fig. 4.

and δ the head resistance in pounds for a wing with an area of 1 sq. ft. and moving at a velocity of 100 m.p.h. Knowing λ and δ, by using equation (2) the values of L and D may be found for any area or any speed. Also, the ratio $\frac{\lambda}{\delta}$ is equal to $\frac{L}{D}$ which is obtained by dividing the L equation by the D equation.

Now, the coefficients λ and δ may assume an entire series of varied values by changing the angle of incidence of the wings. Figs. 3 and 4 show the laws of variation of λ, δ and $\frac{\lambda}{\delta}$ for two different types of wings to which we will refer as wing No. 1 and wing No. 2.

An examination of the diagrams is instructive because it shows how it is possible to build wings which may have totally different values of Lift, the speed being the same for both wings. For example, at an angle of incidence of 3°, wing No. 1 gives $\lambda = 11.8$, while wing No. 2 gives $\lambda = 17.6$; in other words, with equal speeds, wing No. 2 carries a load 49 per cent. greater than wing No. 1.

The laws of variation of λ and δ depend upon the several elements of the wing, namely, the leading edge, top, bottom and trailing edge. Let us consider separately the function of each of these elements:

Actually, the function of the leading edge is to penetrate the air and to deviate it into two streams, one which will pass along the top and the other which will pass along the bottom of the wing. In order to obtain a good efficiency it is necessary that this penetration be made with as little disturbance as possible, in order to prevent eddies. Eddies give rise to considerable head resistance and are therefore great consumers of energy. For that reason, the leading edge should be designed with the same criterions as those adopted in the design of turbine blades. Figs. 5 and 6 show the phenomenon schematically. Due to inertia, the air deviated above the wing tends to continue in its rectilinear path, thus producing a negative pressure or vacuum on top of the wing. This negative pressure exerts a centripetal force on the air molecules, tending to deflect

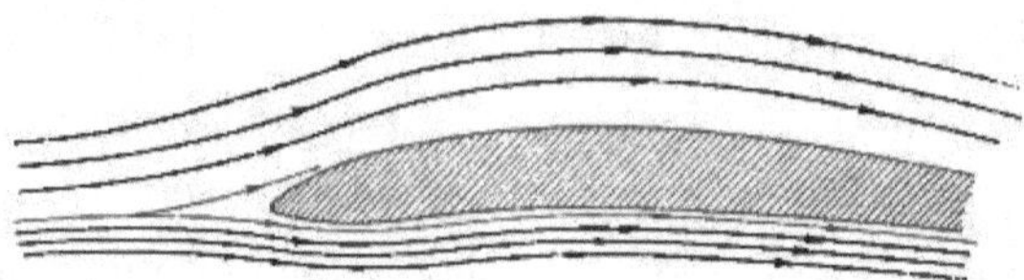

FIG. 5.—Leading edge of good efficiency.

their path downward so as to flow along the top curvature of the wing. A dynamic equilibrium is thereby established between the negative pressure and the centrifugal force of

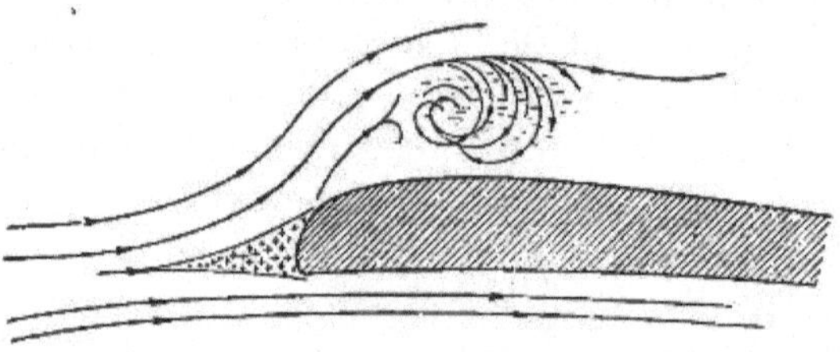

FIG. 6.—Leading edge of poor efficiency.

the various molecules (Fig. 7). It is obvious, then, that the top curvature has a pronounced influence not only upon the intensity of the vacuum, but also on the law of negative pressure distribution along its entire length.

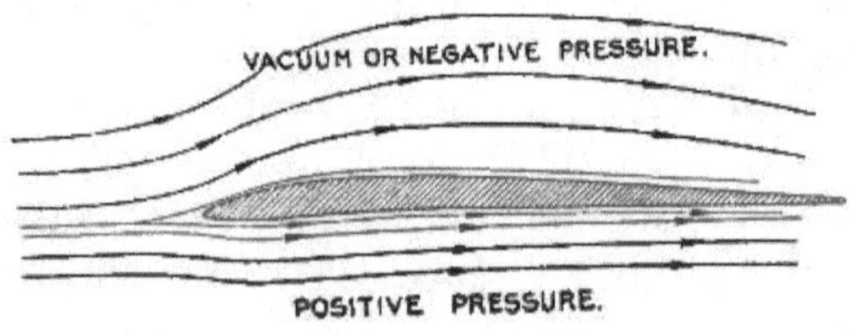

FIG. 7.

The air deviated below the wing tends instead, also due to inertia, to condense, thus producing a positive pressure which forces the air molecules to follow the concairty of the bottom curvature. Because of this change in the direction of velocities, a centrifugal force is developed which is in dynamic equilibrium with the positive pressure produced (Fig. 7).

Curves showing the laws of distribution of the positive and negative pressures are given in Fig. 8. The resultant

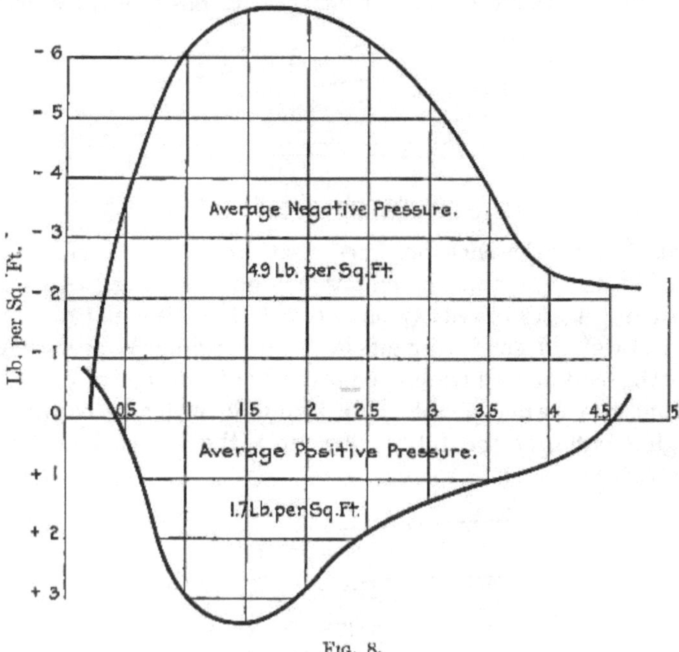

Fig. 8.

of these pressures represents the value $\frac{L}{A}$. It will be noted that the portion of the sustentation due to the vacuum above is much greater than that due to the positive pressure below. In the case under consideration, it is 2.9 times greater, and equal to 74 per cent. of the total Lift. Therefore, the study of the top curvature must be given more careful consideration than that of the bottom curvature, as a wing is not at all defined by the bottom curvature alone. In practice, the means adopted to raise the value of λ is to increase both the convexity of the top and the concavity of the bottom of the wing, thereby increasing the intensities of the negative and positive pressures.

The trailing edge also has its bearing on the efficiency. Its shape must be such as to straighten out the air streamliness when the air leaves the wing, affecting a smooth, gradual decrease in the negative and positive pressures

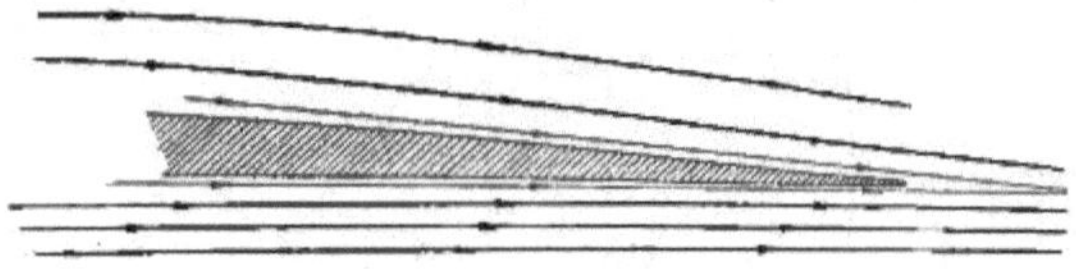

Fig. 9.—Trailing edge of good efficiency.

until their difference becomes zero. In this manner, the formation of a wake or eddies behind the wing, with the resulting losses of energy, is avoided (Figs. 9 and 10).

In brief, for good wing efficiency, it is primarily necessary for the leading and trailing edges to be of a design which will avoid the formation of eddies, and in order to obtain a higher value of the Lift coefficient λ the top and bottom curvatures must be increased.

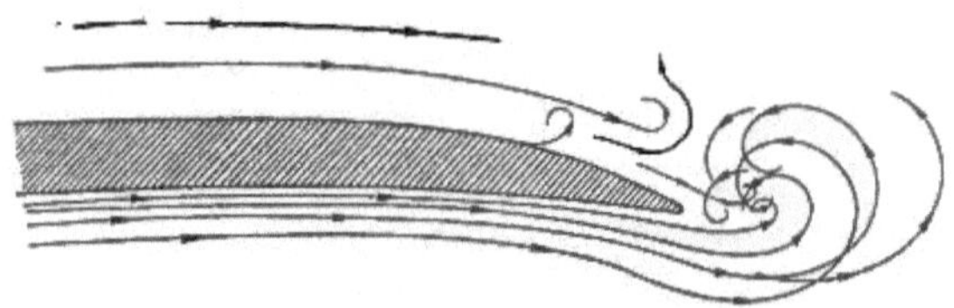

Fig. 10.—Trailing edge of poor efficiency.

From the foregoing it is easy to understand the importance of the ratio $\frac{S}{C}$; that is, the relation between the span S and the chord C of a wing.

Considering the front view of a wing surface, Fig. 11, which represents a section parallel to the leading edge, and shows the mean negative and positive pressure curves for the top and bottom of the wing, it will be seen that while in the central part the curves are represented by lines parallel to the wing, at the wing tips A and B, they suffer serious disruption, for at the end of the wing a short circuit between the compression and depression occurs. This is due to the air under pressure rushing toward the vacuum zone, thus establishing an air flux (the so-called marginal losses), with the result that at the wing tips the average pressure curves come together, and the Lift is decreased considerably, thus lowering the value λ of the wing. It is necessary to

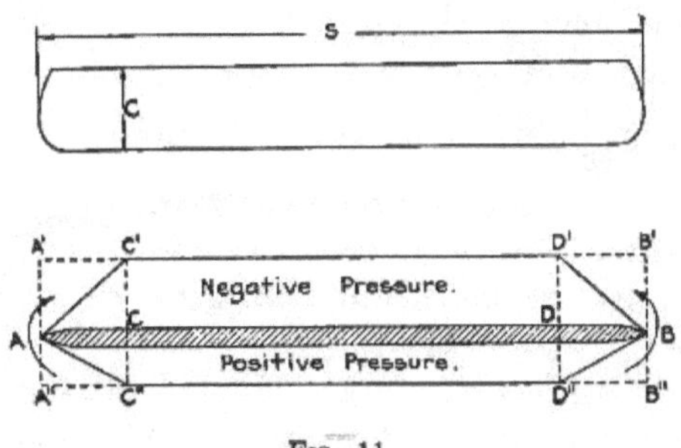

FIG. 11.

reduce the importance of this phenomenon to a minimum, which is done by increasing the ratio of the span to the chord $\left(\frac{S}{C}\right)$.

Assume, as it is sometimes done in practice, that the disruption in the average curves due to marginal losses extends for a distance AC and BD, equal to the chord of the wing; and also that the diagram is modified according to a linear law. This is equivalent to assuming a decrease in the Lift measured by the triangles $AA'C'$, $AA''C''$, $BB'D'$ and $BB''\ D''$. The same result is obtained as though the average λ remained constant and the lifting surface were reduced by the amount c^2, which means that the total surface would be reduced by $s \times c - c^2$. If the product $s \times c$ is kept constant by increasing s and diminishing c correspondingly, the importance of the term c is greatly decreased. The loss is expressed by $\frac{c^2}{s \times c} = \frac{c}{s}$, that is, by the inverse of the ratio

$\frac{\text{span}}{\text{chord}}$. So it is seen that by increasing the ratio $\frac{s}{c}$, the average value of the coefficient of Lift is increased, and it is therefore advantageous to build wings of large spread. In practice, however, there is a limit beyond which this advantage becomes a minimum, and there are also static and structural problems to be considered which limit the value of the ratio $\frac{s}{c}$. In modern machines, this value varies from 5 to 12, and even more.

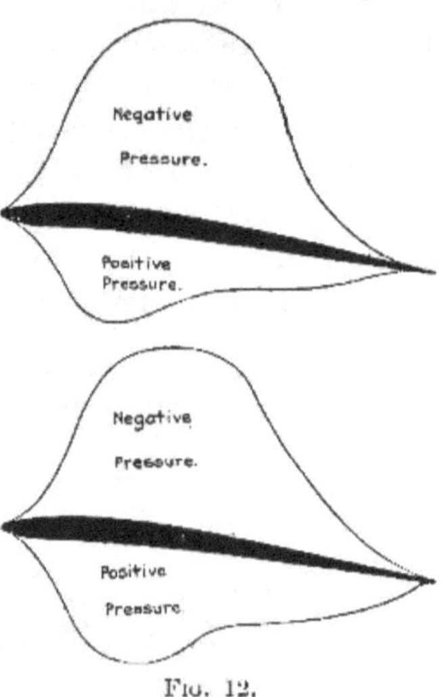

Fig. 12.

In biplanes, triplanes and multiplanes, another very important problem is presented; that of the mutual interference of each plane upon the others. In view of the close arrangement of the surfaces necessitated by structural considerations, and the high values of their negative and positive pressures of air, a confliction of air flow is formed over the entire wing surface, with the result that the value of the Lift coefficient is lowered. Figs. 12 and 13 illustrate this phenomenon for a biplane and triplane respectively. In the case of the biplane, the following effects ensue:

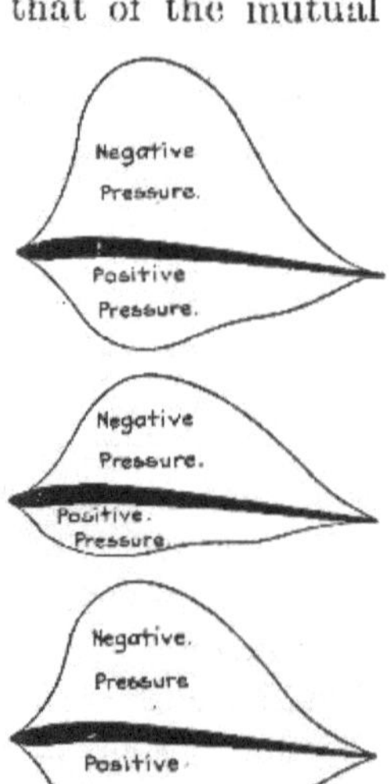

Fig. 13.—Triplane system.

1. Decrease in vacuum on top of lower plane, and

2. Decrease in positive pressures on bottom of upper plane.

In the case of the triplane, the losses are still greater, due to

1. Decrease in vacuum on top of bottom plane,

2. Decrease in positive pressures on bottom of intermediate plane,

3. Decrease in vacuum on top of intermediate plane, and

4. Decrease in positive pressures on bottom of upper plane.

It is thus seen how undesirable, from an aerodynamical point of view, the triplane really is. At the present time, however, the triplane is not a common type of airplane, so the discussion here will be limited to the biplane.

Another important ratio in aeronautics is the unit load on the wings, or the number of pounds carried per square foot of wing surface. Theoretically this value may vary between wide limits; for example, for wing No. 2 set at an angle of 6° and moving at a speed of 150 miles an hour, the ratio is 51 lb. per sq. ft. In practice, however, that value has never been reached. Special racing airplanes have been built whose unit loads were as high as 13 lb. per sq. ft., but the principal disadvantages of such high unit loads are the resulting high gliding and landing speeds, and an appreciable loss in maneuverability. For this reason designers strive to confine the unit load between the limits of 6 and 8 lb. per sq. ft.

Consider a biplane with a chord and gap each of 6 ft. with a unit load equal to 8 lb. per sq. ft. Keeping in mind what has been previously stated (Fig. 8), it can be assumed that the values of positive and negative pressures (vacuum) found at the top and bottom of both wings would be equal to 2 lb. per sq. ft. and 6 lb. per sq. ft. respectively, provided, of course, that the two wing surfaces had no effect on each other. Now, if a difference in pressure of 8 lb. per sq. ft. is produced between two points in the air at a distance of 6 ft. from each other, the air under pressure rushing violently to fill up the vacuum will result in a veritable cyclone in the intervening space.

When a wing is in motion, condensed and rarefied conditions of the air are being constantly produced, so that

δ λ
λ/δ
1.75 35
1.50 30
1.25 25
1.00 20
0.75 15
0.50 10
0.25 5
0
25
22.5
20
17.5
15
12.5
10
7.5
λ/δ
δ
λ
-3 -2 -1 0 1 2 3 4 5 6 7 8 9
Degrees.

Fig. 14.

δ λ
λ/δ
1.75 35
1.50 30
1.25 25
1.00 20
0.75 15
0.50 10
0.25 5
0
25
22.5
20
17.5
15
12.5
10
7.5
λ/δ
δ
λ
-3 -2 -1 0 1 2 3 4 5 6 7 8 9
Degrees.

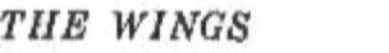

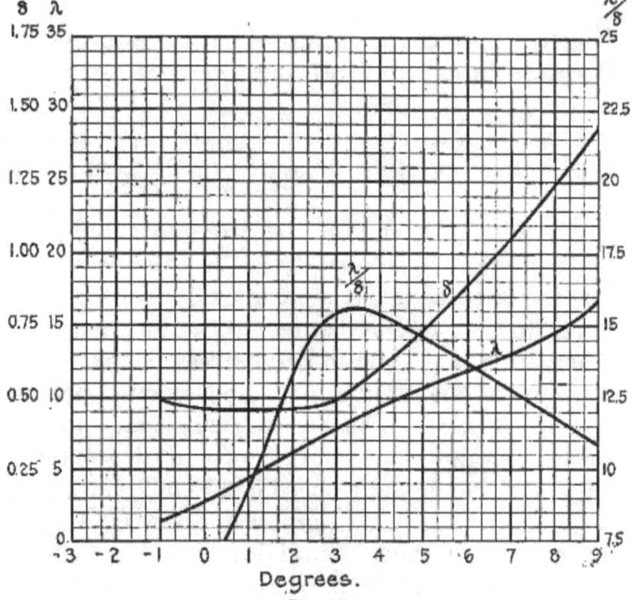

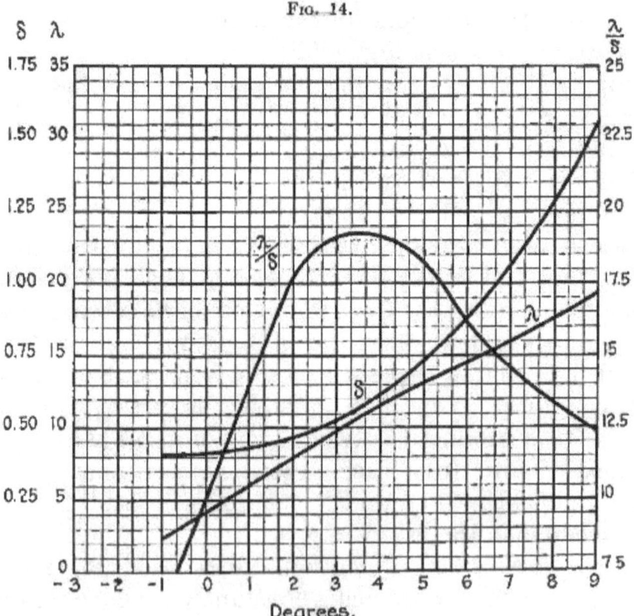

a certain dynamic equilibrium ensues. In order to study the phenomenon more closely, a few brief computations will be made.

Again consider the type of wing curve whose characteristics are given in Fig. 3, and assume that it is to be adopted for a biplane. In such a case, the curves in Fig. 3 are no longer applicable and new curves must be determined experimentally, since the aerodynamical behavior of the wing shown in Fig. 3 will change for every one of the three following conditions:

1. Acting alone, as for a monoplane,
2. Serving as the upper plane of a biplane structure, and
3. Serving as the lower plane of a biplane structure.

Fig. 14 gives the characteristics for wing No. 1 serving as a lower plane. Considered as an upper plane, the aerodynamical curve is practically the same as that in Fig. 3. Fig. 15 gives the characteristics of a complete biplane whose upper and lower planes are similar.

Compare now a monoplane having a wing surface of 200 sq. ft., possessing the type of wing mentioned above, with a biplane also having the same wing section, and whose planes are each 100 sq. ft. in area. Assume each machine to carry a load of 1500 lb. at a speed of 100 miles per hour. The problem then is to find the values of the angles of incidence and the thrust efforts required to overcome the Drag.

From the equation

$$L = \lambda \times A \times \left(\frac{V}{100}\right)^2$$

Since

$$L = 1500 \text{ lb. and}$$
$$A = 200 \text{ sq. ft.,}$$

then

$$\lambda = \frac{1500}{200} = 7.5$$

which value of λ gives, for the monoplane (Fig. 3),

$$i = 1^\circ$$
$$\delta = 0.415$$
$$D = 0.415 \times 200 = 83 \text{ lb.}$$

and for the biplane (Fig. 17),

$$i = 1° 45'$$
$$\delta = 0.450$$
$$D = 0.450 \times 200 = 90 \text{ lb.}$$

In the case of the biplane $\frac{L}{D}$ is seen to be 12 per cent. smaller than in the case of the monoplane. The thrust required is 8 per cent. greater, therefore 8 per cent. more H.P. is required to move the wing surfaces of this biplane than that necessary to move a similar wing in the monoplane structure. However, the final deduction must not be made that a biplane requires 8 per cent. more power than the monoplane of equal area. The power absorbed by the wing system is really only about 25 per cent. of the total H.P. required by the machine, so that the total loss due to the employment of a biplane structure is 8 per cent. of 25 per cent., or 2 per cent.

Of late, the biplane structure has almost entirely supplanted that of the monoplane, due largely to the great superiority, from a structural point of view, offered by a cellular structure over a linear type. For lifting surfaces of equal areas, the biplane takes up much less ground space and is much lighter than the monoplane. Regarding the former, the $\frac{\text{Span}}{\text{Chord}}$ ratio being the same, the span of the biplane is only 0.71 that required by the monoplane.

As to weight, it is to be noted that a wing structure usually consists of two or more main beams called wing spars, running parallel to the span. Wing ribs, constructed to form the outline of the wing section, are fitted to the spars. The junction of the wings to the body or fuselage of a machine is made by means of the spars, which are the main stress-resisting members of the wing. The spars of monoplane wings are fixed or hinged to the fuselage and braced by steel cable rigging (Fig. 16). In the biplane, instead, the corresponding spars of both upper and lower planes are held together by struts and cross bracing, forming a truss (Fig. 17).

For those familiar with the principles of structures it is easy to see the great superiority of the biplane structure over the monoplane structure in stiffness and lightness, and the impossibility of monoplane structure in large machines because of its excessive weight.

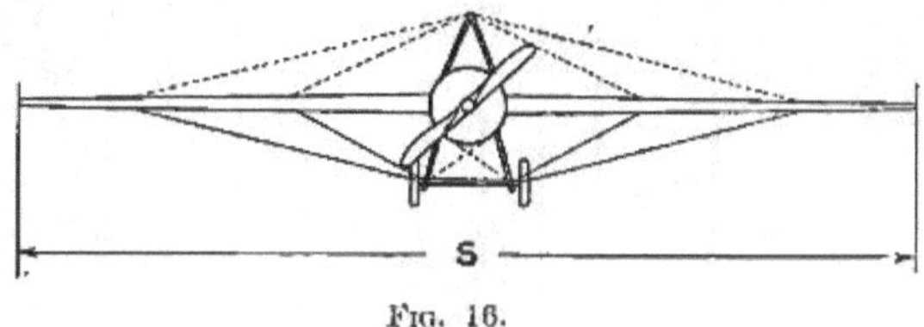

Fig. 16.

Wing structure is becoming more and more uniform for all types of airplanes. As already pointed out, the frame consists of two or more spars on which the ribs are fitted (Fig. 18). A leading edge made of wood connects the front extremities of the ribs, while for the trailing edge a steel wire or wood strip is used. The spars are also held together by wooden or steel tube struts and steel wire cross bracing,

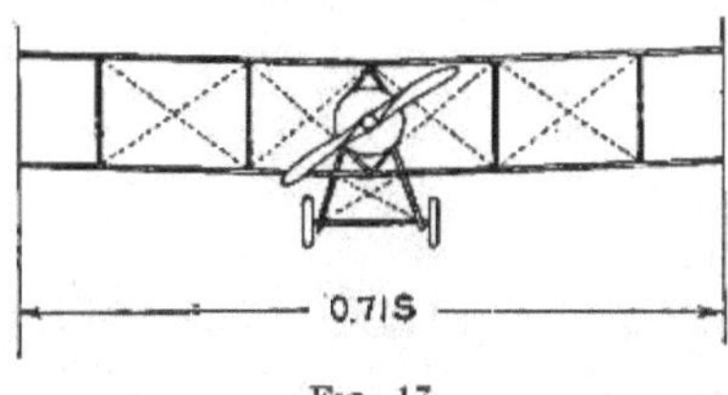

Fig. 17.

the function of which is to stiffen the wing horizontally. The rib is usually built up with a thin veneer web, to which strengthening flanges are glued and nailed or screwed (Fig. 19). The spars are usually of an I,[or box section for lightness (Fig. 20).

The vertical struts between the upper and lower wings of a biplane may be either of wood or steel tubing. In either case, they must have a streamline section to reduce to a minimum their head resistance. Wood struts are often hollowed to obtain lightness. Many different systems of

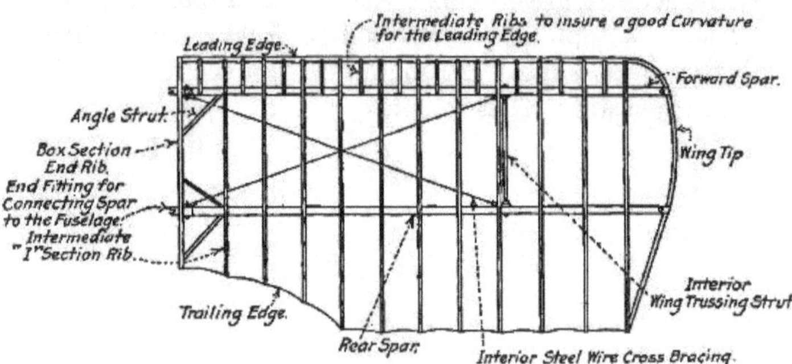

Fig. 18.

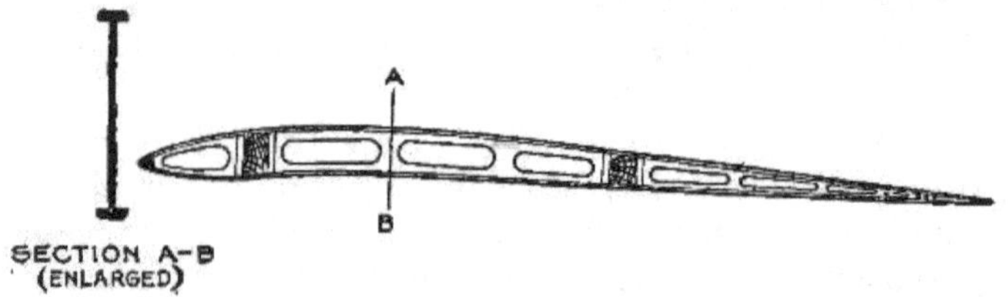

Fig. 19.

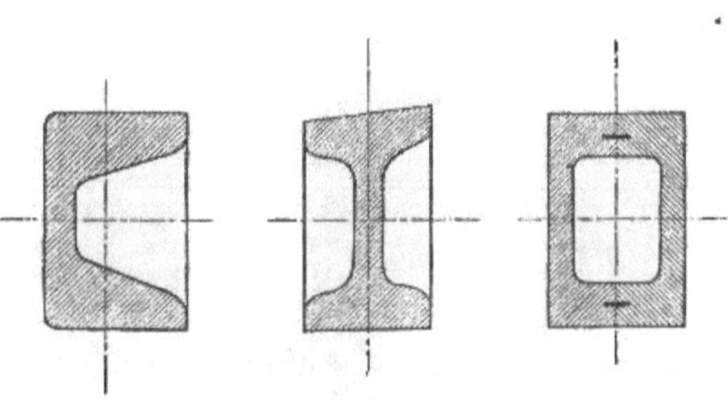

Fig. 20.

attaching the struts and cables to the spars are used, and some of the many possible methods are shown in Fig. 22.

The wing skeleton is covered with linen fabric, attached by sewing it to the ribs, and tacking or sewing it to the leading and trailing edges. It is then given an application of special varnish, called "dope," which stretches it and makes it air tight. The surface is then finished with bright

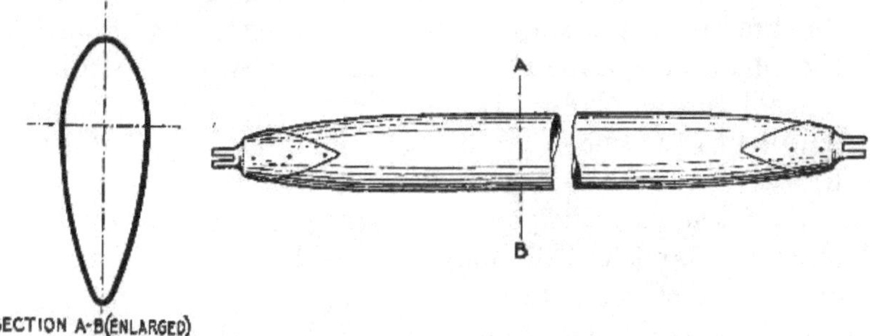

Fig. 21.

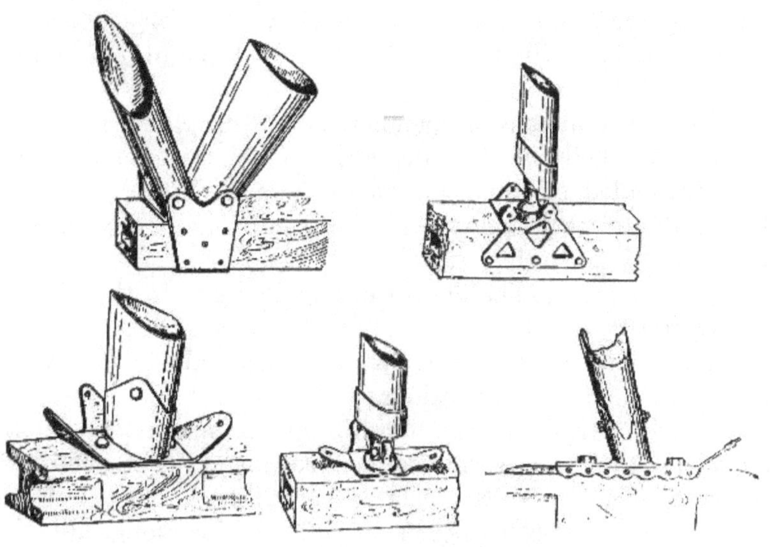

Fig. 22.

waterproof varnish, which leaves the fabric smooth so as to reduce frictional losses to a minimum, thereby detracting as little as possible from the efficiency.

CHAPTER II

THE CONTROL SURFACES

In studying the directional maneuvers of an airplane, reference must be made to its center of gravity (C.G.) and to its three principal axes. Two of the axes are contained in the plane of symmetry of the machine while the third is normal to this plane. One of the two axes in the plane is parallel to the line of flight while the other is perpendicular to it.

By a known principle of mechanics, every rotation of the machine about its C.G. may be considered as the resultant of three distinct rotations, one about each of the three principal axes. On the other hand, if three systems of control are used, each capable of producing a rotation of the airplane about one of its principal axes, any rotation of the machine about its C.G. can be brought about or prevented.

The principal axis perpendicular to the plane of symmetry, is called the pitching axis. Rotations about that axis are called pitching movements. The devices used to bring about, or prevent a pitching movement are called devices of longitudinal stability.

The axis perpendicular to the line of flight, in the plane of symmetry is called the axis of direction of flight. The devices which cause or prevent movements about that axis are called devices of directional stability.

The axis parallel to the line of flight is called the rolling axis, and the devices causing or preventing rolling movements are called devices of lateral stability.

There are usually two surfaces which control longitudinal stability, one fixed, called the stabilizer or tail plane, and the other movable, called the elevator.

The stabilizer or tail plane is a relatively small surface fixed at the rear end of the fuselage. Its function is, first of all, to offset or even completely invert the phenomenon of the inherent instability of curved wings, and secondly, to act as a damper on longitudinal or pitching movements.

The stabilizer may be of various shapes and sections. It may be either lifting or non-lifting, but it must always satisfy the basic condition that its unit loading per sq. ft. be lower than that of the principal wing surface. Under this condition only, will it act as a stabilizer; otherwise it would add to the instability of the wings.

As to the proper dimensions of the stabilizer, they depend on various factors such as the weight of the airplane, its longitudinal moment or inertia, its speed, and the distance the stabilizer is set from the center of gravity of the machine. Moreover, the proportions of the stabilizer with respect to the other parts of the airplane are also dependent on another factor: the type of airplane. For small, swift combat machines which require a high degree of maneuverability, the stabilizer will require relatively less surface than that required for large, heavily loaded machines, such as those used for bombing operations and requiring a much lower degree of maneuverability.

The framework or skeleton of the stabilizer is generally of wood or steel tubing. In general its angle of incidence may be adjusted either on the ground or while in flight. However, that incidence must never be greater than the angle used for the main wing surfaces. Its value is generally 1° to 4° less than that of the wings.

The elevator or movable surface is hinged to the rear edge of the stabilizer, and it may be raised or lowered while in flight.

In normal flight the elevator is set parallel to the air flow so that there is no air reaction on its faces. If it is swung upward or downward the air will strike it, producing a reaction whose direction is upward or downward respectively, thus tending to set the machine for climbing or descending.

The size of the elevator also depends on the weight, moment of inertia, speed of the machine, and on its distance from the center of gravity of the machine; also the type of airplane and the service for which it is intended must be given consideration. However, for quick and responsive machines the elevator must be proportionally larger than

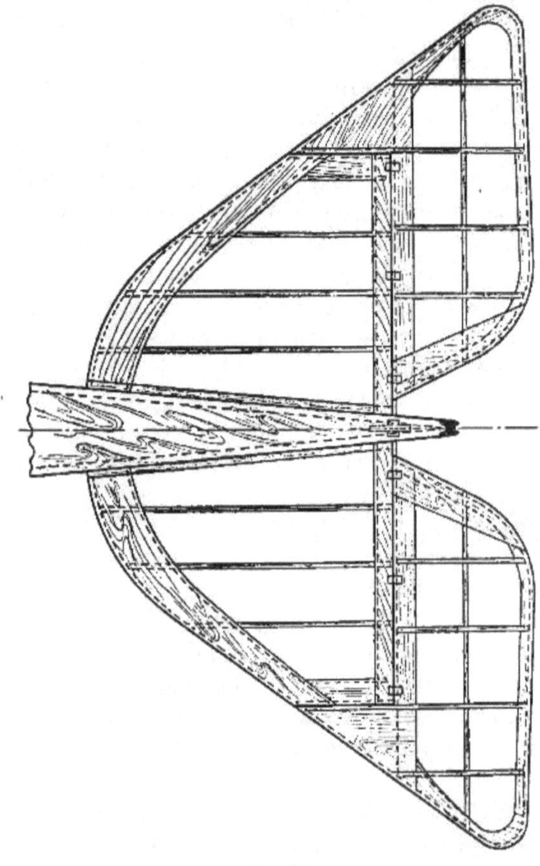

Fig. 23.

for slow machines endowed with a greater degree of stability. In other words, the two proportions vary inversely as those of the stabilizers. However, this will be more easily understood upon considering the functions of the two devices which are in a certain sense, completely opposite.

The function of the stabilizer is to insure longitudinal stability, just as its name implies. The elevators function instead, is to disturb the equilibrium of the machine in order to bring about a change in the normal flying. An outline of a type of stabilizer and elevator system is given in Fig. 23.

A closer study may now be made of the function of these two parts of longitudinal stability. First of all, examination will be made of the mechanism by which the stabilizer, when properly set, exercises its stabilizing property.

When, in an airplane, the incidence of the wing is changed with respect to the air, through which it is progressing, the air reaction will not only vary in intensity but also in location. If the new reaction is such as to antagonize the deviation, the airplane is said to be stable; otherwise it is said to be unstable.

Wings having curved profiles, when acting alone, are unstable. Laboratory experiments have shown that for a wing with a curved profile, the reaction moves forward as the incidence is increased, and vice versa; thus the reaction moves in such a way as to aggravate the disturbance. The point of intersection of the air reaction on the wing chord is called the center of pressure of the wing (Fig. 24). The location of the center of thrust is usually indicated by the ratio $\frac{x}{c}$. The curves for λ and for $\frac{x}{c}$ as functions of the angle of incidence for a given wing section, are shown in Fig. 25. By applying the data from these curves to a wing of 5 ft. chord and 40 ft. span, supposing the normal speed to be 100 m.p.h. and the normal angle of flight 2°, the wing loading will be

$$L = 7.3 \times 200 = 1460 \text{ lb.}$$

and it will be in equilibrium if the center of gravity of the load falls at a distance of 40 per cent. of the chord, or 2 ft. from the leading edge. Suppose now that the incidence is increased from 2° to 4°, then the sustaining force becomes

$$L = 10 \times 200 = 2000 \text{ lb.}$$

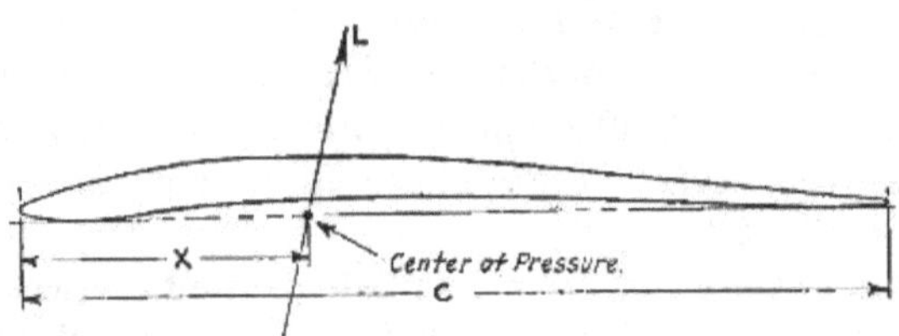

Fig. 24.

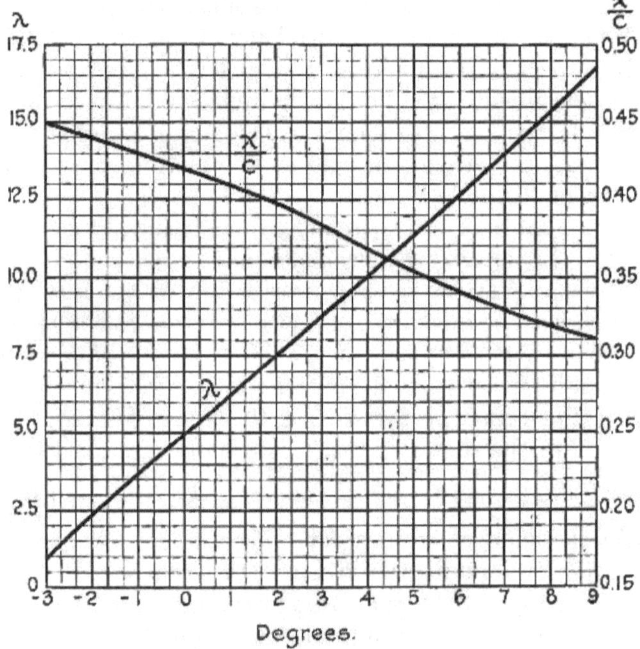

Fig. 25.

and it will be applied at 37 per cent. of the chord, or 1.85 ft. from the leading edge; this result will then produce around the center of gravity, a moment of

$$2000 \times 0.15 = 300 \text{ ft. lb.}$$

and such moment will tend to make the machine nose up; that is, it will tend to further increase the angle of incidence of the wing. Following the same line of reasoning for a case of decrease in the angle of incidence, it will be found in that case that a moment is originated tending to make the machine nose down. Therefore, the wing in question is unstable.

A practical case will now be considered, where a stabilizer is set behind this wing, and constituted of a surface of 15 sq. ft. (2 × 7.5) set in such a manner as to present an angle of −2° with the line of flight when the wing in front presents an angle of + 2°. In normal flight there is

1. The sustaining force of the main wing, equal to

$$L_s = 7.3 \times 200 = 1460 \text{ lb.}$$

2. The center of pressure of the main wing located at

$$0.40 \times 5 = 2 \text{ ft. from the leading edge,}$$

3. The sustaining force of the elevator equal to

$$L_s = 2 \times 15 = 30 \text{ lb., and}$$

4. The center of pressure of the elevator located at

$$0.44 \times 2' = 0.88 \text{ ft. from its leading edge.}$$

Suppose now that the incidence of the machine is increased so that the angle of incidence of the front wing changes from +2° to +5°, then there is

1. The sustaining force of the main wing equal to

$$L_s = 11.30 \times 200 = 2260 \text{ lb.}$$

2. The center of pressure of the main wing located at

$$0.355 \times 5' = 1.78 \text{ ft.,}$$

3. The sustaining force of the elevator equal to

$$L_s = 6.05 \times 15 = 91 \text{ lb., and}$$

4. The center of pressure of the elevator located at

$$0.410 \times 2' = 0.82 \text{ ft. from its leading edge.}$$

With these values, the total resultant of the forces acting in each case is obtained, and it is found that while in normal flight, the moment of total resultant about the c.g. of the machine is equal to zero; when the incidence is increased to 5°, that moment becomes equal to 2351 × (2.71′ − 2.44′) = 645 ft. lb. tending to make the machine nose down; that is, tends to prevent the deviation and therefore is a stabilizing moment (Fig. 26).

In analogous manner it can be shown that if the incidence of the machine is decreased, a moment tending to prevent

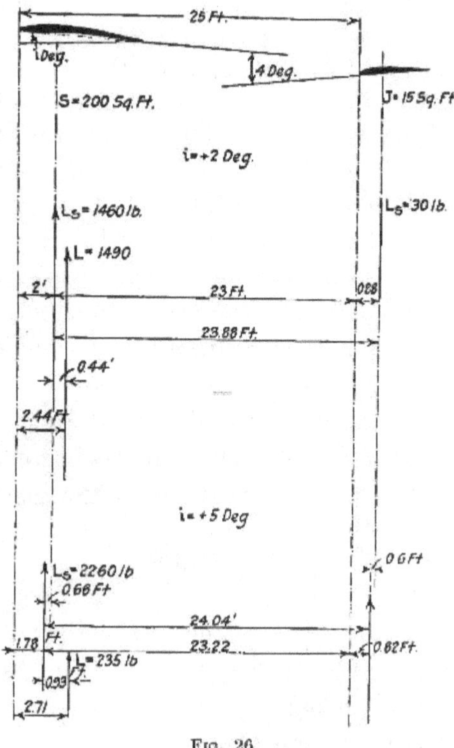

Fig. 26.

that deviation is developed. It is obvious, then, that if the airplane were provided with only a stabilizer and with no elevator, it would fly at only one certain angle of incidence, since any change in this angle would develop a stabilizing moment tending to restore the machine to its original angle. Thus the exact function of the elevator is to produce moments which will balance the stabilizing moments due to the stabilizer. This will allow the machine to assume a complete series of angles of incidence, enabling it to maneuver for climbing or descending.

There are also usually two parts controlling directional stability; one fixed surface called the fin or vertical stabilizer, and one movable surface called the rudder.

Consider, for example, an airplane in normal flight; that is, with its line of flight coincident with the rolling axis

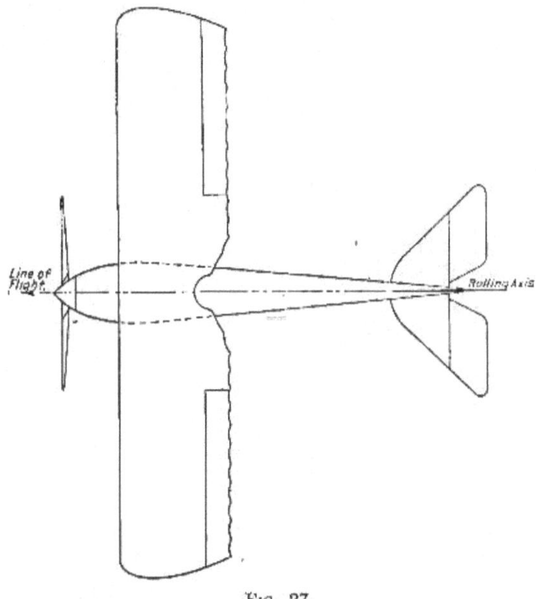

Fig. 27.

(Fig. 27). In this case there is no force of drift, but if for some reason the line of flight is no longer coincident with the rolling axis, a force of drift is developed (Fig. 28), whose point of application is called center of drift. If this center is found to lie behind the center of gravity, the machine tends to set itself against the wind; that is, it becomes endowed with directional stability. If, instead, the center of drift should fall before the center of gravity, normal flight would be impossible, as the machine tends to turn sharply about at the least deviation from its normal course. In practice, since the center of gravity of an airplane is found very close to the front end of the machine, the condition of directional stability is easily attained by the use of a small vertical surface of drift which is set at the extreme rear of the fuselage. This surface is called the fin or vertical stabilizer.

There is, however, a type of airplane called the Canard

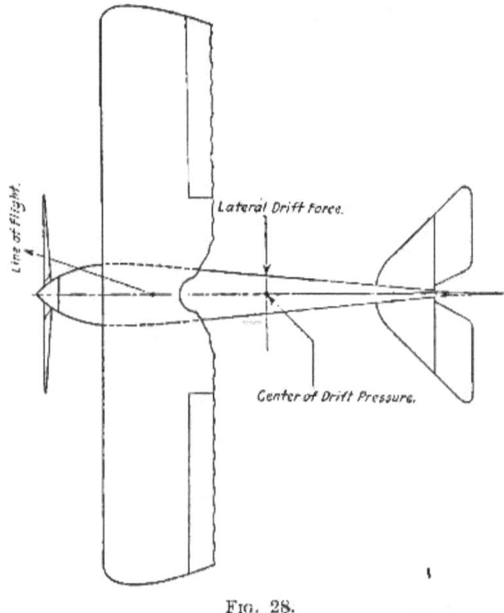

Fig. 28.

type, in which the main wing surface is the one in the rear, (and consequently the c.g. falls entirely in the rear) and in which the problem of directional stability presents considerable difficulty. This type of airplane, however, is not used at the present time.

A machine provided with only a fin would possess good directional stability, but for that very reason it would be impossible for the airplane to change its course. For that reason it is necessary to have a rudder; a vertical movable surface, which, when properly deviated, will produce a balancing moment to overcome the stabilizing moment of the fin, thus permitting a change in the course of the drift.

The phenomenon may now be studied more in detail. Let us suppose that the directing rudder is deviated at an angle; this deviation will then provoke on the rudder a reaction D' (Fig. 29), which will have about the center of gravity a moment $D' \times d'$; as a result, the airplane will

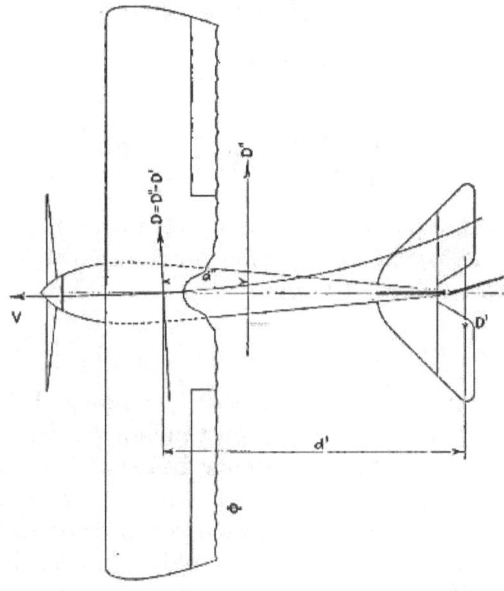

Fig. 29.

rotate about the axis of direction and the line of flight will no longer coincide with the rolling axis; that is, when the airplane starts to drift in its course, a drifting force D'' is originated, which tends to stabilize, and when $D'' \times d'' = D' \times d'$, equilibrium will be obtained. Obviously, then, the line of flight will no longer be rectilinear, since the two forces D'' and D' are unequal, and if transported to the center of gravity they will give a resultant $D = D'' - D'$ other than zero. The equilibrium will be obtained only if the line of flight becomes curvilinear; in fact, a centrifugal force Φ is then developed which will be in equilibrium with the resultant force of drift D. Then equilibrium will be obtained when $\Phi = D$; as

$$\Phi = \frac{W}{g} \times \frac{V^2}{r}$$

where W is the weight of the airplane, g the acceleration due to gravity, V the velocity of the airplane and r the radius of curvature of the line of flight, therefore

$$D = \frac{W}{g} \times \frac{V^2}{r}$$

from which is obtained

$$r = \frac{W}{g} \times \frac{V^2}{D} = \frac{W}{g} \times \frac{V^2}{D'' - D'}$$

From this equation it will be seen that to obtain remarkable maneuverability in turning, the difference $D'' - D'$ must have a large value. Or, since

$$\frac{D''}{D'} = \frac{d'}{d''}$$

it is necessary that the center of drift, although being in the rear of the center of gravity, must be not too far behind it, and it is necessary that the rudder be located at a considerable distance from the center of gravity. In other words, for good maneuverability, an excessive directional stability must not exist. The foregoing applies to what is called a flat turn without banking, which is analogous to that of a ship. The airplane, however, offers the great advantage of being able to incline itself laterally which greatly facilitates turning, as will be shown when reference is made to the devices for transversal stability.

In summarizing the foregoing, it is seen that in addition to the fixed surfaces, stabilizer and fin, whose functions are to insure longitudinal and directional stability, airplanes are provided with movable surfaces, elevator and rudder, which are intended to produce moments to oppose the stabilizing moments of the fixed devices. It will now be better understood that excessive stability is contrary to good maneuverability.

In like manner, for transversal stability, there are two classes of devices opposite in their functions. Some are used to insure stability while others serve to produce moments capable of neutralizing the stabilizing moments.

Let us consider an airplane in normal flight, and suppose that a gust of wind causes the machine to become inclined laterally by an angle α. The weight W and the air reaction L will have a resultant D_α which will tend to make the

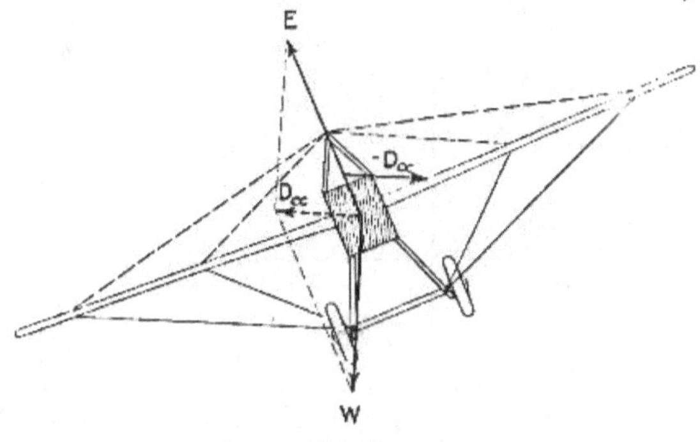

Fig. 30.

machine drift (Fig. 30); this drifting movement will produce a lateral air reaction $-D_\alpha$ acting in the direction opposite to D_α. The resultant of the lateral wind forces acting on the machine is $-D_\alpha$. If this reaction is such as to make with the force D_α a couple tending to restore the machine to its original position, the machine is said to be transversally stable; this is the case shown in Fig. 30. If $-D_\alpha$ has the same axis as D_α, the airplane is said to have an indifferent transversal stability. If, finally, $-D_\alpha$ and D_α form a couple tending to aggravate the inclination of the machine, the latter is said to be transversally unstable.

Consequently, in order to have an airplane laterally stable, conditions must be such that the lateral reaction $-D_\alpha$ together with the force D_α form a stabilizing couple; that is, the point of application of the force $-D_\alpha$ must be situated above the point of application of force D_a, which is the center of gravity. However, the couple of lateral stability must not have an excessive value, as it would decrease the maneuverability to such an extent as to make the machine dangerous to handle, as will now be explained.

It has been explained before how a turning action may be obtained by merely narrowing the rudder, and how

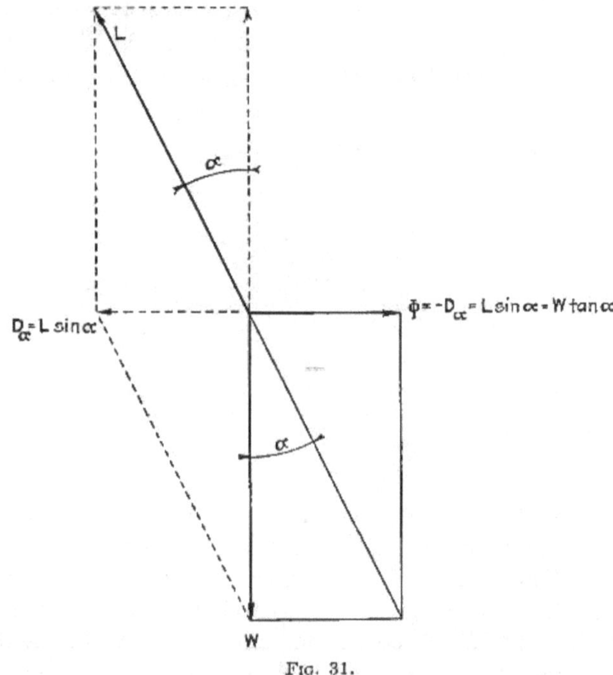

FIG. 31.

this cannot be actually done in practice since there is a possibility of the machine banking while turning. Now, when the airplane "banks," the forces L and W will admit a lateral resultant D_α which tends to deviate laterally the line of flight. A centrifugal force Φ is thereby developed, tending to balance the force D_α and equilibrium will obtain when $\Phi = D_\alpha$ (Fig. 31); that is, when

$$\Phi = \frac{W}{g} \times \frac{V^2}{r}$$

where r is the radius of curvature of the line of flight; therefore

$$D_a = \frac{W}{g} \times \frac{V^2}{r}$$

which will give

$$r = \frac{W}{g} \times \frac{V^2}{D_a}$$

As $D_a = W \tan \alpha$, we obtain

$$r = \frac{1}{g} \times \frac{V^2}{\tan \alpha}$$

This equation shows that the turn can be so much sharper as the speed is decreased, and the angle α of the bank is increased. This explains why pilots desiring to turn sharply, make a steep bank and at the same time nose the machine upward in order to lose speed.

Now the angle of bank may be obtained in two ways; by operating the rudder or by using the ailerons which are the controls for lateral stability. In using the rudder, it has been observed that the machine assumes an angle of drift. If the force of drift $D = D'' - D'$ (Fig. 29) passes through the center of gravity, a flat turn without banking will result. If force D passes below the center of gravity, the airplane will incline itself so as to produce a resultant D_a of L and W, in a direction opposite to force D. Then the total force of drift is equal to $D - D_a$. This case is of no practical interest, since it corresponds to the case of lateral instability, which is to be avoided. If, instead, force D passes above the center of gravity, then the angle of bank α is such that D_a is of the same direction as D. Therefore, the total force of drift is $D + D_a$.

Now if force D_a had its point of application too far above the center of gravity, the result would be that with a slight movement of the rudder, a strong overturning moment would develop which would give the machine a dangerous angle of bank. Therefore it is evident that an excessive stabilizing moment must be avoided.

The ailerons are two small movable surfaces located at the wing ends (Fig. 32). Let us now observe what happens when they are operated.

The ailerons are hinged along the axes AA' and BB', and are controlled in such a manner that when one swings upward the other swings downward. With this inverse movement, the equity of the sustaining force on both the

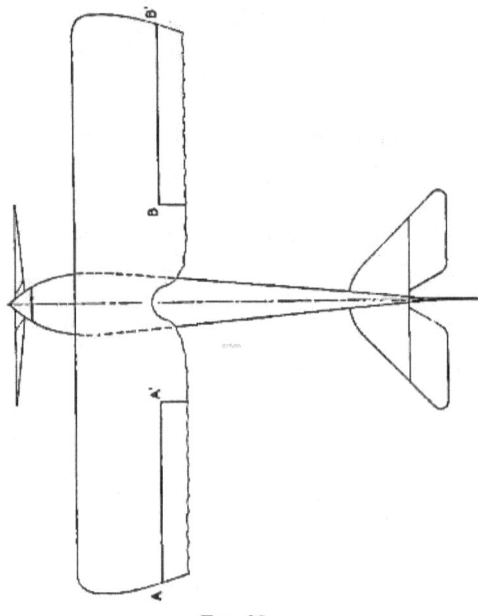

Fig. 32.

right and left wings, is broken. Thus a couple is brought into play which tends to rotate the machine about the rolling axis. Since it is possible to operate the ailerons in either direction, the pilot can bank his machine to the right or to the left.

Supposing that the pilot operates the ailerons so that the machine banks to the right; let α be the angle of bank; then, a force D_α is produced, which, in a laterally stable machine will tend to oppose the banking movement caused by the ailerons. The rapidity of turning, and consequently the mobility of the machine, will increase in proportion as the rapidity of the banking movement increases. Now, all other conditions being similar, the rapidity with which the machine banks is proportional to the difference of the couple due to the actions of the ailerons, and the couple due to the force of drift α; if the value of the latter is very large (that

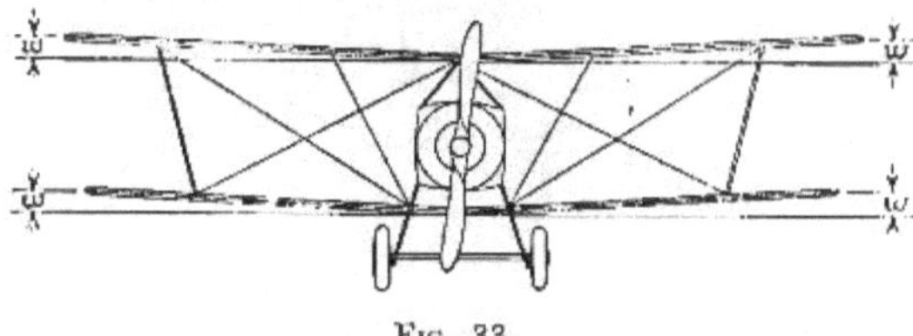

Fig. 33.

is, if D_α is applied very far above the center of gravity) the maneuver will be slow. Therefore for good mobility of the airplane, the force D_α must not be too far above the center of gravity.

The foregoing considerations show the close interdependency existing between the problems of directional stability and those of transversal stability. It is practically possible

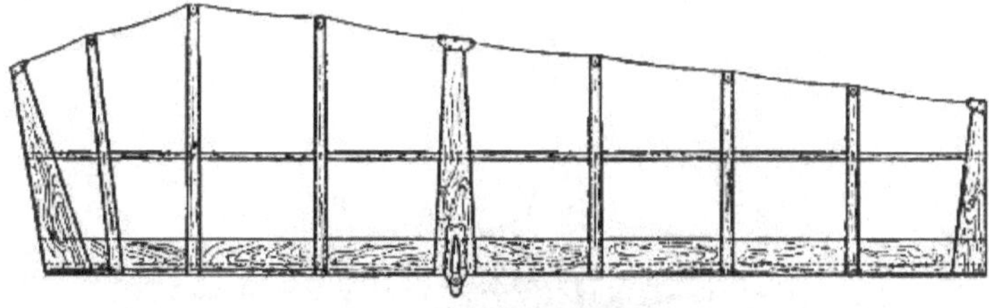

Fig. 34.

to control directional stability by means of the lateral controls, and vice versa. For example, birds possess no means of control for directional stability alone, but use the motion of their wings for changing the direction of their flight.

To raise the force D_α with respect to the center of gravity, we may either install fins above the rolling axis, or, better still, give the wings an upward inclination from the center to the tip of the wing, the so-called dihedral angle (Fig. 33). The effect of this regulation is that when the machine takes an angle of drift, the wing on the side toward which the machine drifts, assumes an angle of incidence greater than the inci-

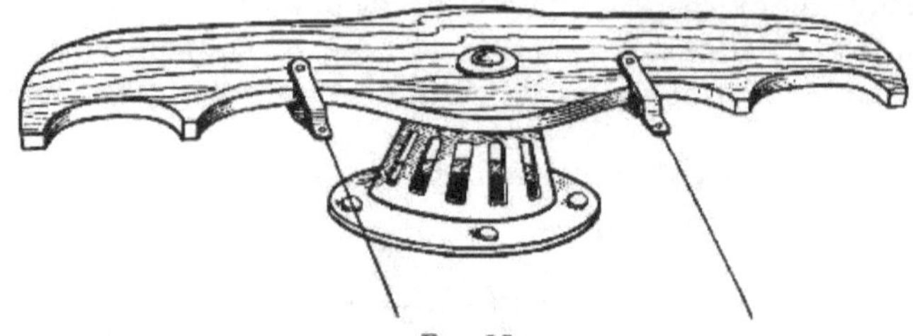

Fig. 35.

dence of the opposite wing, thereby developing a lateral couple which is favorable to stability.

The framework of the ailerons is usually of wood, steel tubing or pressed steel members. An outline of wood ailerons is given in Fig. 34.

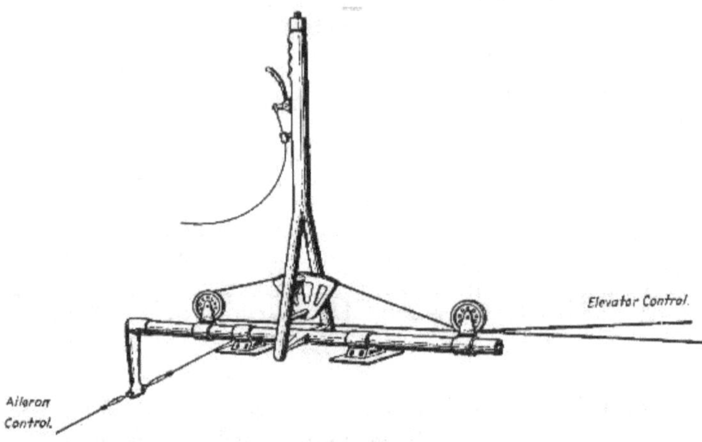

Fig. 36.

Concluding to be relatively safe and controllable at the same time, an airplane must be provided with devices which will produce stabilizing couples for every deviation from the position of equilibrium; but these couples must not be of excessive magnitude, for the machine would then be too slow in its maneuvers, and consequently dangerous in many cases. These stabilizing couples must be of the same magnitude as the couples which can be produced by the controlling devices. In this manner the pilot always has control of the machine and it will answer readily and effectively to his will.

The system of control of maneuvering by the pilot usually consists of a rudder-bar operated by the feet, and a hand-controlled vertical stick (called the "joy stick") piv-

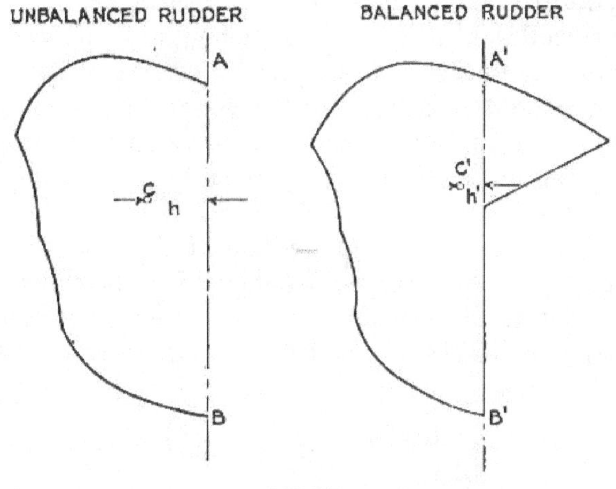

Fig. 37.

oted on a universal joint, moved forward and backward to lower and raise the elevator, and from left to right to move the ailerons (Figs. 35 and 36).

Balanced rudders are found on some of the high-powered machines, as they reduce, to a slight degree, the muscular effort of the pilot. The effort required to move a control surface depends on the distance h (Fig. 37) between the center of pressure C and the axis AB of rotation. If axis AB is moved to $A'B'$, the value of h is reduced to h', and therefore the required effort for the maneuver is decreased.

CHAPTER III

THE FUSELAGE

The fuselage or body of an airplane is the structure usually containing the engine, fuel tanks, crew and the useful load. The wings, landing gear, rudder and elevator are all attached to the fuselage. The fuselage may assume any one of various shapes, depending on the service for which the machine is designed, the type of engine, the load, etc. In general, however, the fuselage must be designed so as to have, as nearly as possible, the shape of a solid offering a minimum head resistance. In the discussion on wings, it was observed that the air reaction acting on them is generally considered in its two components of Lift and Drag. For a fuselage moving along a path parallel to its axis, the Lift component is zero, or nearly so; the Drag component is predominant, and must be reduced to a minimum in order to minimize the power necessary to move the fuselage through the air.

Let S indicate the major section of the fuselage, and V the velocity of the airplane. Laboratory experiments have shown that head resistance is proportional to S and V^2. Assuming our base speed as 100 m.p.h. for a given fuselage, then

$$R = K \times S \times \left(\frac{V}{100}\right)^2 \qquad (1)$$

therefore, if $S = 1$ and $V = 100$, then $R = K$. Thus the coefficient K is the head resistance per square foot of the major section of the fuselage, when $V = 100$ m.p.h. This is called the coefficient of penetration of the fuselage. The lower K is, the more suitable will be the fuselage, as the corresponding necessary power will be decreased.

Equation (1) shows two ways of decreasing the necessary power;

(*a*) By reducing the major section of the fuselage to a minimum, and (*b*) by lowering the value of coefficient K as much as possible.

In order to solve problem (*a*) it is necessary first to adapt the section of the fuselage to that of the engine. The

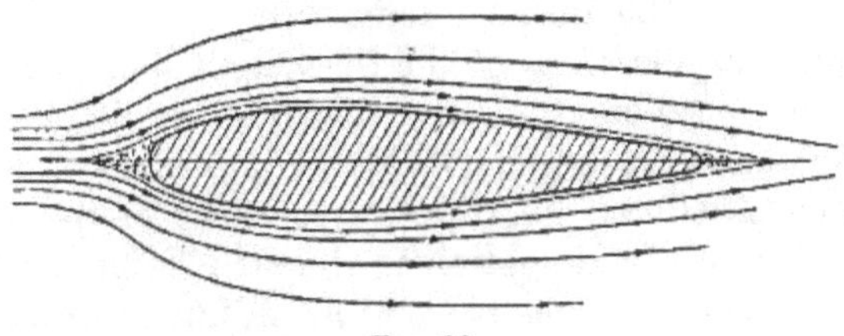

Fig. 38.

fuselage may be of circular, square, rectangular, triangular, etc., section, so designed that its major section follows the form of the major section of the engine. In the second place, it is good practice, when other reasons do not prevent it, to arrange the various masses constituting the load (fuel, pilot, passengers, etc.) one behind the other, so as to keep the transversal dimension as small as possible.

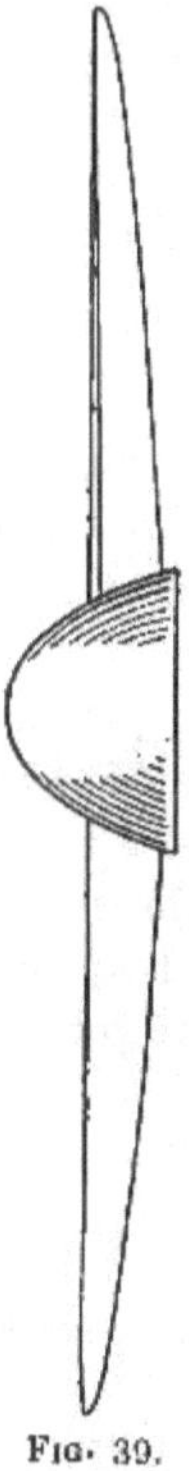

To decrease the coefficient of head resistance, the shape of the fuselage must be carefully designed, especially the form of the bow and stern. Analogous to that of the wings, the phenomenon of head resistance of the fuselage is due to the resultant of two positive and negative pressure zones, developing on the forward and rear ends respectively (Fig. 38). Whatever be the means employed to reduce the importance of those zones, the value of K will be lowered, thus improving the penetration of the fuselage.

To improve the bow, it must be given a shape which will as nearly as possible approach that of the nose of a dirigible. This is easily affected with engines whose contours are circular, but the problem presents greater difficulties with vertical types of engines, or V types without reduction gear. Sometimes a bullet-nosed colwing is fitted over the propeller hub, fixed to and rotating with

Fig. 39.

the propeller. Its form is then continued in the front end of the fuselage contour, its lines gradually easing off to meet those of the fuselage (Fig. 39).

To improve the stern of the fuselage it must be given a strong ratio of elongation, and the shaping with the rest of the machine must be smoothly accomplished. A special advantage is offered by the reverse curve of the sides; in fact, in this case, a deviation in the air is originated in the zone of reverse curving (Fig. 40) tending to decrease the pressure, and consequently increasing the efficiency.

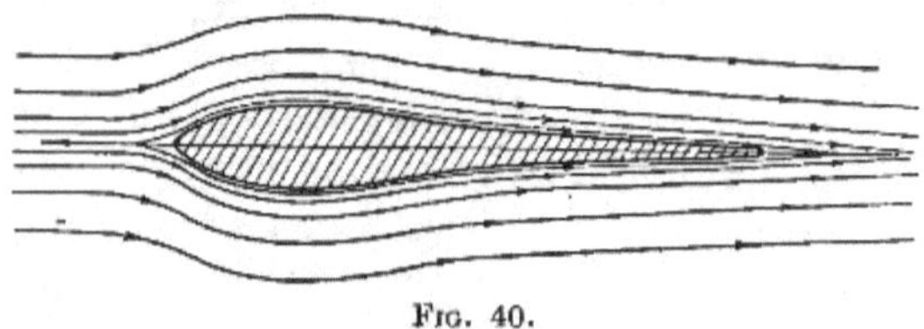

Fig. 40.

The value of coefficient K varies from 7 (for the usual types of fuselage) to 2.8 (for perfect dirigible shapes). It is interesting to compare such values with the coefficient of head resistance of a flat disc 1 sq. ft. in area, which is equal to 30. To move the above disc at a speed of 100 m.p.h. we must overcome a resistance of 30 lb., while in the case of the fuselage of equal section, but having a perfect streamline shape, we must overcome a resistance of only 2.8 lb., or less than one-tenth the head resistance of the disc. Practically, a well-shaped fuselage has a coefficient of about 6, so if its major section is, for instance, 12 sq. ft., the resistance to be overcome at a speed of 150 m.p.h. is

$$6 \times 12 \times \left(\frac{150}{100}\right)^2 = 162 \text{ lbs.}$$

which will theoretically absorb about 66 H.P.

Fuselages may be divided into three principal classes, depending on the type of construction used:

(*a*) Truss structure type,

(*b*) Veneer type, and

(*c*) Monocoque type.

Main Longerons.
Strut.
Transverse Strut.
Motor Supports.
Motor Supporting Beams.

FIG. 41.

Tail Strut.
Bulkheads.
Veneer
Panel
Stiffening Wood Cross-Bracing.
Main Spars
Motor Supports.
Motor Supporting Beams.

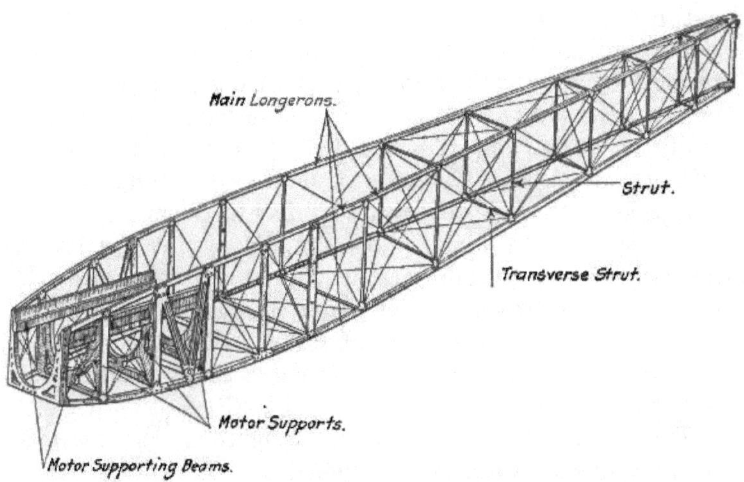

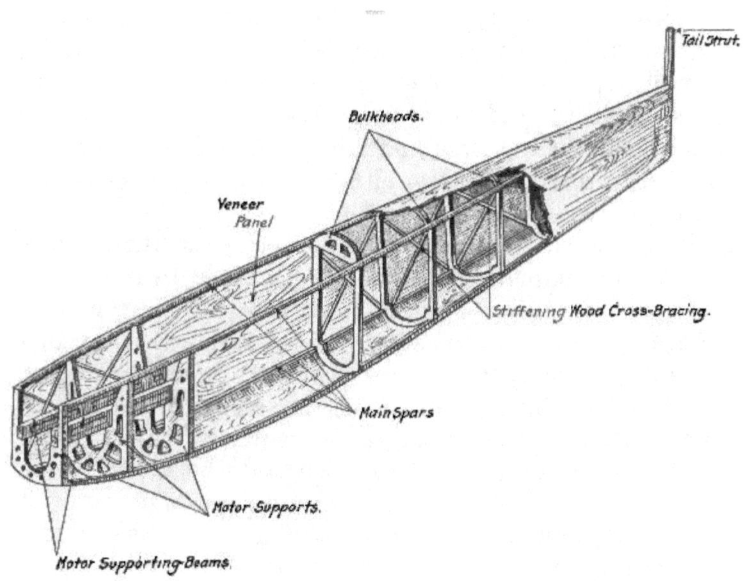

FIG. 42.

The truss type generally consists of 4 longitudinal longerons, held together by means of small vertical and horizontal struts and steel wire cross bracing (Fig. 41). The whole frame is covered in the forward part with veneer and aluminum and in the rear with fabric. The longerons are generally of wood, and the small struts are often of wood, although sometimes they are made of steel tubing.

Fuselages built of veneer are similar to the truss type as they also have 4 longitudinal longerons, but the latter, instead of being assembled with struts and bracing, are held in place by means of veneer panels glued and attached by nails or screws. By the use of veneer, which firmly holds the longerons in place along their entire length, the section of the longerons can be reduced (Fig. 42).

The monocoque type has no longerons, the fuselage being formed of a continuous rigid shell. In order to insure the necessary rigidity, the transverse section of the monocoque is either circular or elliptical. The material generally used for this type is wood cut into very thin strips, glued together in three or more layers so that the grain of one ply runs in a different direction than the adjacent plies. This type of construction has not come into general use because of the time and labor required in comparison with the other two types, although it is highly successful from an aerodynamical point of view.

Whatever the construction of the fuselage be, the distribution of the component parts to be contained in it does not vary substantially. For example, in a two-seater biplane (Fig. 43), at the forward end we find the engine with its radiator and propeller; the oil tank is located under the engine, and directly behind the engine are the gasoline tanks, located in a position corresponding to the center of gravity of the machine. It is important that the tanks be so located, as the fuel is a load which is consumed during flight, and if it were located away from the center of gravity, the constant decrease in its weight during flight would disturb the balance of the machine.

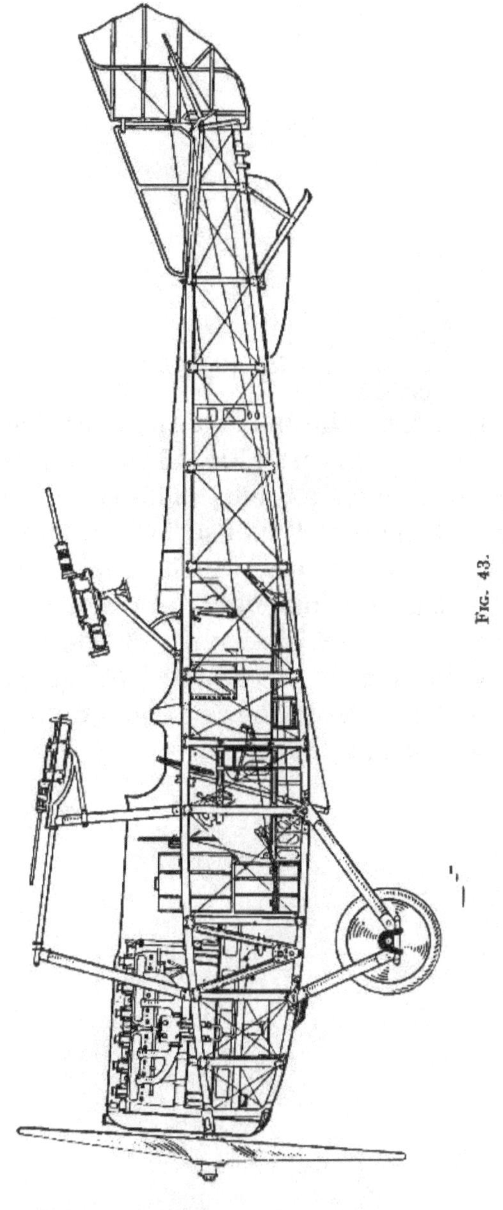

Fig. 43.

Directly behind the tanks is the pilot's seat, and behind the pilot is the observer. Fig. 43 shows the positions of the machine-guns, cameras, etc. The stabilizing longitudinal surfaces and the directional surfaces are at the rear end of the fuselage. The wings, which support the entire weight of the fuselage during flight, are attached to that part on which the center of gravity of the machine will fall. Under the fuselage is placed the landing gear. Its proper position with respect to the center of gravity of the machine will be dealt with later on.

CHAPTER IV

THE LANDING GEAR

The purpose of the landing gear is to permit the airplane to take off and land without the aid of special launching apparatus.

The two principal types of landing gears are the land and marine types. There is a third, which might be called the intermediate type, the amphibious, which consists of both wheels and pontoons, enabling a machine to land or "take

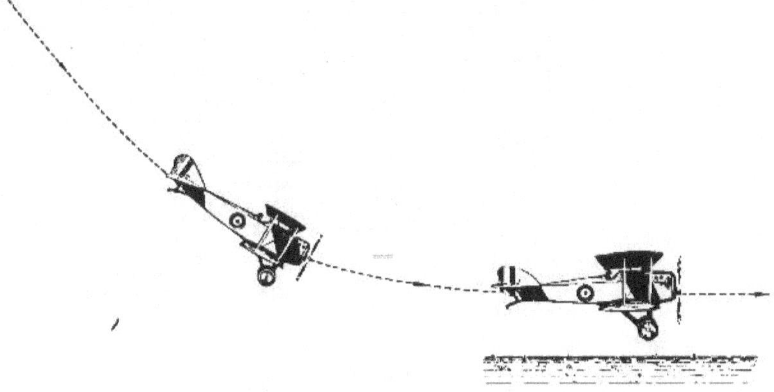

Fig. 44.

off" from ground or water. This discussion will be devoted solely to wheeled landing gears, the study of which pertains especially to the outlines of the present volume.

The "take off" and landing, especially the latter, are the most delicate maneuvers to accomplish in flying. Even though a large and perfectly levelled field is available, the pilot when landing must modify the line of flight until it is tangent to the ground (Fig. 44); only by doing this will the kinetic force of the airplane result parallel to the ground, and only then will there be no vertical components capable of producing shocks.

In actual practice, however, the maneuvers develop in a rather different manner. First, the fields are never perfectly level, and secondly, the line of flight is not always exactly parallel to the ground when the machine comes in contact with the ground. The landing gear must therefore be equipped with shock absorbers capable of absorbing the force due to the impact.

The system of forces acting on an airplane in flight is generally referred to its center of gravity, but for an air-

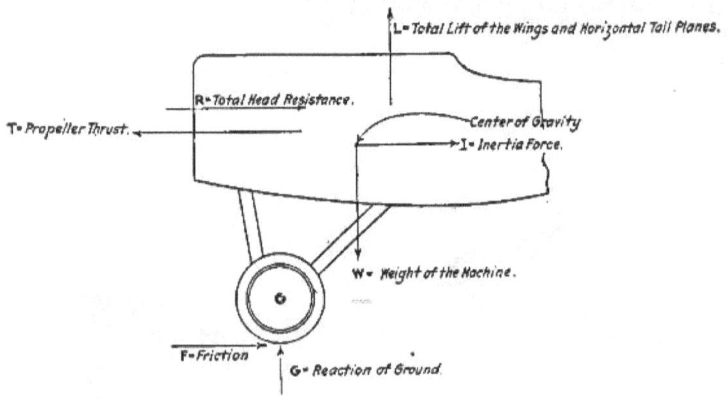

FIG. 45.

plane moving on the ground, the entire system of the acting forces must be referred to the axis of the landing wheels. Such forces are (Fig. 45),

T = propeller thrust,
W = weight of airplane,
L = total lift of wing surfaces,
R = total head resistance of airplane,
I = inertia force,
F = friction of the landing wheels, and
G = reaction of the ground.

The moments of these forces about the axis of the landing gear may be divided into four groups:

1. Forces whose moments are zero (the reaction of the ground, G),

2. Forces whose moments will tend to make the machine sommersault (forces T and F),

3. Forces whose moments tend to prevent sommersaulting (forces W and R), and

4. Forces whose moments may aid or prevent sommersaulting (forces L and I).

In group 4, the moment of the force L may be changed in direction at the pilot's will, by maneuvering the elevator; force I prevents sommersaulting when the machine accelerates in taking off, and aids sommersaulting in landing when the machine retards its motion.

In practice it is possible to vary the value of these moments by changing the position of the landing gear, placing it forward or backward.

By placing the landing gear forward, the moment due to the weight of the machine is particularly increased, and it may be carried to a limit where this moment becomes so excessive that it cannot be counterbalanced by moments of opposite sign. Then the airplane will not "take off," for it cannot put itself into the line of flight.

By placing the landing gear backward, the moment due to the weight is decreased, and this may be done until the moment is zero, and it can even become negative; then the machine could not move on the ground without sommersaulting. Consequently it is necessary to locate the landing gear so that the tendency to sommersault will be decreased and the "take off" be not too difficult. In practice this is brought about by having an angle of from 14° to 16° between the line joining the center of gravity of the machine to the axis of the wheels, and a vertical line passing through the center of gravity.

Let us examine the stresses to which a landing gear is subjected upon touching the ground. Assume, in this case, an abnormal landing; that is, a landing with a shock. (In fact, in the case of a perfect landing, the reaction of the ground on the wheels is equal to the difference between the weight W and the sustaining force L, and assumes a maximum value when $L = 0$; that is, when the machine is standing.) In the case of a hard shock, due either to the encounter of some obstacle on the ground, or to the fact that the line of flight has not been straightened out, the kinetic energy of the machine must be considered. That kinetic energy is equal to

$$\frac{1}{2} \times \frac{W}{g} \times V^2$$

where g is the acceleration due to gravity, and V the velocity of the airplane *with respect to the ground.* The foregoing is the amount of kinetic energy stored up in the airplane.

Naturally, it would be impossible to adopt devices capable of absorbing all the kinetic energy thus developed, as the weight of such devices would make their use prohibitive. Experience has proven that it is sufficient to provide shock absorbers capable of absorbing from 0.5 per cent. to 1 per cent. of the total kinetic energy. Then the maximum kinetic energy to be absorbed in landing an airplane of weight W and velocity V, is equal to

$$0.0025 \text{ to } 0.0050 \times \frac{W}{g} \times V^2$$

For example, for an airplane weighing 2000 lb., moving at a velocity of 100 m.p.h. (146 ft. per sec.), assuming 0.004, it will be necessary that the landing gear be capable of absorbing a maximum amount of energy equal to

$$0.004 \times \frac{2000}{32.2} \times 146^2 = 5300 \text{ ft.-lb.}$$

The parts of the landing gear intended to absorb the kinetic energy of an airplane in landing, are the tires and shock absorbers. Fig. 46 gives the work diagrams for a wheel. The wheel is capable of absorbing 900 ft.-lb. with a deformation of 0.25 ft. Fig. 47 gives the diagram of the work referred to per cent. elongation for a certain type of elastic cord. The work absorbed by n ft. of elastic cord under a per cent. elongation of x is equal to the product of $\frac{n}{100}$ times the area of the diagram corresponding to x per cent. elongation. Supposing, for instance, to have a shock absorbing system 32 ft. long, allowing an elongation of 150 per

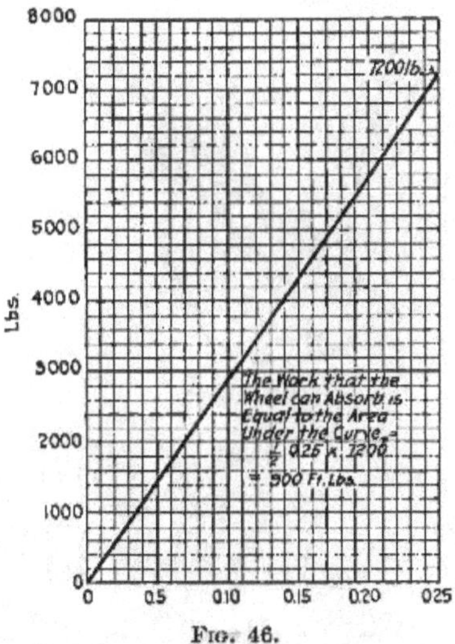

Fig. 46.

cent.; the work that it can absorb is equal as shown in the diagram to 1800 ft.-lb. As this gives a total of 2700 ft.-lb.,

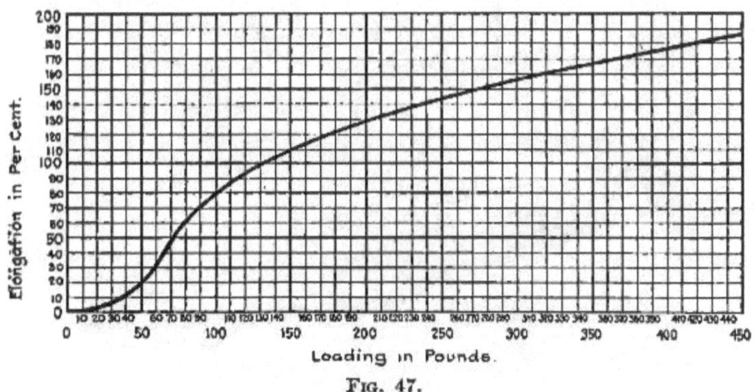

Fig. 47.

two wheels and two shock absorbers of such type will be sufficient for the airplane in question.

Rubber cord shock absorbers, which perform work by their elongation, have proven to be the lightest and most

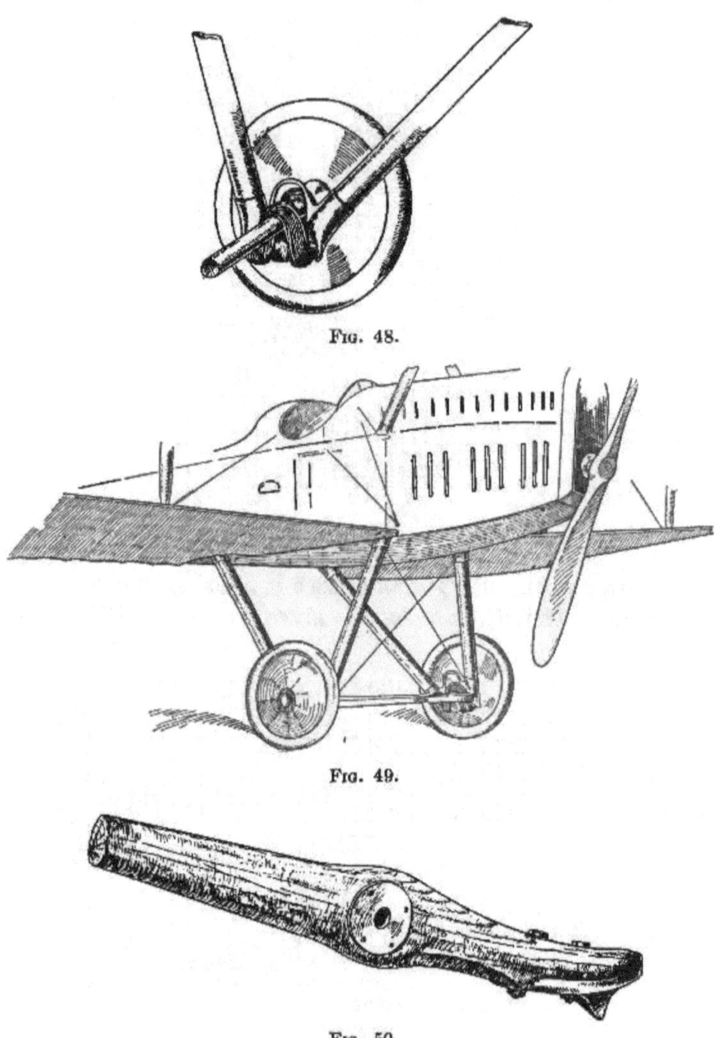

Fig. 48.

Fig. 49.

Fig. 50.

practical. Experiments have been made with other types, such as the steel spring, hydraulic and pneumatic, but the results have shown these types to possess but little merit. Fig. 48 illustrates an example of elastic cord binding. Fig. 49 shows the outline of a landing gear.

Up to this point, our discussion has been only on the vertical component of the kinetic energy. Consideration must also be given the horizontal component, whose only effect is to make the machine run on the ground for a certain distance. When the available landing space is limited, the machine must be slowed down by means of some braking device, in order to shorten the distance the machine has to roll on the ground. Friction on the wheels, head resistance and the drag all have a braking effect, but it often happens that these retarding forces are not sufficient. The practice therefore prevails of providing the tail skid with a hook, which, as it digs into the ground, exerts on the machine an energetic braking action (Fig. 50). On some machines, a short arm, with a small plow blade at its lower end, is attached to the middle of the landing gear axle, which can be caused to dig into the ground and produce a braking effect.

Similar to the landing gear, the tail skid is also provided with a small elastic cord shock absorber to absorb the kinetic energy of the shock.

On certain airplanes, use is made of aerodynamical brakes consisting of special surfaces which normally are set in the line of flight, and consequently offering no passive resistances, but when landing they can be maneuvered so as to be disposed perpendicularly to the line of motion, producing an energetic braking force.

CHAPTER V

THE ENGINE

The engine will be dealt with only from the airplane designers point of view. For all the problems peculiar to the technique of the subject, special texts can be referred to.

There are various types of aviation engines—with rotary or fixed cylinders, air cooled or water cooled, and of vertical, V, and radial types of cylinder disposition. Whatever the type under consideration, there exist certain fundamental characteristics which enable one to judge the engine from the point of view of its use on the airplane. Such characteristics may be grouped as follows:

1. Weight of engine per horsepower,
2. Oil and gasoline consumption per horsepower per hour,
3. Ratio between the major section of the engine and the number of horsepower developed,
4. Position of the center of gravity of the engine with respect to the propeller axis, and
5. Number of revolutions per minute of the propeller shaft.

In order to judge the light weight of an engine, it is not sufficient to know only its weight and horsepower; it is also essential to know it specific fuel consumption. If we call E the weight of the engine, P its power, C the total fuel consumption per hour (gasoline and oil), and x the number of hours of flight required of the airplane, then the smaller the value of the following equation, the lighter will be the motor:

$$y = \frac{E}{P} + x \times \frac{C}{P} \qquad (1)$$

For a given engine, equation (1) gives the linear relation between y and x, which can be translated into a simple, graphic, representation. Let us consider two engines, A and B, having the following characteristics:

TABLE 1

Engine	E lbs.	P H.P.	$e = \frac{E}{P}$ lbs. per H.P.	C lbs.	$c = \frac{C}{P}$ lbs. per H.P.
A	600	300	2	180	0.6
B	750	300	2.5	144	0.48

For engine A, equation (1) will give $y = 2 + 0.6x$.
For engine B, equation (1) will give $y = 2.5 + 0.48x$.

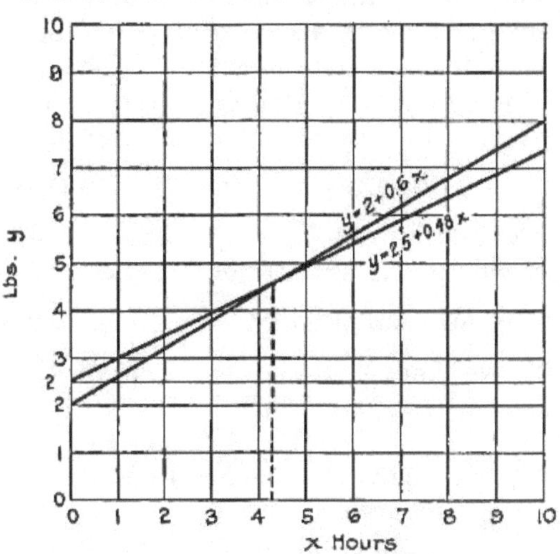

FIG. 51.

Translating these equations into diagrams (Fig. 51), we see that engine A is lighter than engine B, for flights up to 4 hours 10 minutes beyond which point, B is the lighter.

If $$x = 10 \text{ hours},$$
then
$$y_A = 8 \text{ lb.}$$
$$y_B = 7.3$$
that is, B has an advantage of 0.7 lb. per H.P.; since $P =$ 300 H.P. the total advantage is 270 lb.

Practically, for engines of the same general types, the value of the specific consumption $c = \frac{C}{P}$, varies around the same values. In that case, only the weight per horsepower, $e = \frac{E}{P}$ is of interest. In fact, that ratio is so important that it may often be convenient to adopt an engine of lower power in comparison with another of high power, for the sole reason that for the latter the above ratio is higher.

Let us suppose that we wish to build an airplane of given horizontal and climbing speed characteristics, capable of carrying fuel for a flight of three hours and a useful load of 600 lb. (pilot, observer, arms, ammunition, devices, etc.). Fixing the flying characteristics is equivalent to fixing the maximum weight per horsepower, of the machine with its complete load. In fact, we shall see further on in discussing the efficiency of the airplane, that the lower the ratio $\frac{W}{P}$ between the total weight W, and the power P of the motor, the better will be the flying characteristics of the machine.

Supposing for example, that $\frac{W}{P} = 10$ lb.. Analyzing the weight W, we find it to be the sum of the following components:

W_A = weight of airplane without engine group and accessories,
W_P = weight of the complete engine group,
W_C = weight of oil and gasoline,
W_U = useful load,

We can then write

$$W = W_A + W_P + W_C + W_U$$

Generally $W_A = \frac{1}{3} W$; $W_P = eP$. In this case (assuming 4 hours of flight) $W_C = 4CP$, where C is the specific consumption per horsepower which can be assumed to be equal to 0.55; this gives $W_C = 2.2P$; furthermore $W_U = 600$ lb.

We shall then have

$$W = \frac{1}{3} W + eP + 2.2P + 600$$

that is, since $\frac{W}{P}$ must be equal to 10

$$P = \frac{600}{4.46 - e}$$

and consequently

$$W = \frac{600}{0.446 - 0.1e}$$

In Fig. 52 these relations have been translated into curves, and it is seen that there are innumerable couples of values e, P, which satisfy the conditions necessary for the construction of the airplane under consideration.

Let us examine the extreme values for $e = 2$ lb. per H.P. and $e = 3$ lb. per H.P. We see that

if $e = 2$; $P = 246$ H.P. and $W = 2460$ lb.
if $e = 3$; $P = 416$ H.P. and $W = 4160$ lb.

From these it is obvious then, that although using an engine of 70 per cent. more power, the same result is obtained, plus the disadvantage of having an airplane of which the surface (and consequently the required floor space), is 70 per cent. greater.

However, in practice it often happens that an engine of higher power than another, not only *does not* possess higher weight per horsepower, but on the contrary, has a lower weight per horsepower. It is only necessary to note the importance of this matter.

Another important consideration is the bulk of the engine. Of two engines having the same power, but different major sections, we naturally prefer the engine of lesser major section, because it permits the construction of fuselages offering less head resistance. An example will make the point clearer. Supposing we have two engines, each of 300 H.P., whose characteristics with the exception of their bulk, are absolutely similar. Suppose that one of these engines has a major section of 6 sq. ft., and the other of 9 sq. ft., the head resistance of the fuselage of the second engine is 50 per cent. greater than that of the first. Let

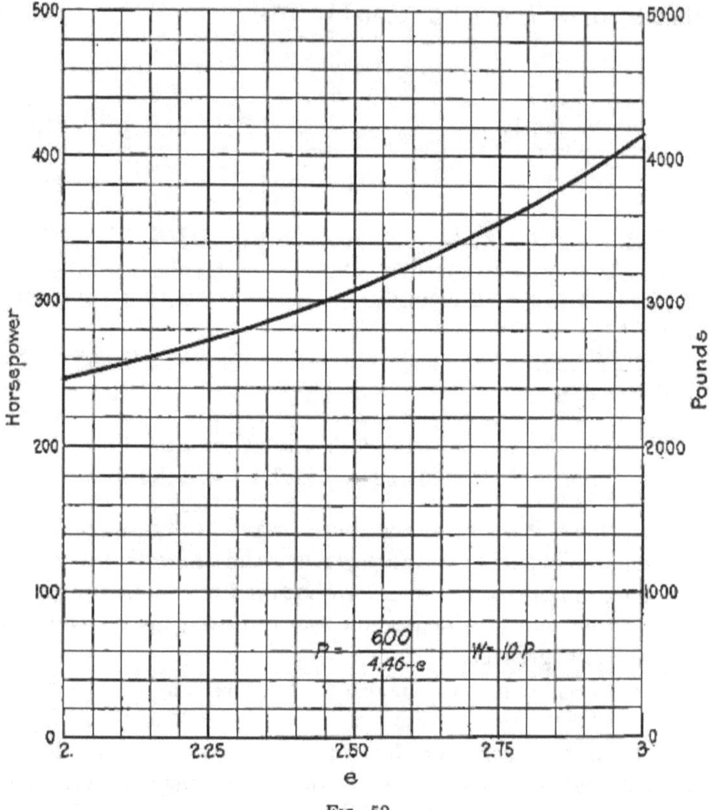

Fig. 52.

us assume that the power developed is used up in the following manner:

30 per cent. for the resistance of the wing surface,
40 per cent. for the resistance of the fuselage, and
30 per cent. for the resistance of all the other parts.

The result is that with the second engine, a machine can be constructed whose head resistance will be 20 per cent.

greater, thereby losing about 7 per cent. of the speed, due to the relations between the various head resistances and the speeds, as we shall see in the discussion on the efficiency of the airplane.

The position of the center of gravity with respect to the propeller axis, has a great importance in regard to the installation of an engine in the airplane. An ideal engine should have its center of gravity below, or at the most, coincident with the line of thrust. This last condition is true for all rotary and radial engines. Instead, for engines with vertical or V types of cylinders, the center of gravity is generally found above the line of thrust, unless the propeller axis is raised by using a transmission gear. In speaking of the problems of balancing, we shall see the great importance of the position of the center of gravity of the machine with respect to the axis of traction, and the convenience there may be in certain cases, of employing a transmission gear in order to realize more favorable conditions.

Furthermore, the transmission gear from the engine shaft to the propeller shaft, may in some cases prove very convenient in making the propeller turn at a speed conducive to good efficiency. In the following chapter we shall see that the propeller efficiency depends on the ratio between the speed of the airplane and the peripheral speed of the propeller; since the peripheral speed depends on the number of revolutions, this factor consequently becomes of vast importance for the efficiency.

Let us see now which criterions are to be followed in installing an engine in an airplane, and let us discuss briefly, the principal accessory installations such as the gasoline and oil systems, and the water circulation for cooling.

As has been pointed out before, in the type of machine most generally used today, *the tractor biplane*—the engine is installed in the forward end of the fuselage—on properly designed supports, usually of wood, to which it is firmly bolted. The supports, in turn, are supported on transverse fuselage bridging and are anchored with steel wires which take up the propeller thrust (Fig. 53).

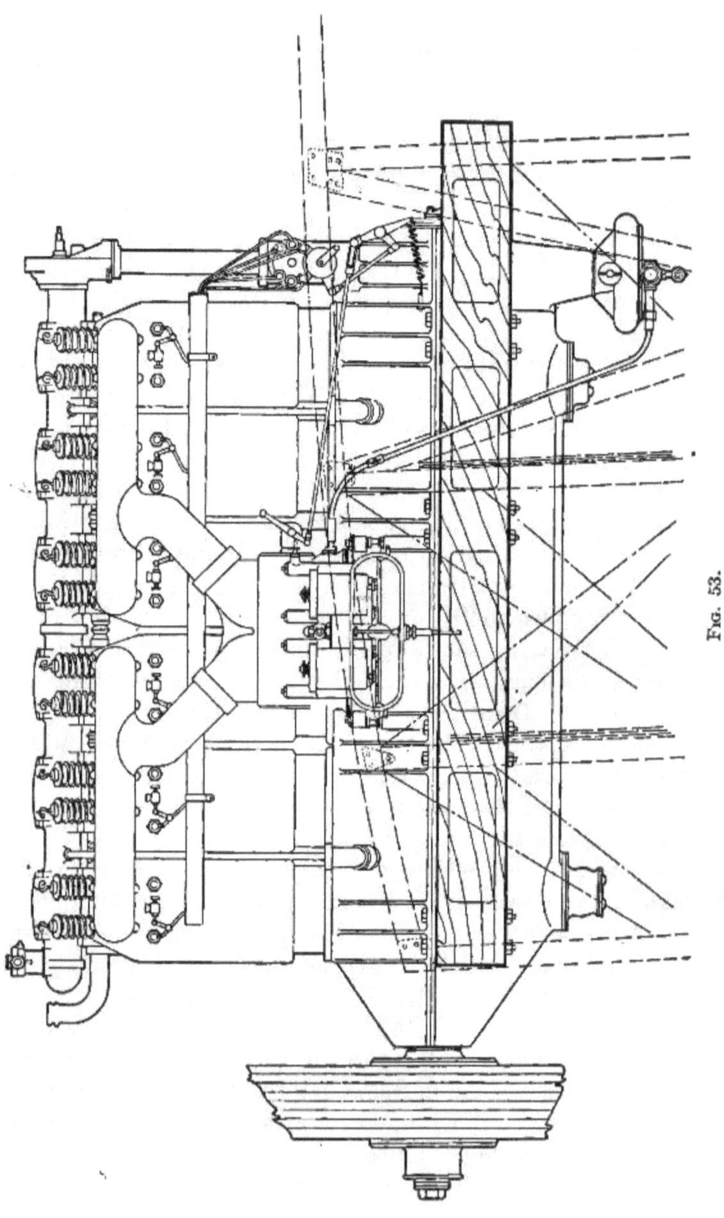

Fig. 53.

The oil tank is generally situated under the engine, so as to reduce to a minimum the piping system. There are two pipe lines—one leading from the bottom of the tank and which is used for the suction, the other, for the return and leading into the top of the tank (Fig. 54). The oil tank is usually made of copper or leaded steel sheets; it generally weighs from 10 per cent. to 12 per cent. as much as the oil it contains.

It is easy to place all the oil in one tank, as the oil consumption per horsepower is about $\frac{6}{100}$ of the gasoline

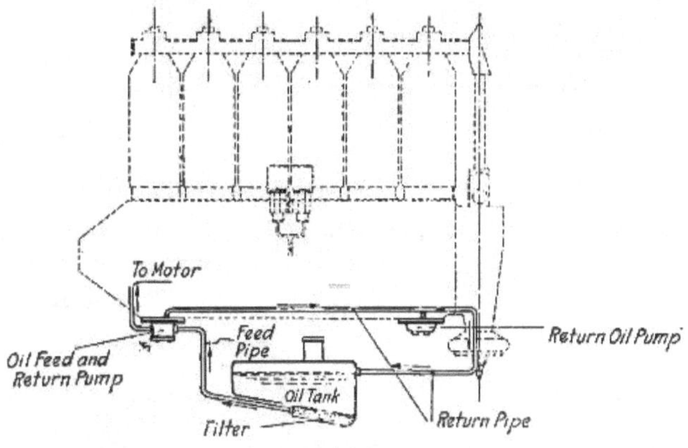

Fig. 54.

consumption, but it is a difficult matter to contain all the required gasoline in a single tank, especially for powerful engines. Therefore, multiple tanks are used. As the gasoline must be sent to the carburetor which is generally located above the tanks, it is necessary to resort to artifices to insure the feeding. The principal artifices are

a. Air pump pressure feed,

b. Gasoline pump feed.

The general scheme of the pressure feed is shown in Fig. 55. The motor *M*, carries a special pump which compresses the air in tank *T*; the gasoline flowing through cock *i*, goes to carburetor *C*. Cock *i* enables the opening or closing of the flow between tank *T* and the carburetor. Furthermore, it allows or stops a flow between the carburetor and a small auxiliary safety tank *t*, situated above the level of the carburetor, so that the gasoline may flow to the carbu-

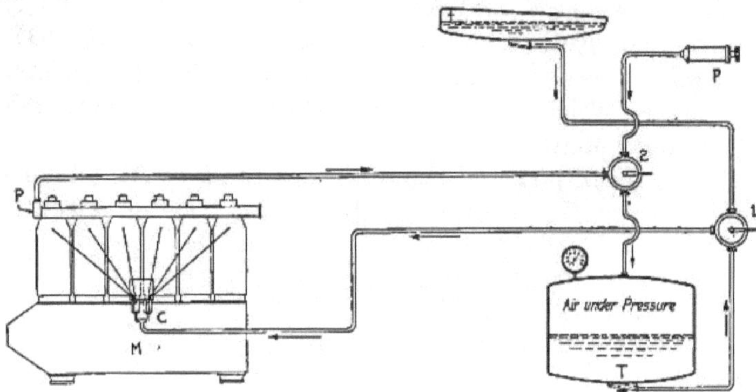

Fig. 55.—Gasoline pressure feed system.

retor by gravity; the gasoline in this tank is used in case the feed from the main tank should cease to operate. Fi-

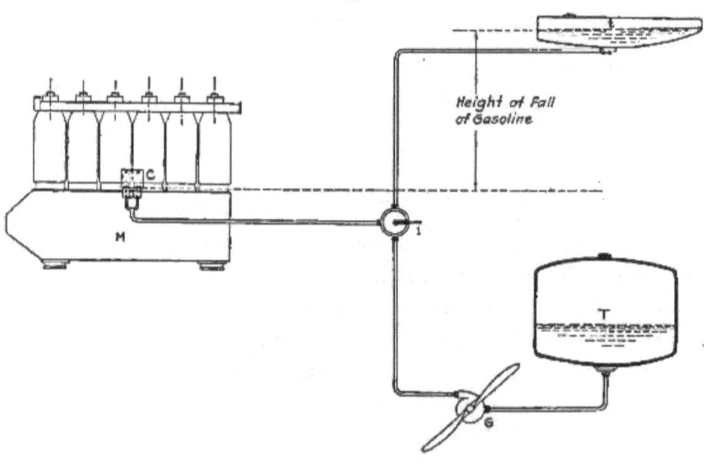

Fig. 56.

nally, cock *i* also enables a flow between the main tank *T* and the auxiliary tank *t*, in order that the latter may be replenished. The scheme of circulation is completed by a hand pump p, which serves to produce pressure in the tank before starting the engine; cock 2 establishes a flow between tank T and either or both of the pumps P and p, or excludes them both.

Fig. 56 shows the scheme of circulation by using the gasoline pump feed. The gasoline in the main tank T flows to a pump G, which sends it to the carburetor. Cock i permits or stops a flow between tank T and the carburetor, or between tank t and the carburetor, or between T and t. Pump G may be operated by a special small propeller or by the engine.

In the schemes of Figs. 55 and 56, an example of only one main tank is shown. If there are two or more tanks the conception of the schemes remains the same, the cocks only changing so as to allow simultaneous or single functioning of each of the tanks.

Gasoline pump feed is much more convenient than pressure feed because it is more reliable. It does not use compressed air, is less tiresome for the pilot, as it requires of him only the maneuver of opening or closing a cock, and finally, because the tanks can be much lighter as they do not have to withstand the air pressure.

As a matter of interest, a tank operating under pressure weighs from 14 per cent. to 18 per cent. as much as the gasoline it contains, while a tank operating without pressure weighs from 10 per cent. to 13 per cent.

We shall note finally, that it is necessary to install proper metallic filters or strainers in the gasoline feed system, in order to prevent impurities existing in the gasoline, from clogging up the carburetor jets.

The piping systems for gasoline and oil are made of copper. The joints are usually of rubber. As to the diameter of the piping system, it must be comparatively large for the oil, in order to avoid obstruction due to congealing. For the gasoline, the diameter must be such that the speed of gasoline flow does not exceed 1 ft. to 1.5 ft. per second; thus for instance, supposing an engine to consume 24 gallons an hour (that is, 0.00666 gallon a second) the inside diameter of the gasoline pipe must be from 5/16 in. to 3/8 in.

It is often necessary to resort to special radiators to cool the oil. On the contrary, in order to avoid freezing, in winter, it is necessary to insulate the tank with felt.

The water circulation exists only in water-cooled engines. Fig. 57 shows the principle of the water-cooling system. The engine is provided with a water pump *P*, which pumps the water into the cylinder jackets; after it has been

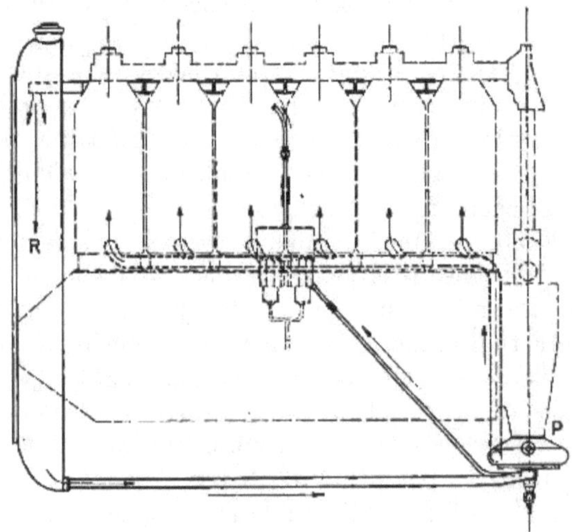

Fig. 57.—Water-cooling system.

warmed by contact with the cylinders, it flows to the radiator *R*, which lowers its temperature. Finally, from the radiator, the water flows back to the pump, and the circuit is completed.

The gasoline consumption of the engines varies from 0.45 to 0.55 lb. per H.P. per hour. Assuming an average of 0.5 lb. per H.P., and since the heat of the combustion of gasoline is about 18,600 B.t.u. per lb., then for 1 H.P. per hour, 9300 B.t.u. are necessary. Now, the thermal equivalent of 1 H.P. per hour is 2550 B.t.u., therefore only $\frac{2550}{9300} = 27.5$ per cent. of the heat of combustion of the gasoline is utilized in useful work; the rest, 72.5 per cent. or 6550 B.t.u. are to be eliminated through exhaust gases or through the cooling water. The B.t.u. taken up by the exhaust, compared with those taken up by the cooling water, vary not only for each engine, but even for each type of exhaust system. On the average, we can assume the water to absorb about 30 per cent. of the B.t.u., or about 2800 B.t.u. for every horsepower per hour; the quantity of B.t.u. to be absorbed by the cooling water of an engine of power P, is consequently equal to $(2800P)$ B.t.u.

This quantity of heat must naturally be given up to the air, and the radiator is used for that purpose.

From the standpoint of its application to the airplane, the radiator must possess two fundamental qualities, which are:

First, it must be as light as possible, and

Second, It must absorb the minimum power to move it through the air.

Since the weight also involves a loss of power, suppose that, as we have indicated, the flying characteristics depend on the weight per horsepower, we may then say that the lower the percentage of power absorbed the more efficient will be the radiator. It is possible to determine experimentally the coefficients which classify a given type of radiator according to its efficiency, with respect to its application to the airplane.

Before all, it must be remembered that a radiator is nothing more than a reservoir in which the water circulates in such a way as to expose a large wall surface to the air which passes conveniently through it. There are two main types of radiators: the water tube type, and the air tube or *honeycomb* type. In the first, the water passes through a great number of small tubes, disposed parallel to, and at some distance from each other; the air passes through the gaps between the tubes. In the air tube radiators (also called *honeycomb* radiators because of their resemblance to the cells of a beehive), the water circulates through the interstices between the tubes, while the air flows through the tubes. For the present great flying speeds, the latter type of radiator has proven much more suitable, and therefore is more generally used.

To compare two types of *honeycomb* radiators, we will take into consideration a cubic foot of radiator, and study its weight, water capacity, cooling surface, head resistance, and cooling coefficient. The first three are geometrical elements which can be defined without uncertainties.

The head resistance depends not only on the speed of the airplane, but also on its position in the machine, and frontal area.

Finally, the cooling coefficient beside depending on the type of radiator, depends on the velocity of water flow and air flow, and the initial temperatures of the air and water. As one can see, there are many factors which would be difficult to condense into one single formula. We must therefore content ourselves with studying separately, the influence of each of the above factors.

In the following table are given the values of the weight W_R, water capacity W_W, and radiating surface Σ per cubic foot, of radiator for certain types of radiators; also let us call α the ratio between the weight of 1 cu. ft. of radiator including the water, and its radiating surface.

TABLE 2

Type of radiator	Weight W_R lb. per cu. ft.	Water capacity W_W lb. per cu. ft.	Total weight W lb. per cu. ft.	Radiating surface Σ sq. ft. per cu. ft.	Total weight of radiating surface lbs. per sq. ft. α
Circular tubes with hexagonal sides...	34.8	20.5	55.3	97.8	0.4660
Square tubes.......	38.9	9.3	48.2	188.5	0.2653
Square tubes........	42.8	8.8	51.6	161.5	0.3095
Hexagonal tubes....	29.7	12.9	42.6	132	0.3227

The power absorbed by the head resistance of 1 cu. ft. of the radiator, may assume the following expression:

$$\beta \times S \times V^3,$$

where S is the frontal area of the radiator, and V is the speed of the machine in feet per second.

Let us call d the depth of the radiator core; $S \times d = 1$ or $S = \frac{1}{d}$; thus the preceding expression becomes

$$\beta \times \frac{1}{d} \times V^3 \qquad (1)$$

The coefficient β varies not only with the different types of radiators, but with the same radiator, depending on whether it is placed in the front of the fuselage, or whether it is completely surrounded by free air.

Equation (1) shows that to decrease the head resistance it is convenient to augment the depth of the radiator d. This increase, however, is limited by the fact that it is advisable to keep at a maximum the difference in the water and air temperatures; then if the depth of the radiator tubes is greatly increased, the air is excessively heated, thus decreasing the difference in temperature between it and the water. For this reason the depth d may become greater as the air flow v through the tubes is increased in velocity. The following is a practical formula that may be used in determining d:

$$d = 8 \times l \times \sqrt{v} \qquad (2)$$

where l is the diameter of the tubes in feet, and v the velocity of the air flow through the tubes in feet per second.

The quantity of heat radiated by 1 cu. ft. of radiator, not only depends on the type, but on the difference between the temperature t_w of the water, and t_a of the air, on the velocity of water flow, on the velocity v of air flow through the tubes, and on the radiating surface Σ per cubic foot of the given radiator.

Assuming the velocity of water flow to be constant, the quantity of B.t.u. may be expressed by

$$\gamma \times (t_w - t_a) \times v\Sigma \qquad (3)$$

where γ is the cooling coefficient, varying with the type of radiator.

Now, if the engine has power P, the radiator must take

care of 2800P calories. Therefore the volume C of the radiator must be such that

$$C \times \gamma \times (t_w - t_a) \times v \times \Sigma = 2800P$$

or,

$$C = \frac{2800P}{\gamma \times (t_w - t_a) \times v \times \Sigma} \qquad (4)$$

The weight of the radiator will be $C \times W$, and the power absorbed by its head resistance will be

$$C \times \beta \times \frac{1}{d} \times V^3 = \frac{C \times \beta \times V^3}{8 \times l \times \sqrt{v}}$$

If we call $\frac{L}{D}$ the ratio $\frac{\text{Lift}}{\text{Drag}}$, the power required to carry $C \times W$ lb. will be (in ft. lbs.),

$$C \times W \times \frac{L}{D} \times V$$

Therefore the total power absorbed by the cooling system will be

$$P_R = \frac{C \times \beta \times V^3}{8 \times l \times \sqrt{v}} + C \times W \times \frac{L}{D} \times V$$

and by equation (4)

$$P_R = P \times \frac{2800}{\gamma \times (t_w - t_a) \times v \times \Sigma} \times \left[\frac{\beta V^3}{8l\sqrt{v}} + W \times \frac{L}{D} \times V\right] \qquad (5)$$

We can further simplify the preceding expression. First of all we will note that v (the velocity of air flow inside of the tubes), is proportional to the speed of the airplane; we can then write

$$v = \delta \times V$$

The temperature t_w is usually taken at 176°F. (80°C.); it is not convenient to increase it, as the airplane must be able to fly at considerable altitude, where due to the atmospheric depression, the boiling point of water is lowered. For the air temperature t_a, we must take the maximum annual value of the region in which the machine is to fly; in cold seasons, the cooling capacity of the radiator becomes excessive, and therefore, special devices are resorted to, for cutting off part, or all of the radiator.

In very warm climates, we may take for example $t_a = 104°$, then the result is

$$t_w - t_a = 176° - 104° = 72°\text{F}.$$

As to the dimension l (the diameter of the tube through which the air passes), experiments have shown that to diminish W, and increase Σ, l must be kept around 0.396 in. = 0.033 ft. Finally, the ratio $\frac{L}{D}$ for a good wing, varies around 15. Then letting p = ratio $\frac{P_o}{P}$, where P_o is the power absorbed by the radiator, and P the total power, equation (5), remembering that $\frac{W}{\Sigma} = \alpha$, by the proper reductions, becomes

$$p = \frac{149\ \beta}{\gamma \times \Sigma\delta^{3/2}} \times V^{3/2} + \frac{583 \times \alpha}{\gamma \times \delta} \tag{6}$$

where the coefficients have the following significance:

$p = \frac{P_R}{P}$ = percentage of power absorbed by the radiator,

$\alpha = \frac{W}{\Sigma}$ = weight of radiator per square foot of radiating surface,

β = coefficient of head resistance,

γ = cooling coefficient of the radiator,

$\delta = \frac{v}{V}$ = coefficient of velocity reduction inside the tubes, with respect to the speed of the airplane, and

Σ = radiating surface per cubic foot of radiator.

Similarly, if we call $c = \frac{C}{P}$ the volume of radiator required per horsepower, and simplifying as before, equation (4) gives

$$c = \frac{38.9}{\gamma \times \Sigma \cdot \delta} \times \frac{1}{V} \tag{7}$$

The two equations (6) and (7), allow one to solve the problem of determining the volume of the radiator and the power absorbed. For a given type of radiator, α, β, δ, and Σ are constants, then one can write

$$\frac{149\beta}{\gamma \times \Sigma \times \delta^{3/2}} = B; \quad \frac{583 \times \alpha}{\gamma \times \delta} = C; \quad \frac{38.9}{\gamma \times \delta \times \Sigma} = A$$

and therefore equations (6) and (7) become, respectively,

$$\begin{cases} c = A \dfrac{1}{V} \\ p = B \times V^{3/2} + C \end{cases} \tag{8}$$

Naturally, such relations can be used within the present limits of airplane speeds (80 m.p.h. to 160 m.p.h.). They state that the volume of the radiator is inversely proportional to the speed, and the power required is proportional to the 3/2 power of the speed.

Before leaving the discussion on radiators, we will briefly discuss the systems of reducing the cooling capacity. There are two general methods; to decrease the speed of water circulation, or to decrease the speed of air circulation. The second is preferable, and is today more generally adopted. It is effected by providing the front face of the radiator with shutters which can be more or less closed until the air passage is completely obstructed.

Mufflers have not as yet been extensively adopted for aviation engines, principally because they entail a direct loss of power amounting to from 6 per cent. to 10 per cent.; and because of their bulk and weight. Ordinary exhaust tubes are used, exhausting singly for each cylinder, or joined together, the point being, to convey the gases away from those parts of the machine that might be damaged by them.

Before concluding this chapter, it is desirable to note the functioning of the engine at high altitudes. Modern airplanes have attained heights up to 25,000 ft.; battleplanes carry out their mission at heights varying from 10,000 to 20,000 ft., therefore it is necessary to study the actions of the engine at such altitudes.

Since the density of the air decreases as one rises above the ground, according to a logarithmic law; let H be the height in feet, at some point in the atmosphere above sea level, and μ the ratio between the density at height H, and that at ground level; then

$$H = 60{,}720 \log \frac{1}{\mu}$$

Fig. 58 shows the diagram for μ as a function of H, constructed on the basis of the preceding formula.

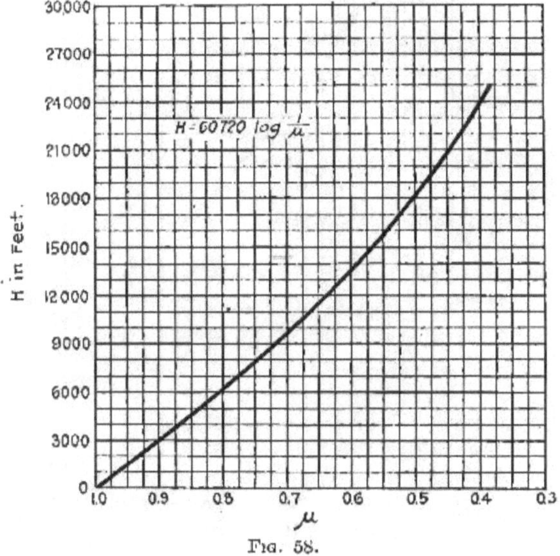

Fig. 58.

In practice, however, it happens that the temperature of the air also decreases as one rises above the ground. Then at a given height H, the density μ with respect to the ground level, is greater than the value given by the above formula. In the following discussion, which is primarily qualitative in nature, we will not take into account this decrease in temperature, in order not to complicate the treatment of the subject.

With this foreword, let us remember that the moving

power P, is equal to the product of the angular velocity ω by the engine torque M.

$$P = \omega \times M$$

At height H, the engine torque M is proportional to the

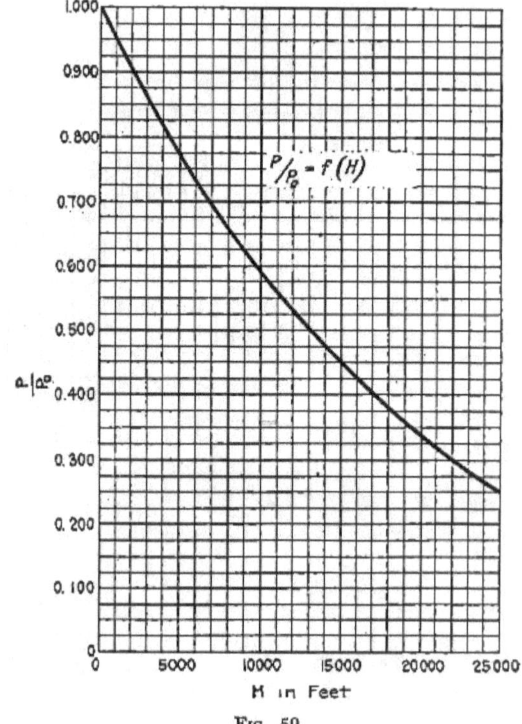

FIG. 59.

mass of oxygen burned in one unit of time, or to the density of the air. Therefore

$$P = \mu M_o \omega = \mu \times P_o \times \frac{\omega}{\omega_0} \qquad (1)$$

where

$$P_o = \omega_o M_o = \text{power at sea level.}$$

It is obvious then, that as the machine climbs, the power of the engine decreases.

In Fig. 59, a diagram is given for the reduction in percentage $\frac{P}{P_o}$ of the power, corresponding to the increase of H.

In one of the following chapters will be shown the influence that the decrease in the air density exerts on the power required for the sustentation of the machine. It will be readily perceived, that if a machine is to climb 25,000 ft., it must be able to maintain itself in the air with 0.251 of the power of the engine; in other words, it must carry an engine which will develop $\frac{1}{0.251} = \approx 4$ times the minimum power strictly necessary for its sustentation. In practice, these are the actual means chosen by designers to attain high altitudes. That is, the machines are equipped with engines of such excess power, as to be sufficient to maintain flight even after the strong reduction of power mentioned above.

Such a method is evidently irrational, since at ground level the airplane employs a useless excess of power, while at high altitudes it is overloaded with a weight of engine entirely out of proportion to the power actually developed.

To eliminate this loss of efficiency, two solutions present themselves. One provisional solution (but of inestimable value in augmenting the efficiency of engines as they are actually conceived and constructed) consists of providing the engine with an air compressor which will feed the carburetor. In this way, the mass of gas mixture taken in by the engine at each admission stroke, is greater than the amount which would be sucked in from the atmosphere directly, and as a result, the engine torque is increased.

Two types of compressors have thus far been experimented with; the turbo compressor designed by Rateau (France), actioned by means of the exhaust gases, and the centrifugal multiple compressor designed by Prof. Anastasi (Italy), actioned by the engine shaft.

The latter type, for example, with an increase in weight of less than 10 per cent., allows a complete recuperation of the power at 13,000 ft., or it recuperates 50 per cent. of the power. Since it absorbs 10 per cent. of the power in operation, the actual power recuperated is 40 per cent.

These compressors have not yet been adopted for practical use, because of reasons inherent to the operation of the propeller, which will be seen in the following chapter.

The second solution (the one toward which engine technique must inevitably direct itself in order to open a way for further progress), consists in predisposing the engines so that the compression of air at high altitudes may be effected without the aid of external compressors.

CHAPTER VI

THE PROPELLER

The propeller is the aerial propulsor universally adopted in aviation.

Its scope is to produce and maintain a force of traction capable of overcoming the various head resistances of the wings and other parts of the airplane.

Calling T the propeller traction in pounds, and V the velocity of the airplane in feet per second, the product $T \times V$ measures the useful work in foot pounds per second accomplished by the propeller. If P_o is the power of the engine in horsepower, the propeller efficiency is expressed by

$$\rho = \frac{TV}{550 \times P_o} \qquad (1)$$

Every effort must of course be used in making the propeller efficiency as high as possible. In fact, equation (1) may also be written as

$$P = \frac{TV}{550 \times \rho}$$

which means that having assumed a given speed and a given head resistance, the power required for flight will be so much greater as the value of ρ is smaller. Suppose for example that $T = 500$ lb. and $V = 200$ ft. per sec., then

$$\text{for } \rho_1 = 0.70 \qquad P_1 = 260 \text{ H.P.}$$
$$\text{for } \rho_2 = 0.80 \qquad P_2 = 227 \text{ H.P.}$$

and P_2 is 13 per cent. less than P_1.

A propeller is defined by a few geometric elements, and by its operating characteristics.

The geometric elements of a propeller are the number of blades, the diameter, the pitch, the maximum width of the blades and their profile.

Propellers are built with 2, 3, and 4 blades. The type most commonly used is the 2-blade propeller, especially when quick-firing guns with synchronized devices for firing through the propeller, are mounted on the airplane. On machines that have their propellers in front, the problem of firing directly forward is solved by equipping the machine guns with special automatic devices operated by the engine

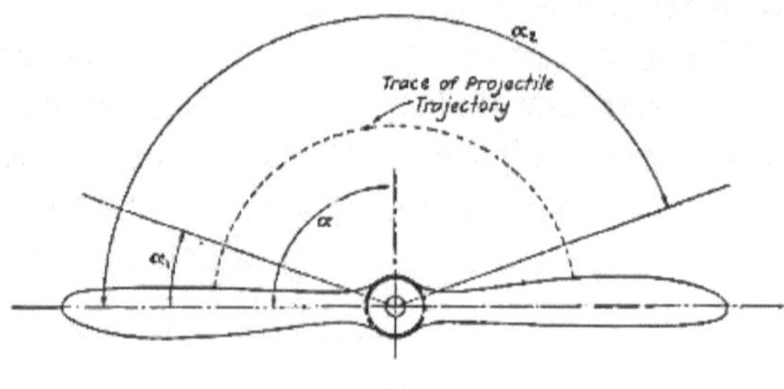

FIG. 60.

(devices called *synchronizers*), which release the projectiles at the instant the propeller blades have passed in front of the machine gun muzzle; in other words, the projectile is fired through the plane of rotation of the propeller when the blade has rotated by an angle α (Fig. 60). Angle α is not fixed, but varies with the number of revolutions of the propeller, which is easily understood if one considers that

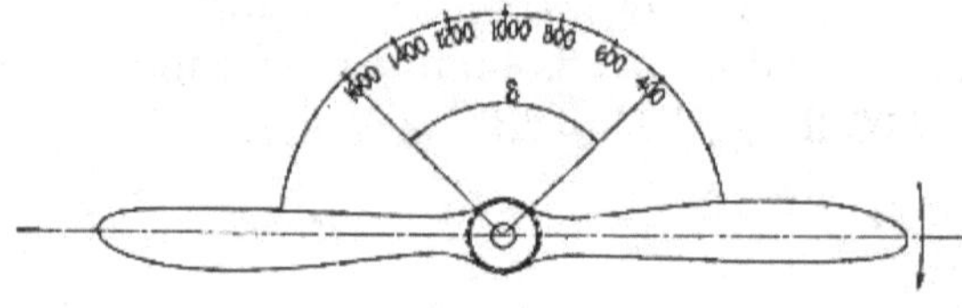
FIG. 61.

the velocity of the projectile remains constant, while the angular velocity of the propeller varies. Thus, as the number of revolutions change, there is a dispersion of projectiles; these fall in a sector δ, which is called the angle of dispersion of the synchronizer (Fig. 61). Now, if this angle is greater than 90°, as it often happens, it is impossible to use 4-bladed propellers, altho in certain cases, 4-bladed propellers may be convenient for reasons of efficiency, as will be observed further on.

The diameter of the propeller depends exclusively upon the power the propeller has to absorb, and upon its number of revolutions.

The *pitch* of the propeller, from an aerodynamical point of view, should be defined as "*the distance by which the propeller must advance for every revolution in order that the traction be zero.*" In practice, however, the pitch is measured by the tangent of the angle of inclination of the propeller blade with respect to its plane of rotation; if θ is the angle for a cross section AB of the propeller, at a distance r from

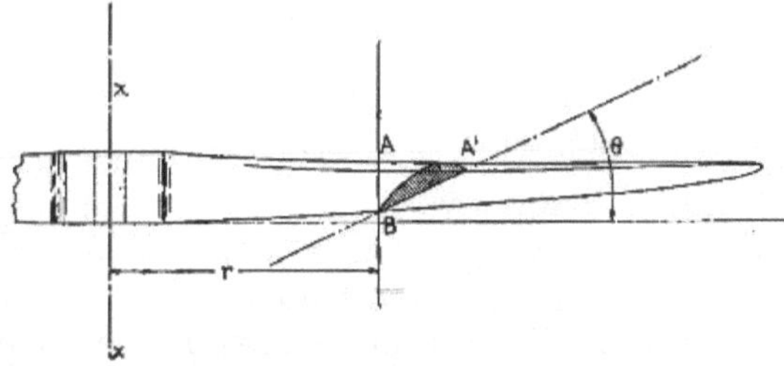

Fig. 62.

the axis XX (Fig. 62), the pitch of the propeller at that section will be

$$p = 2\pi r \text{ tang } \theta$$

Practically, propellers are made with either a constant pitch for all sections, or a more or less variable one. Figs. 63 and 64 illustrate respectively, two examples of propellers, one with constant pitch, the other with variable pitch.

The width of the blade is not important as to its absolute value, but is important with respect to the diameter. Since the propeller blade may be considered as a small wing moving along an helicoidal path, it is evident that to increase the efficiency, it is convenient to reduce the width of the blades to a minimum with respect to the diameter. However, it is not possible to reduce the blade width below a certain limit, for reasons of construction and resistance of the propeller. Practically, it oscillates from 8 to 10 per cent. of the diameter.

The profile of a propeller, although varying from section to section, characterizes the type of the propeller. It bears a great influence on the characteristics of a propeller.

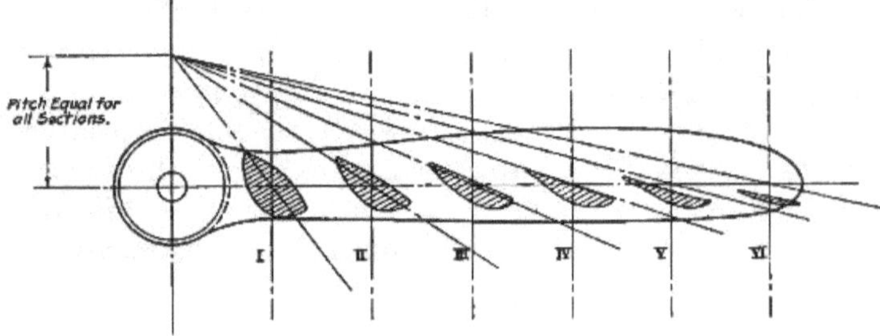

Fig. 63.

All propellers having the same type of profile, are said to belong to the same family.

Numerous laboratory experiments on propellers, by Colonel Dorand, have demonstrated that there exist certain well-determined relations between the elements of

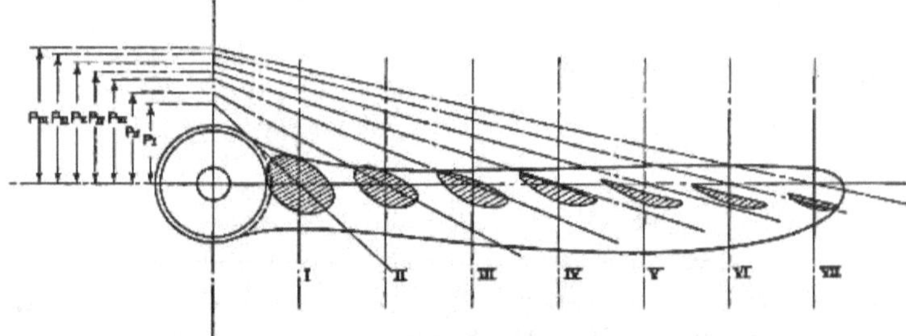

Fig. 64.

propellers that are of the same family and geometrically similar, so that once the coefficients of these relations are known, it is easily possible to obtain all the data for the design of the propeller. Let

D = the diameter of the propeller in feet,

p = the pitch of the propeller in feet,

P_o = the power absorbed by the propeller on the ground,

N = number of revolutions per second,
V = the speed of the machine in feet per second, and
ρ = the efficiency of the propeller,

than the relations binding the preceding parameters are

$$P_o = \alpha n^3 D^5 \quad (1)$$

$$\alpha = f_1\left(\frac{V}{\pi n D}\right) \quad (2)$$

$$\rho = f_2\left(\frac{V}{\pi n D}\right) \quad (3)$$

Equation (2) states that the coefficient α of equation (1) is not a constant, but depends on the ratio $\frac{V}{\pi n D}$. Let us examine the graphical interpretation of this ratio.

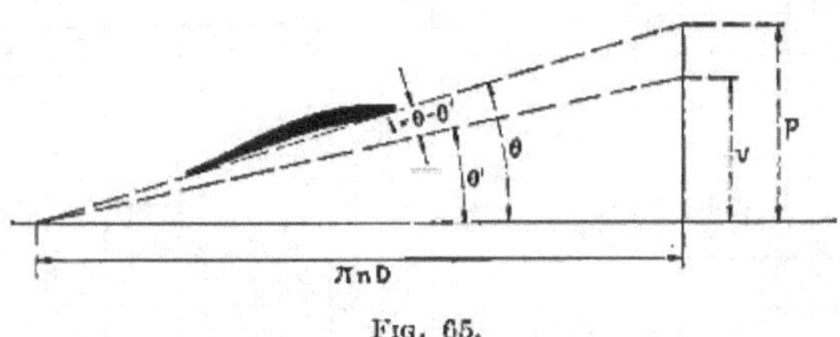

FIG. 65.

Since $\pi n D$ is the peripheral speed of the blade tip, $\frac{V}{\pi n D}$ measures the angle θ that the path of the blade tip makes with the plane of rotation of the propeller (Fig. 65). Now, the angle of incidence i of the blade with respect to its path, is measured exactly by the difference $\theta - \theta'$; as θ is fixed, i varies with the variation of θ'; this explains why as tangent $\theta' = \frac{V}{\pi n D}$ varies, the power absorbed by the propeller varies, and consequently coefficient α varies. This also explains equation (3), which shows that the propeller efficiency is dependent upon $\frac{V}{\pi n D}$; in fact, as in the case of a wing, the efficiency of a propeller blade varies with the variation of the angle of incidence i.

Returning to equation (1), and assuming a given value for α, for instance, $\alpha = 3 \times 10^{-8}$, then that equation becomes

$$P_o = 3 \times 10^{-8} n^3 D^5$$

and states

1. For a propeller of a given diameter, the power required to rotate it, increases as the 3d power of n. In Fig. 66 a curve is drawn illustrating that law, assuming $D = 10$ ft.; the curve is a cubic parabola.

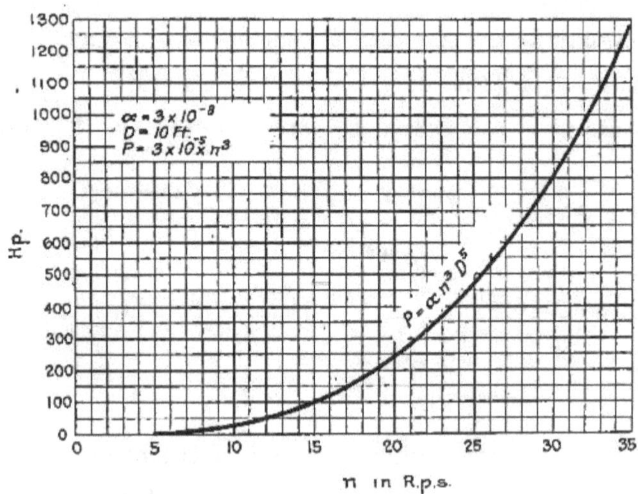

Fig. 66.

2. For a given number of revolutions, the power required to rotate a propeller, increases as the 5th power of the diameter.

In Fig. 67 the curve is drawn illustrating that law for $n = 25$ r.p.s. = 1500 R.P.M. It is a parabola of the 5th degree.

3. Assuming the power, the diameter to be given to the propeller is inversely proportional to the $\frac{3}{5}$ power of the number of revolutions. The curve for that law is drawn in Fig. 68. It is an hyperbola.

α=3×10⁻⁸
n= 25
P= 3×10⁻⁸×25⁻³×D⁵
P=αn³D⁵
Hp.
Diameter, Ft.
0 1 2 3 4 5 6 7 8 9 10
0 100 200 300 400 500

Fig. 67.

α=3×10⁻⁸
P= 300 Hp.
300=3×10⁻⁸×n³×D⁵
D, Ft.
n R.p.s.
0 5 10 15 20 25 30 36
0 10 20 30 40

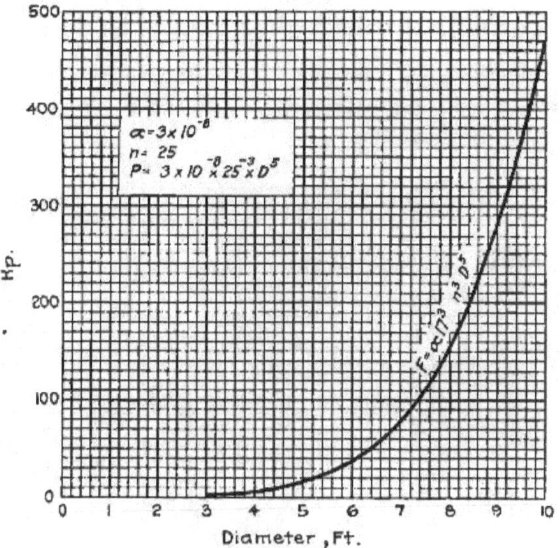

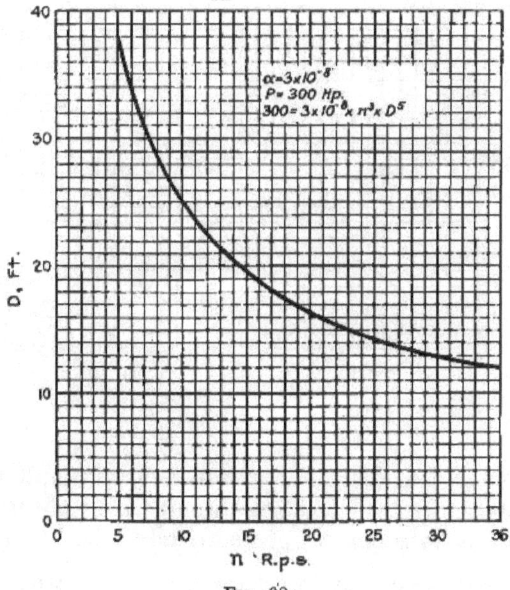

Fig. 68.

Equation (2), which gives α as a function of $\frac{V}{\pi n D}$, is of interest only inasmuch as it is necessary to know the value of α for equation (1). Therefore, we shall not pause in examination of it.

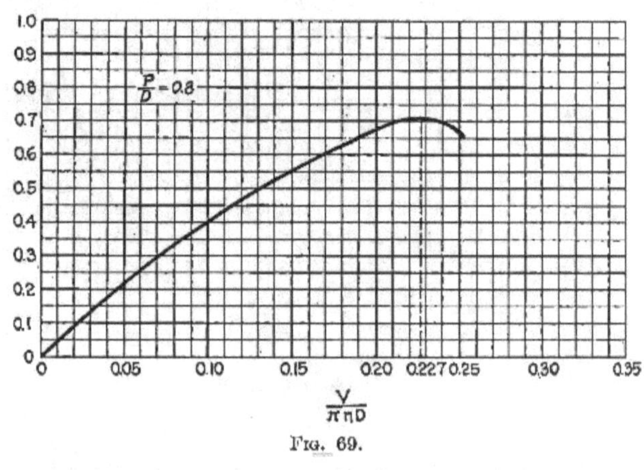

Fig. 69.

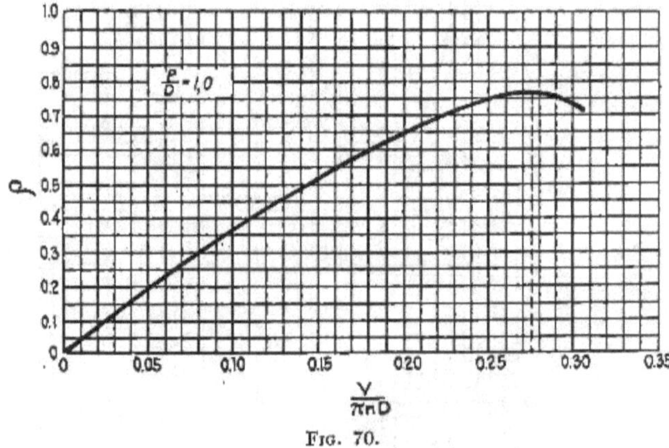

Fig. 70.

On the contrary, it is of maximum interest to examine equation (3), which gives the efficiency of the propeller.

Let us consider all geometrically similar propellers of the same family, having diameter D and pitch p, so that $\frac{p}{D}$

= 0.8; Fig. 69 gives the diagram $\rho = f_2\left(\frac{V}{\pi nD}\right)$ for such propellers. The diagram shows that ρ increases and reaches a maximum value $\rho_{max} = 0.71$ corresponding to the value $\frac{V}{\pi nD} = 0.227$.

Let us now consider a group of propellers also of similar profile, but having $\frac{p}{D} = 1.0$, and let us draw the efficiency diagram (Fig. 70). This will be similar to the preceding one in shape, but will reach a value $\rho_{max} = 0.77$ corresponding to a value of $\frac{V}{\pi nD} = 0.275$.

If this experience is repeated for various values of $\frac{p}{D}$, it will be observed that the maximum efficiency obtainable from a propeller of certain profile, varies with the variation of that ratio; it is easy to construct a diagram giving all the values of φ_{max} as functions of $\frac{p}{D}$. Such a diagram shows that a propeller of a certain type, gives its maximum efficiency when $\frac{p}{D} = 1.20$. Naturally this condition does not suffice, as the propeller must rotate at a number of revolutions n, such that the ratio $\frac{V}{\pi nD}$ will be the one at which the propeller actually attains the maximum efficiency.

Fig. 71 gives the values of $\frac{p}{D}$, α, and ρ, as functions of $\frac{V}{\pi nD}$, for the best propellers actually existing.

The use of these diagrams requires a knowledge of all the aerodynamical characteristics of the machine for which the propeller is intended. However, even a partial study of them is very interesting for the results that can be attained.

First, we see that for $\frac{p}{D} = 1.18$ and $\frac{V}{\pi nD} = 0.32$, the maximum efficiency ρ reaches a value of 82 per cent. Obviously that is very high, especially when the great simplicity of the aerial propeller is considered. But unfortunately, it often occurs in practice, that that value of efficiency cannot be attained because there are certain

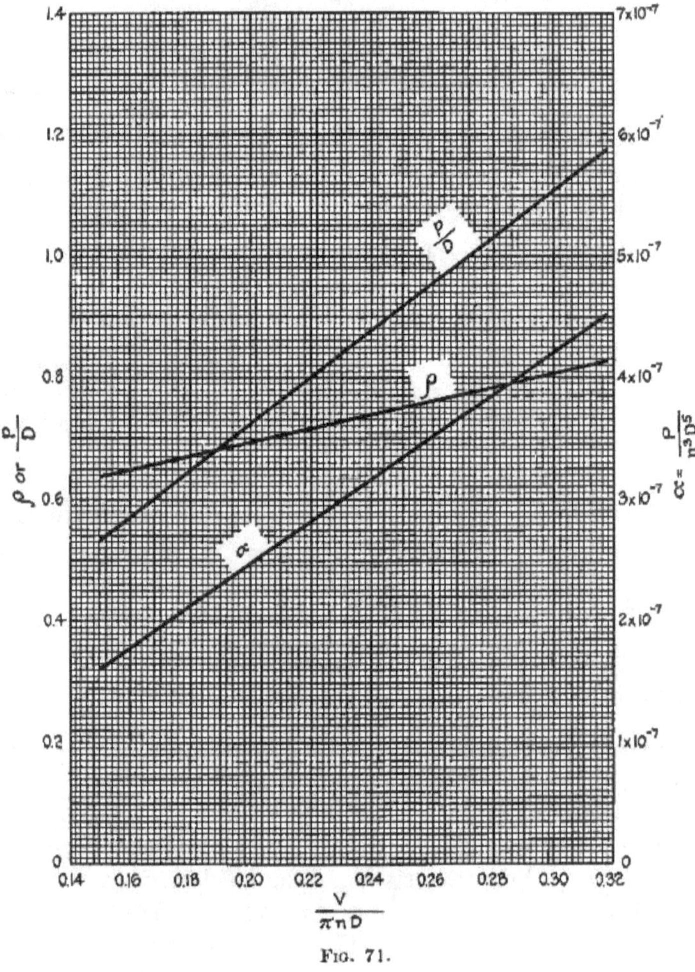

Fig. 71.

parameters which it is impossible to vary. An example will illustrate this point.

Let us assume that we have at our disposition an engine

developing 300 H.P., while its shaft makes 25 r.p.s., and let us assume that we wish to adopt such an engine on two different machines, one to carry heavy loads and consequently slow, the other intended for high speeds. Let the speed of the first machine be 125 ft. per sec., and that of the second 200 ft. per sec. We shall then determine the most suitable propeller for each machine.

For the first machine, as $n = 25$, and $V = 125$, the expression $\frac{V}{\pi n D}$ becomes equal to $\frac{1.60}{D}$. We must choose a value of D, such that together with the value of α corresponding to $\frac{1.60}{D}$, (Fig. 71), it will satisfy the equation

$$300 = \alpha n^3 D^5$$

or, for $n = 25$

$$\alpha \times D^5 = 0.0192$$

Now the corresponding values of α and D satisfying those equations are

$\alpha = \sim 1.4 \times 10-7$ and $D = 10.6$; in fact, for this value of D, $\frac{D}{\pi n D} = \frac{125}{3.14 \times 25 \times 10.6} = \sim 0.15$, to which corresponds

$\alpha = 1.4 \times 10^{-7}$; the corresponding value of ρ is ~ 0.62, that is, our propeller will have an efficiency of 62 per cent.; its pitch will be $0.48 \times 10.6 = 5.1$ ft.

For the second machine instead—$n = 25$, and $V = 200$—the expression $\frac{V}{\pi n D}$ becomes $\frac{200}{3.14 \times 25 \times D} = \frac{2.55}{D}$ and $\alpha \times D^5 = 0.0192$; the two values satisfying the desired conditions are

$$\frac{V}{\pi n D} = \frac{200}{3.14 \times 25 \times 8.6} = 0.296; \ \alpha = 4.1 \times 10^{-7},$$

and corresponding to these values $\rho = 0.79$. The pitch results equal to 9.3 ft.

We can see then, that the propeller for the second machine, has an efficiency of 79 per cent.; that is $\frac{79}{62} = \sim 1.27$ more than that of the first machine. It would be possible to improve the propeller efficiency of the first machine by using a reduction gear to decrease the number of revolutions of the propeller. In this case, it would even be possible by properly selecting a reduction gear, to attain the maximum efficiency of 82 per cent.

But this would require the construction of a propeller of such diameter, that it could not be installed on the machine. Consequently we shall suppose a fixed maximum diameter of 14 ft. Then it is necessary to find a value of n, such that value α corresponding to $\frac{V}{\pi n D}$ gives

$\alpha \times n^3 \times D^5 = 300$. That value is $n = 12.4$ r.p.s., for which $\frac{V}{\pi n D} = 0.23$ and $\rho = 0.72$. We see then that in this case, the reduction gear has gained $\frac{0.72}{0.62} = 1.16$ or 16 per cent. of the power, which may mean 16 per cent. of the total load; and if we bear in mind that the useful load is generally about ⅓ of the total weight, we see that a gain of 16 per cent. on the total load, represents a gain of about 50 per cent. on the useful load; this abundantly covers the additional load due to the reduction gear.

From the preceding, we see that in order to obtain good efficiency, modern engines whose number of revolutions are very high, must be provided with a reduction gear when they are applied to slow machines. On the contrary, for very fast machines, the propeller may be directly connected, even if the number of revolutions of the shaft is very high.

Concluding we can say, that it is not sufficient for a propeller to be well designed in order to give good efficiency, but it is necessary that it be used under those conditions of speed V and number of revolutions n, for which it will give good efficiency.

Until now we have studied the functioning of the propeller in the atmospheric conditions at sea level. Let us see what happens when it operates at high altitudes. The equation of the power then becomes

$$P = \mu \times \alpha \times n^3 \times D^5$$

where μ is the ratio between the density at the height under consideration and that on the ground (see Chapter 5). This means that the power required to rotate the propeller decreases as the propeller rises through the air, in direct proportion to the ratio of the densities.

As to the number of revolutions, the preceding equation gives

$$n^3 = \frac{P}{\mu\alpha\ D^5}$$

Theoretically, the power of the engine varies proportionally to μ, that is

$$P = \mu P_o$$

so that theoretically we should have

$$n^3 = \frac{\mu P_o}{\mu\alpha D^5} = \frac{P_o}{\alpha \times D^5}$$

and this would mean that the number of revolutions of the propeller would be the same at any height as on the ground. Practically, however, the motive power decreases a little more rapidly than proportionally to μ (see Chapter 5), and consequently the number of revolutions slowly decreases as the propeller rises in the air.

If instead, by using a compressor or other device, the power of the engine were kept constant and equal to P_o, then the number of revolutions would increase inversely as $\sqrt[3]{\mu}$. So for instance, at 14,500 ft., where $\mu = 0.5$ the revolutions should be $\frac{n}{\sqrt[3]{0.5}} = \frac{n}{0.79} = 1.26\ n$. A propeller making 1500 revolutions on the ground, would make 1900 revolutions at a height. This, then, is one of the principal difficulties that have until now opposed the introduction of compressors for practical use. In fact, as it is unsafe that an engine designed for 1500 revolutions make 1900, it would practically be necessary for the propeller to brake the engine on the ground, so as not to allow a number of revolutions greater than $1500 \times 0.79 = 1180$. In this way, however, the engine on the ground could not develop all its power, and therefore the characteristics of the machine would be considerably decreased.

To eliminate such an inconvenience, there should be the solution of adopting propellers whose pitch could be variable in flight, at the will of the pilot; thus the pilot would be enabled to vary the coefficient of the formula

$$P = \alpha \times n^3 \times D^5$$

and consequently could contain the value of n within proper limits. Today, the problem of the variable propeller has not yet been satisfactorily solved; but tentatives are being made which point to positive results.

The materials used in the construction of propellers, the stresses to which they are subjected, and the mode of designing them, will be dealt with in Part IV of this book.

PART II

CHAPTER VII

ELEMENTS OF AERODYNAMICS

Aerodynamics studies the laws governing the reactions of the air on bodies moving through it.

Very little of these laws can be established on a basis of theoretical considerations. This can only give indications in general; the research for coefficients, which are definitely those of interest in the study of the airplane, cannot be completed except in the experimental field.

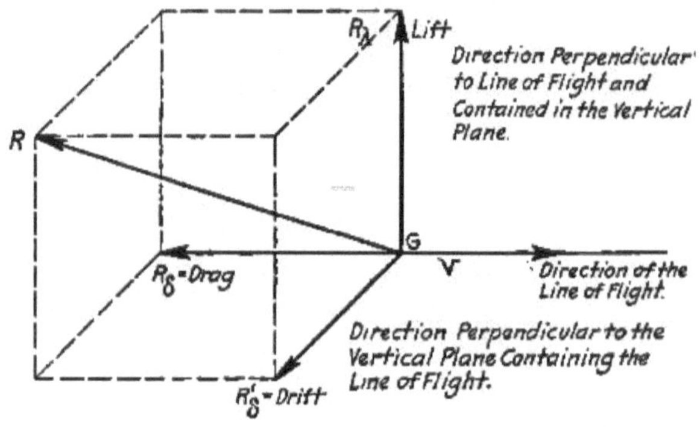

FIG. 72.

For these reasons, we shall consider aerodynamics as an "Applied Mechanics" and we shall rapidly study the experimental elements in so far as they have a direct application to the airplane.

Let us consider any body moving through the air at a speed V, and let us represent the body by its center of gravity G (Fig. 72). Due to the disturbance in the air, positive and negative pressure zones will be produced on the various surfaces of the body, and in general, the resultant

R of these pressures, may have any direction whatever. Let us resolve that resultant into three directions perpendicular to one another, the first in the sense of the line of flight, the second perpendicular to the line of flight and lying in the vertical plane passing through the center of gravity, and the third perpendicular to that plane.

These components R_λ, R_δ, and R'_δ, shall be called respectively:

R_λ, the Lift component,

R_δ, the Drag component,

R'_δ, the Drift component,

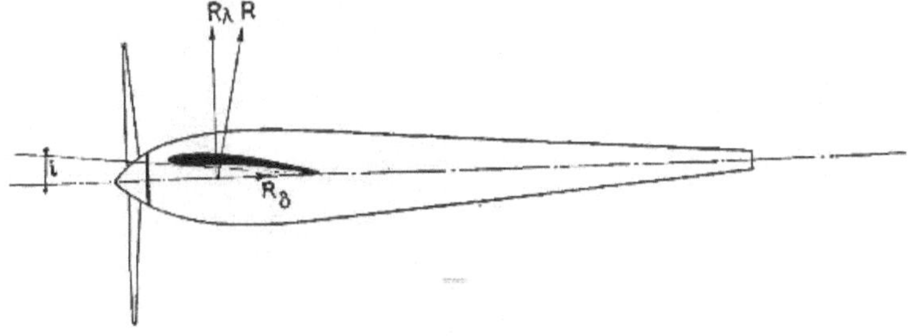

Fig. 73.

If we wished to make a complete study of the motion of the body in the air it would be necessary to know the values, of R_λ, R_δ, and R'_δ, for all the infinite number of orientations that the body could assume with respect to its line of path; practically, the most laborious research work of this kind would be of scant interest in the study of the motion of the airplane.

Let us first note that the airplane admits a plane of symmetry, and that its line of path is, in general, contained in that plane of symmetry; in such a case, the component $R'_\delta = 0$. This is why the study of components R_λ and R_δ is made by assuming the line of path contained in the plane of symmetry, and referring the values to the angle i that the line of path makes with any straight line contained in the plane of symmetry and fixed with the machine.

In general, this reference is made to the wing chord (Fig.

73), and i is called the *angle of incidence;* as to the force of drift, usually the study of its law of variation is made by keeping constant the angle i between the chord and the projection of the line of path on the plane of symmetry, and varying only the angle δ between the line of path and the plane of symmetry (Fig. 74); the angle δ is called the *angle of drift.*

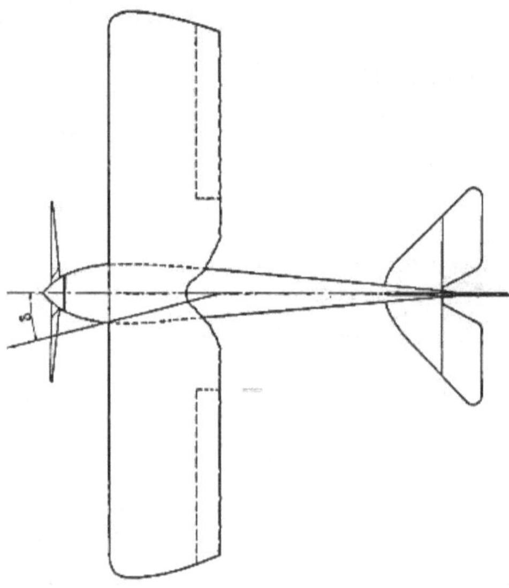

FIG. 74.

Summarizing, the study of components R_λ, R_δ, and R'_δ, is usually made in the following manner:

1. To study R_λ and R_δ, considering them as functions of the angle of incidence i.

2. To study R'_δ by considering it as a function of the angle of drift δ.

For the study of the air reactions on a body moving through the air, the *aerodynamical laboratory* is the most important means at the disposal of the aeronautical engineer.

The equipment of an aerodynamical laboratory consists of a special tube system of more or less vast proportions, inside of which the air is made to circulate by means of special fans (Fig. 75). The small models to be tested are

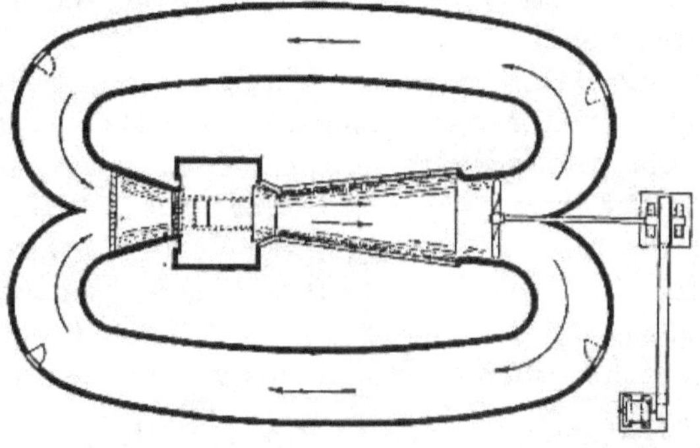

Fig. 75.

suspended in the air current, and are connected to instruments which permit the determination of the reactions

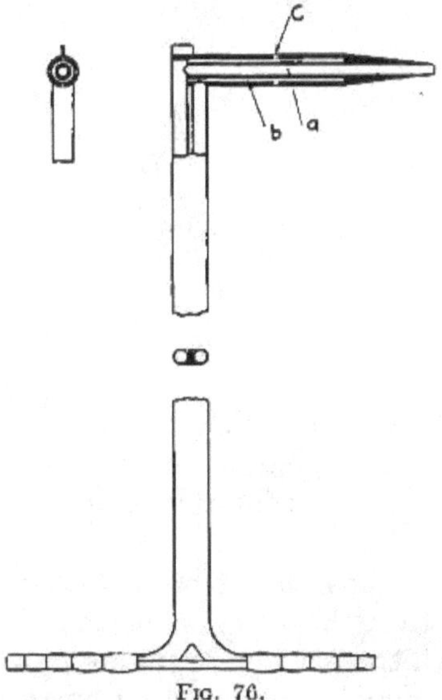

Fig. 76.

provoked upon them by the air. The section in which the models are tested is generally the smallest of the tube system, and a room is constructed corresponding to it, from which the tests may be observed. The speed of the air current may easily be varied by varying the number of revolutions of the fan.

The velocity of the current may be measured by various systems, more or less analogous. We shall describe the Pitot tube, which is also used on airplanes as a speed indicator. The Pitot tube (Fig. 76), consists of two concentric tubes, the one, internal tube *a* opening forward against the wind, the other external tube *b*, closed on the forward end but having small circular holes. These tubes are connected with a differential manometer. The pressure transmitted by tube *a* is equal to $P + \frac{dV^2}{2g}$; the pressure transmitted by tube *b* is equal to P; thus, the differential manometer will indicate a pressure h in feet of air, equal to

$$P + \frac{dV^2}{2g} - P$$

that is

$$h = \frac{dV^2}{2g}$$

consequently

$$V = \sqrt{\frac{2gh}{d}}$$

as $g = 32.2$, the result will be

$$V = \sim 8 \times \sqrt{\frac{h}{d}}$$

d represents the specific weight of the air. The preceding formula consequently gives us the means of graduating the manometer so that by using the Pitot tube it will read air speed directly.

With this foreword, let us note that experiments have demonstrated that the reaction of the air R, on a body moving through the air, and therefore also its components R_λ, R_δ and R'_δ, may be expressed by means of the formula

$$R = \alpha \frac{d}{g} \times A \times V^2$$

where

α = coefficient depending on the angle of incidence or the angle of drift,

d = the specific weight of the air,

g = is the acceleration due to gravity (which at the latitude of 45° = 32.2),

A = the major section of model tested (and defined as will be seen presently), and

V = the speed.

As a matter of convenience we shall give the coefficients assuming that the specific weight of the air is the one corresponding to the pressure of one atmosphere (33.9 ft. of water), and to the temperature of 59°F. Furthermore the coefficients will be referred to the speed of 100 m.p.h.

Then the preceding formula can be written

$$R = K \times A \times \left(\frac{V}{100}\right)^2 \qquad (1)$$

and knowing K, it gives the reaction of the air on a body similar to the model to which K refers, but whose section is equal to A sq. ft., and the speed to V m.p.h.

It is of interest to know the value of coefficient K, when the pressure and the temperature of the air are no more 1 atmosphere and 59° F., but have respectively any value h whatsoever (in feet of water), and $t°$ (degrees F.). The value of the new coefficient K_{ht} is then evidently given by

$$K_{ht} = K \times \frac{h}{33.9} \times \frac{460° + 59°}{460° + t°}$$

This equation will be of interest in the study of flight at high altitudes.

Interpreted with respect to the speed, formula (1) states that the reaction of the air on a body moving through it, is proportional to the square of the speed of translation. This is true only within certain limits. In fact, we shall soon see that in some cases, coefficient K determined by equation (1) changes with the variation of the speed, although the angle of incidence remains constant.

From the aerodynamical point of view, the section of the parts which compose an airplane may be grouped in three main classes which are:

1. Surfaces in which the Lift component predominates,
2. Surfaces in which the Drag component predominates, and
3. Surfaces in which the Drift component predominates.

The first are essentially intended for sustentation. Among them, the elevator is also to be considered, of which the aerodynamical study is analogous to that of the wings.

The second, surfaces in which the component of head resistance exists almost solely, are the major sections of all those parts, as the fuselage, landing gear, rigging, etc., which although not being intended for sustentation, form essential parts of the airplane.

Finally, the last surfaces are those in which the air reaction equals zero until the line of path is contained in the plane of symmetry of the airplane, but manifests itself as soon as the airplane drifts.

In Chapter I, we have spoken diffusely enough of the criterions followed for the aerodynamic study of a wing. Consequently, we shall repeat briefly what has already been said.

Let us consider a wing which displaces itself along a line of path which makes an angle i with its chord; a certain reaction will be borne upon it which may be examined in its two components R_λ and R_δ respectively perpendicular and opposite to the line of path, and which shall be called Lift and Drag, indicating them respectively by the symbols L and D.

We may then write,

$$L = \lambda \times A \times \left(\frac{V}{100}\right)^2$$

$$D = \delta \times A \times \left(\frac{V}{100}\right)^2$$

Where the coefficients λ and δ are functions of the angle of incidence, and define a type of wing, and A is the total surface of wing. The wing efficiency is given by

$$\frac{L}{D} = \frac{\lambda}{\delta}$$

and measures the number of pounds the wing can sustain for each pound of head resistance.

In Chapter I, we have given the diagrams for λ, δ and $\frac{\lambda}{\delta}$, as functions of i for two types of wings; consequently, it is unnecessary to record further examples.

For a complete aerodynamical study of a wing, it is necessary to determine in addition to the diagrams of λ, δ and $\frac{\lambda}{\delta}$, as functions of i, also the diagram of the ratio $\frac{x}{C}$ as a function of i, which defines the position at the center of thrust (see Chapter II). Knowledge of the law of variation of $\frac{x}{C}$ as a functon of i, is necessary to enable the study of the balance of the airplane.

In the reports on aerodynamical experiments conducted in various laboratories, American, English, Italian, etc., the reader will find a vast amount of experimental material which will assist him in forming an idea of the influence borne on the coefficients λ and δ, not only by the shape and relative dimensions of the wings, as for instance the ratios $\frac{\text{span}}{\text{chord of the wing}}$ and $\frac{\text{thickness of the wing}}{\text{chord of the wing}}$, but also by the relative positions of the wings with respect to each other; such as multiplane machines with superimposed wings, with wings in tandem, etc.

In the study of coefficients of resisting surfaces, in general, solely the knowledge of the component R_δ is of interest; the sustaining component R_λ is equal to zero, or is of a negligible value as compared with that of R_δ. We then have

$$R_\delta = K \times A \times \left(\frac{V}{100}\right)^2$$

where K is a function of i, and A measures the surface of the major section of the form under observation, taken

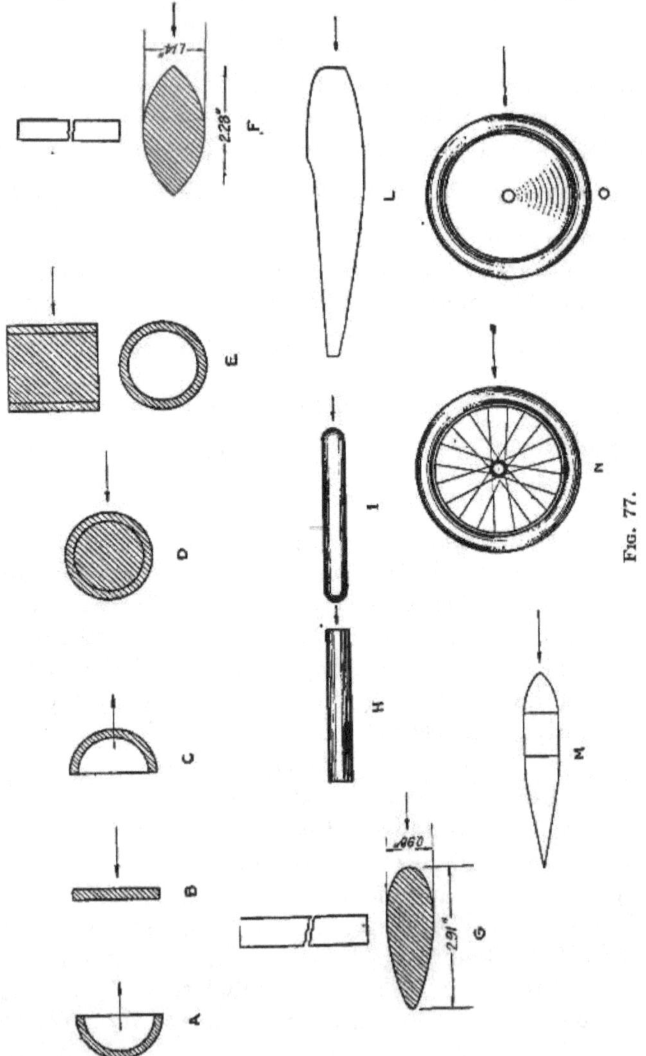

FIG. 77.

perpendicular to the axis of symmetry of the body, or to the axis parallel to the normal line of path.

In general, the head resistance is usually determined for only one value of i, that is, for the value corresponding to normal flight. In fact, it should be noted that an airplane normally varies its angle of incidence within very narrow limits, from 0° to 10°; now, while for wings such variations of incidence bring variations of enormous importance in the values of L as well as in those of D, the variation of coefficient K for the resistance surfaces is relatively small. Consequently, in laboratories, only one value is found. Nevertheless, exception is made for the wires and cables, which are set on the airplanes at a most variable inclination, and therefore it is interesting to know coefficient K for all the angles of incidence.

A table is given below compiled on the basis of Eiffel's experiments, which gives the value of K for the following forms (Fig. 77), and for a speed of 90 feet per second:

A = Half sphere with concavity facing the wind,
B = Plain disc perpendicular to the wind,
C = Half sphere with convexity facing the wind,
D = Sphere,
H = Cylinder with ends having plain faces, with axis parallel to the wind,
I = Cylinder with spherical ends, with axis parallel to the wind,
E = Cylinder with axis perpendicular to the wind,
F = Airplane strut—fineness ratio ½,
G = Airplane strut—fineness ratio ⅓,
L = Airplane fuselage with radiator in front,
M = Dirigible shape,
N = Airplane wheel without fabric, and
O = Wheel covered with fabric.

TABLE 3

A	B	C	D	E	F	G	H	I	L	M	N	O
43.5	28.6	7.8	4.1	8.7	15.6	3.5	22.6	6.1	8	3	28	14

In the above table, one is immediately impressed by the very low value for the dirigible form. Its resistance is about 10 times less than that of the plain disc.

The preceding table contains values corresponding to a speed of 90 ft. per sec. If the law of proportion to the square of the speed were exact, these values would also be available for other speeds. On the contrary, at different speeds these

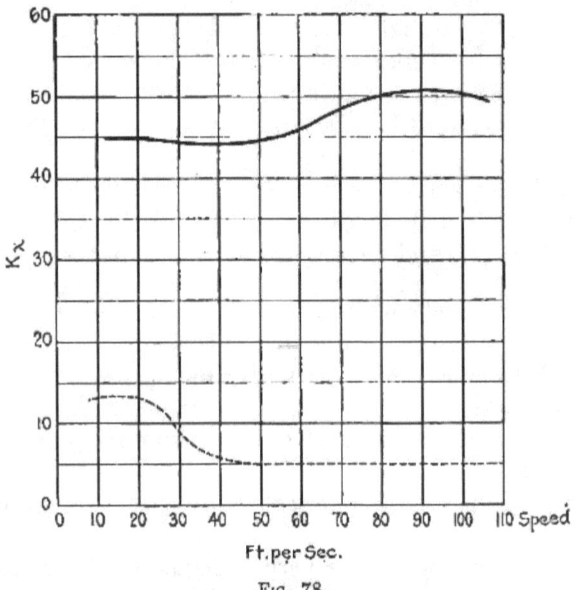

Fig. 78.

values vary. An example will better illustrate this point. In Fig. 78 diagrams are given of the variation of K for the forms A and D, and for the speed of from 13 to 100 ft. per sec. (Eiffel's experiments). We see that coefficient K of form A, increases with the speed, while that of D decreases. These anomalies can be explained by admitting that the various speeds vary the vortexes which are formed behind the bodies in question, therefore varying the distribution of the positive and negative pressure zones, and consequently the coefficients of head resistance.

Figs. 79 and 80 give the diagrams of the coefficient K, for the wires and cables (Eiffel); for the wires, coefficient K first decreases, then increases; for the cables instead, the value of K shows an opposite tendency. Finally,

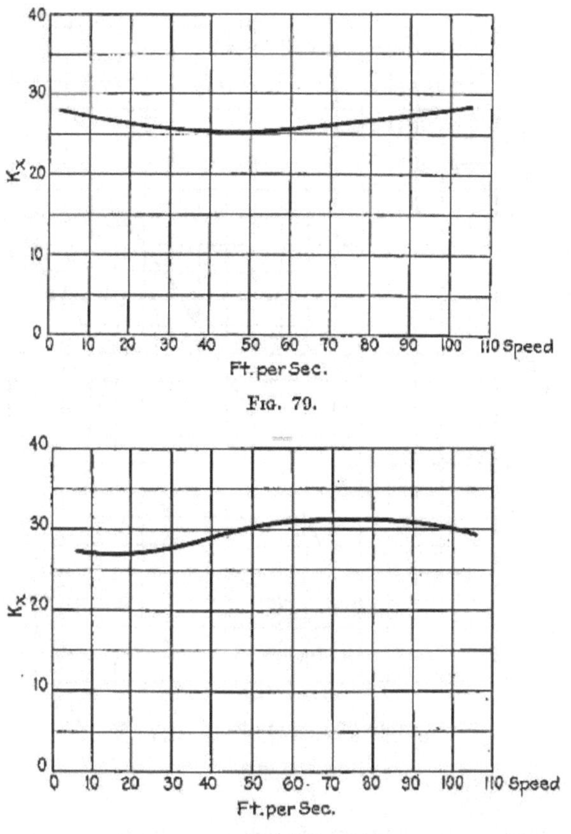

Fig. 79.

Fig. 80.

Fig. 81 gives the diagram showing how coefficient K varies for the wires and cables when their angle of incidence varies from 0° to 90°.

In studying the airplane, it is more interesting to know the total head resistance than that of the various parts;

if we call A_1, A_2, . . . and A_n the major sections of the various parts constituting the airplane and which produce a head resistance, (fuselage, landing gear, wheels, struts, wires, radiators, bombs, etc.), and K_1, K_2, . . . and K_n,

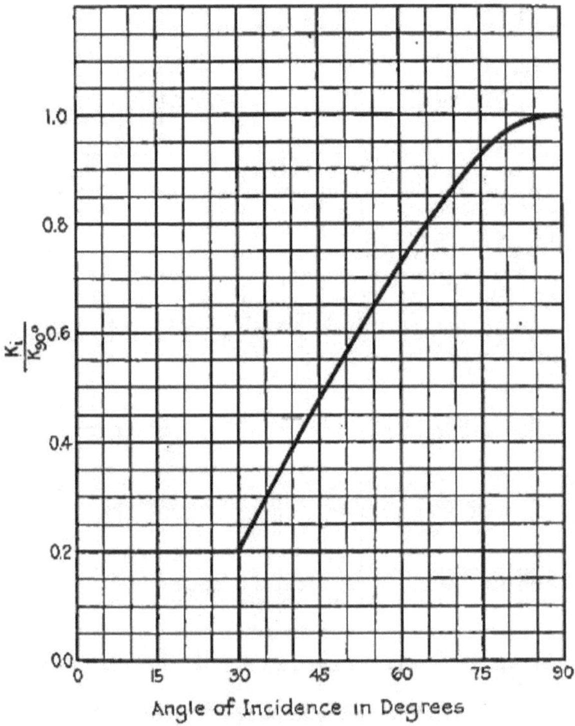

Fig. 81.

the respective coefficients of head resistance, the total head resistance R_s of the airplane will be

$$R_s = K_1A_1\left(\frac{V}{100}\right)^2 + K_2A_2\left(\frac{V}{100}\right)^2 + \ldots K_nA_n\left(\frac{V}{100}\right)^2$$

$$= (K_1A_1 + K_2A_2 + \ldots K_nA_n)\left(\frac{V}{100}\right)^2 = \sigma\left(\frac{V}{100}\right)^2$$

where $\sigma = K_1A_1 + K_2A_2 + \ldots K_nA_n$ and is called the *total coefficient of head resistance* of the airplane.

As to the study of the drift surfaces, it is accomplished by taking into consideration only the drift component, and not the component of head resistance, as the latter is negligible with respect to the former. Furthermore, in this study it is interesting to know the center of drift at various angles of drift, in order to determine the moments of drift and their efficaciousness for directional stability. When the line of path lies out of the plane of symmetry, all the parts of the airplane can be considered as drift surfaces. Nevertheless, the most important are the fuselage, the fin, and the rudder. From the point of view of drift forces, the fuselages without fins and without rudders, may be unstable; that is, the center of drift may be situated before the center of gravity in such a way as to accentuate the path in drift when this has been produced for any reason whatsoever.

For what we have already briefly said in speaking of the rudder and elevator, and for what we shall say more diffusely in discussing the problems of stability, it is opportune to know both of the coördinates of the center of drift, which define its position on the surface of drift.

Finally, we shall make brief mention of the aerodynamical tests of the propeller.

Let us suppose that we have a propeller model rotating in the air current of an experimental tunnel. By measuring the thrust T of the propeller, its number of revolutions n, the power P absorbed by the propeller, and the velocity V of the wind, it is possible to draw the diagrams of T, P, and the efficiency ρ. Numerous experiments by Colonel Dorand have led to the establishing of the following relations;

$$T = \alpha' n^2D^2$$
$$P = \alpha n^3D^5$$
$$\rho = \frac{TV}{P} = \frac{\alpha'}{\alpha} \times \frac{V}{nD}$$

where D is the diameter of the propeller, and α' and α are numerical coefficients which vary with the variation of $\frac{V}{nD}$. This ratio is proportional to the other

$$\frac{V}{\pi nD} = \frac{\text{velocity of translation}}{\text{peripheral velocity}}$$

which defines the angle of incidence of the line of path with respect to the propeller blade.

Knowing the values of α' and α as functions of $\frac{V}{nD}$, it is possible to obtain those of T, P, and ρ, thereby possessing the data for the calculation of the propeller.

CHAPTER VIII

THE GLIDE

Let us consider an airplane of weight W, of sustaining surface A, and of which the diagrams for λ, δ and the total head resistance σ, are known.

Let us suppose that the machine descends through the air with the engine shut off; that is gliding. Suppose the pilot keeps the elevator fixed in a certain position maintaining the ailerons and the rudder at zero. Then if

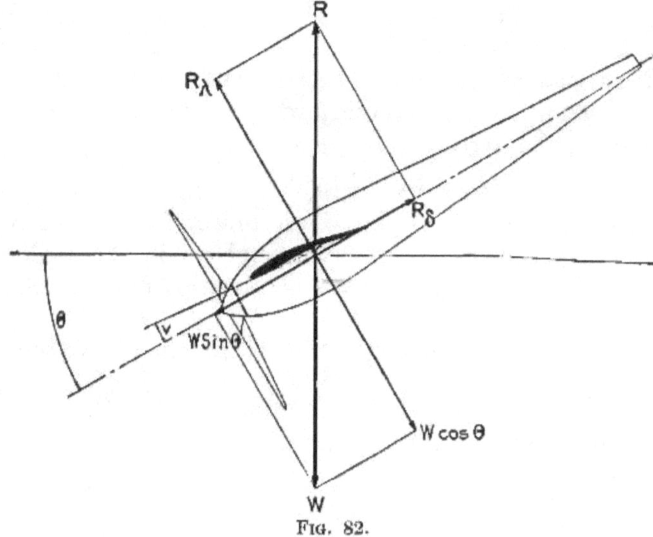

Fig. 82.

the airplane is well balanced, it will follow a sloping line of path θ (Fig. 82), which will make a well-determined angle of incidence i, with the wing; in fact, if this angle should vary, some restoring couples (see Chapter II), tending to keep the machine at incidence i, would be produced.

Let us study the existing relations among the parameters W, A, λ, δ, σ, θ and V. When the machine has reached its normal gliding speed (that is, V = constant), the forces acting on it are reduced only to the weight W, and the total air reaction R. By a known theorem of mechanics, all the forces acting on a body in ùniform rectilinear motion, balance each other; that is, in this case force R is equal and of opposite direction to W; that is,

$$R + W = 0$$

Let us consider the two components R_λ and R_δ of R (on the line of the path and perpendicular to the line of path). The preceding equation can then be divided into two others

$$R_\delta + W \sin \theta = 0 \qquad (1)$$
$$R_\lambda + W \cos \theta = 0 \qquad (2)$$

Let us express the components R_λ and R_δ as function of λ, δ, σ and γ. Remembering what we have said in the preceding chapters,

$$R_\lambda = 10^{-4}\ \lambda A V^2$$

Where R_λ is expressed in lb., A in sq. ft., V in m.p.h. and λ is a coefficient which depends upon the angle of incidence and of which the law of variation must be found experimentally.

As to R_δ its expression results from the sum of two terms, one due to the wings $\delta \times A \times \left(\frac{V}{100}\right)^2$ and the other due to parasite resistances σ of the form

$$\sigma \left(\frac{V}{100}\right)^2$$

Thus we shall have

$$R_\delta = 10^{-4}\ \delta A V^2 + 10^{-4}\sigma V^2$$

The equations (1) and (2) become

$$10^{-4}\ (\delta A + \sigma)\ V^2 = -\ W \sin \theta \qquad (3)$$
$$10^{-4}\ \lambda A V^2 = -\ W \cos \theta \qquad (4)$$

We have immediately, by squaring and by adding the preceding equations

$$10^{-4}\ V^2 = \frac{W}{A} \quad \frac{1}{\sqrt{\left(\delta + \frac{\sigma}{A}\right)^2 + \lambda^2}} \qquad (5)$$

and dividing (3) by (4)

$$\frac{\delta}{\lambda} + \frac{\sigma}{\lambda A} = \tan \theta \qquad (6)$$

As, once the angle of incidence i is fixed, the values λ and δ are fixed, equations (5) and (6) enable us to find, corresponding to each value of i, a couple of values θ and V. Thus all the elements of the problem are known.

Equation (5) enables us to state the following general principles:

1. Other conditions being equal, the gliding speed is directly proportional to the ratio $\frac{W}{A}$, that is, to the unit load on the wings.

2. Other conditions being equal, the gliding speed is inversely proportional to the coefficient λ; therefore with wings having a heavily curved surface and consequently of great sustaining capacity, the descending speed is much lower than with wings having a small sustaining capacity.

3. Other conditions being equal, the gliding speed is inversely proportional to the value of sum $\left(\delta + \frac{\sigma}{A}\right)$, which represents $\frac{R_\delta}{A}$ for $V = 100$ m.p.h.

Equation (6) enables us to state the following general principles:

4. Other conditions being equal, the angle of glide θ is inversely proportional to the ratio $\frac{\lambda}{\delta}$, that is, to the efficiency of the wing.

5. Other conditions being equal, the angle of glide θ is directly proportional to the ratio $\frac{\sigma}{A}$ between the coefficient of parasite resistance and the surface of the wings. This ratio is also usually called *coefficient of fineness*.

6. The angle θ of volplaning is independent from the weight of the airplane. This weight doesn't influence but the speed. In other words, by increasing the load, the gliding speed will increase but the angle of descent will not change.

With this premise we propose, following a method suggested by Eiffel, to draw a special logarithmic diagram which will enable us to study all the relations existing among the variable parameters of gliding.

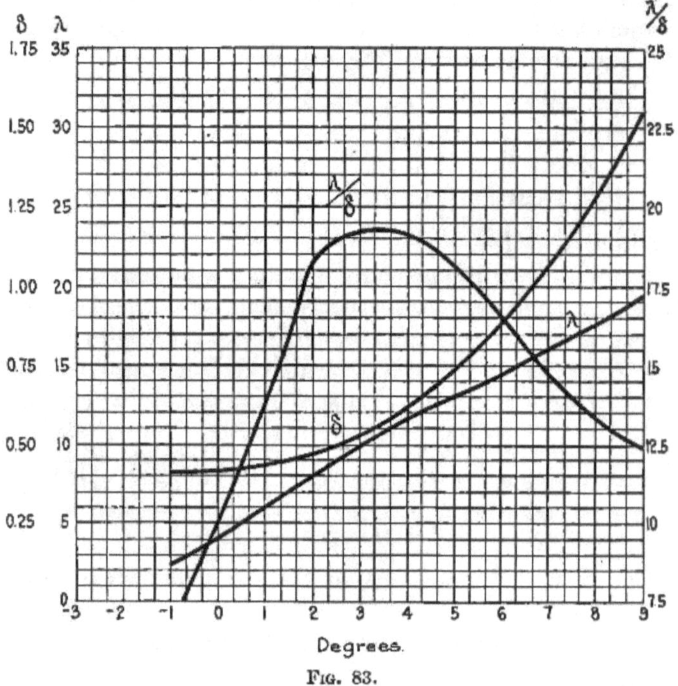

Fig. 83.

Let us go back to formulas (3) and (5) and write them in the following form

$$\frac{-W \sin \theta}{V^2} = 10^{-4}\,(\delta A + \sigma)$$

$$\frac{W}{V^2} = \sqrt{[10^{-4}\,(\delta A + \sigma)]^2 + (10^{-4}\,\lambda A)^2}$$

Furthermore let us assume

$$\Lambda = 10^{-4}\,\lambda A \tag{7}$$

$$\Delta = 10^{-4}\,(\delta A + \sigma) \tag{8}$$

Then the preceding equations become

$$\frac{W}{V^2} = \sqrt{\Delta^2 + \Lambda^2} \tag{9}$$

$$\frac{-\sin\theta \cdot W}{V^2} = \Delta \tag{10}$$

Now, as for each value of the angle of incidence i, δ, λ and σ, are known, and as A is constant, we can, by means of equations (7) and (8), determine a couple of values of Λ and Δ and consequently of $\sqrt{\Delta^2 + \Lambda^2}$ and Δ corresponding to each value of i; it will be then possible to draw the logarithmic diagram of $\sqrt{\Delta^2 + \Lambda^2}$ as function of Δ. A numerical example will better explain this.

Let us consider an airplane having the following characteristics:

W = 2700 lb.
A = 270 sq. ft.
σ = 160 (average value between $i = 0°$ and $i = 9°$).
λ, δ functions of i as from the diagram of Fig. 83.

We can then compile the following table:

TABLE 4

i	0°	1°	2°	3°	4°	5°	6°	7°	8°	9°
λ	4.11	6.03	8.08	9.70	11.8	13.00	14.40	16.10	17.75	19.40
δ	0.41	0.44	0.45	0.51	0.62	0.73	0.89	1.06	1.31	1.16
λA	1110	1630	2180	2620	3180	3510	3890	4340	4780	4240
$\delta A + \sigma$	271	278	281	297	327	358	400	446	514	582
$\sqrt{\Delta^2 + \Lambda^2}$	0.11	0.16	0.22	0.26	0.32	0.35	0.39	0.43	0.48	0.52
Δ	0.0271	0.0278	0.0281	0.0297	0.0326	0.0358	0.0398	0.0445	0.0512	0.0580

Thus we have a certain number of pairs of corresponding values of $\sqrt{\Lambda^2 + \Delta^2}$ and Δ which enable us to draw the diagram of $\sqrt{\Lambda^2 + \Delta^2}$ as a function of Δ.

Now, instead of drawing this diagram on paper graduated to cartesian coördinates, let us draw it on paper with logarithmic graduation (Fig. 84). We shall have a logarithmic diagram which gives

$$\sqrt{\Delta^2 + \Lambda^2} = f(\Delta)$$

or

$$\frac{W}{V^2} = f\left(\frac{-W \sin\theta}{V^2}\right)$$

Let us consider any part whatever of this curve for instance the point A; the abscissa OX of this point is

$$OX = \log \frac{-W.\sin\theta}{V^2}$$

Now $\log \frac{-W.\sin\theta}{V^2} = \log W + \log(-\sin\theta) - 2\log V$

Therefore we can consider OX as the algebraic sum of the segments $\log W$, $\log(-\sin\theta)$ and $-2\log V$.

Analogously the ordinate of point A is

$$OY = \log \frac{W}{V^2}$$

and as $\log \frac{W}{V^2} = \log W - 2\log V$, we can consider OY as the algebraic sum of the two segments $\log W$ and $-2\log V$.

Thus in order to pass from the origin O to the point A of the diagram it is sufficient to sum the segments $\log W$, $\log(-\sin\theta)$ and $-2\log V$, following the axis of the abscissæ and $\log W$ and $-2\log V$, following the axis of the ordinates.

As evidently, the segments can be summed in any order whatever, we can sum them in the following order:

1. Log W parallel to OX.
2. Log W parallel to OY.
3. $-2\log V$ parallel to OX.
4. $-2\log V$ parallel to OY.
5. Log $(-\sin\theta)$ parallel to OX.

Now, it is evident that the two segments corresponding to W, can be replaced by a single oblique segment of inclination 1/1 on the axis OX and of lengths $\sqrt{2}\log W$. Similarly the two segments corresponding to V can be replaced by a single segment also inclined by 1/1 on OX and of length $-\sqrt{2^2 + 2^2}\log V$. Thus we can pass from the origin O to the point A of the diagram by drawing 3 segments, two parallel to an axis of inclination 1/1 on OX and one parallel to OX, and which measure W, V and $-\sin\theta$ in the respective scales. The condition necessary and suf-

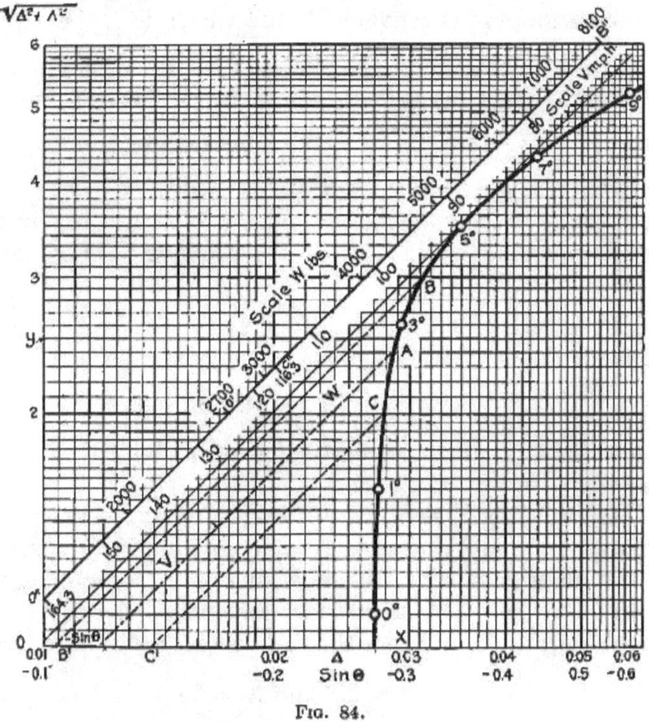

FIG. 84.

ficient in order that a system of values of W, V and $-\sin\theta$ be realizable with the given airplane, is evidently that the three corresponding segments, summed geometrically starting from the origin, end on the diagram.

The units of measure selected for drawing the diagram of Fig. 84, are the following:

W in lb.
V in m.p.h.

In order to determine the relation between the scales of $\sqrt{\Delta^2 + \Lambda^2}$ and Λ^2 and the scales of W, V and $-\sin\theta$, it is first of all necessary to fix the origin of the scale of W and V.

It is convenient to select W equal to the weight of the airplane, in our case $W = 2700$ lb.

Furthermore it is convenient that the ratio $\frac{W}{V^2}$ be equal to any one whatever of the values 1×10^x, where x is a whole positive or negative number; thus we have from the equation $\Delta = -\sin\theta . \frac{W}{V^2}$ that the same scale of Δ, if divided by 10^x, gives the scale of $-\sin\theta$.

It would be convenient to make $x = -1$ in order to keep the scale of $-\sin\theta$ within the drawing. Then from

$$\frac{2700}{V^2} = 1 \times 10^{-1}$$

We have

$$V^2 = 27{,}000 \text{ and } V\ 164.3 \text{ m.p.h.}$$

The scale of $-\sin\theta$ is equal to that of Δ divided by 10^{-1}, that is, multiplied by 10.

Then, making $V \doteq 164.3$ the corresponding segment is zero and we pass from the origin O to a point of the diagram by summing geometrically the segments corresponding to $-\sin\theta$ and W. Let us consider any point whatsoever B of the diagram, for instance the point whose coördinates are:

$$\sqrt{\Delta^2 + \Lambda^2} = 0.3 \text{ and } \Delta = 0.031$$

For this point and for $V = 164.3$, the weight W is represented by the segment BB'; because

$$\sqrt{\Delta^2 + \Lambda^2} = \frac{W}{V^2}$$

substituting the preceding values of $\sqrt{\Delta^2 + \Lambda^2}$ and V, we have

$$W = 8100$$

that is $$BB' = 8100$$

Let us make now $W = 2700$; then the corresponding segment is zero and in order to pass from the origin O to a point of the diagram it would be sufficient to sum geometrically the segment corresponding to $-\sin\theta$ and V. Let us take any other point whatsoever C on the diagram, for instance that whose coördinates are:

$$\sqrt{\Delta^2 + \Lambda^2} = 0.2 \text{ and } \Delta = 0.0278$$

For this point and for $W = 2700$ we shall have, as it is demonstrated with an analogous process, that $CC' = V =$ 116.3 m.p.h.

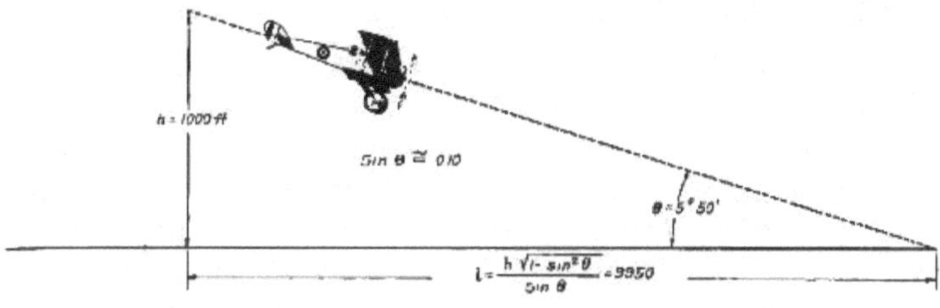

FIG. 85.

Taking BB' to $O'B''$ on the scale of W and marking 2700 lb. in O' and 8100 lb. in B'', the scale of weights will be individuated. Analogously taking CC' to $O''C''$ on the scale of V and marking 164.3 on O'' and 116.3 on C'', the scales of speed will be individuated.

With the preceding scales and for the airplane of our example weighing 2700 lb., the diagram of Fig. 84 gives immediately the pair of corresponding values of $-\sin\theta$ and V. In fact for any value whatsoever of $-\sin\theta$ for instance, from the point C' correspondent to $-\sin\theta =$ 0.139, it is sufficient to draw a parallel to the scale of speeds until it meets the diagram in C; the segment $C'C$, read on the scale of the speeds gives the value of the speed V corresponding to $-\sin\theta$; in our case $C'C = 116.3$. From the diagram we see that by increasing the angle of incidence, the angle θ decreases to a minimum, after which it increases again. This means that the line of path raises its inclination up to a limit which in our case is equal to about 0.1 corresponding to the incidence of 5° to 6°; if our airplane was descending for instance from the height of 1000 ft. it could reach any point whatsoever, situated within a radius of 9950 ft. (Fig. 85).

Our example, however, is referred to an exceptional case; in practice with the present airplanes, the minimum value of sin θ is between 0.12 and 0.14. Furthermore the diagram shows the law variation of the speed of the airplane with a variation of the angle of incidence. It is seen that it is not safe to decrease too much the angle of incidence in order not to increase excessively the speed.

In practice the pilots usually dispose the machine even vertical but for a very short time, so as not to give time to the airplane to reach dangerous speeds. On the other hand one has to look out not to increasing excessively the angle of incidence in order not to fall in the opposite inconvenience of reducing excessively the speed, which causes a strong decrease in the sensibility of the controlling devices and consequently in the control of the machine by the pilot.

The use of speedometers, today much diffused, is a very good caution in order that the pilot, while gliding may keep the speed within normal limits, keeping it preferably slightly below the normal speed which the machine has with engine running.

Until now we have treated the rectilinear glide. It is necessary to take up also the spiral glide which is today the normal maneuver for the descent.

The spiral descent is accomplished by keeping the machine turning during the glide. We have seen that a centrifugal force is then originated

$$\Phi = \frac{W}{g} \cdot \frac{V^2}{r}$$

equal and opposite to the centripetal force R'_s which has provoked the turning (Fig. 86). This force R'_s can be produced by the inclination of the airplane or by the drifting course of the airplane or by both phenomena. When this

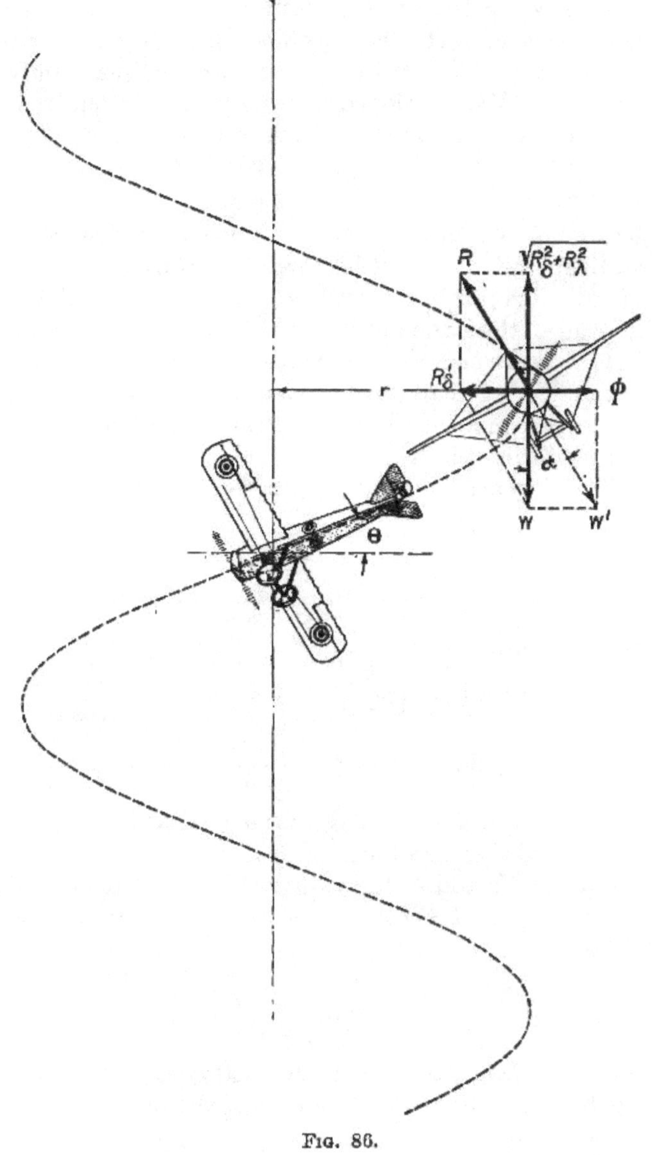

Fig. 86.

force is provoked solely by the inclination of the airplane, that is, when the angle of drift is zero, we say the spiral descent is correct, the machine then doesn't turn flat; as in practice this is the normal case, we shall study only this case. We developed the discussion for this case as if the weight were increased from W to W' where

$$W' = \frac{W}{\cos \alpha}$$

Therefore we can apply the formulæ of the rectilinear gliding, but we shall be careful to consider the angle θ' of the line of path, with a plane perpendicular to W' instead of the angle θ of the line of path with the horizontal; in fact, as we consider the fictitious weight W' instead of the weight W, we shall have to consider a fictitious horizontal plane perpendicular to W' instead of the horizontal plane perpendicular to W.

Then equations (3) and (4) become

$$10^{-4} (\delta A + \sigma) V^2 = \frac{-W}{\cos \alpha} \sin \theta' \qquad (11)$$

$$10^{-4} \lambda A V^2 = \frac{-W}{\cos \alpha} \cos \theta' \qquad (12)$$

from which

$$10^{-2} V = \sqrt{\frac{W}{A}} \sqrt[4]{\frac{1}{\lambda^2(\delta A + \sigma)^2}} \frac{1}{\sqrt{\cos \alpha}}$$

$$\sin \theta' = 10^{-4} \frac{(\delta A + \sigma) V^2}{W} \cos \alpha$$

If we make $\alpha = 0$, we have $\cos \alpha = 1$, and we fall back to the formula for rectilinear gliding.

Calling V_o and θ_o the values of V and θ for $\alpha = 0$, and calling V_α and θ'_α the values for the angle α, we have

$$V_\alpha = V_o \frac{1}{\sqrt{\cos \alpha}}$$

$$\sin \theta'_\alpha = \sin \theta_o$$

From known theorems of geometry, calling θ_α the angle of the line of path with the horizontal, we have

$$\sin \theta'_\alpha = \sin \theta_\alpha \, . \, \cos \alpha$$

from which

$$\sin \theta_\alpha = \frac{\sin \theta_o}{\cos \alpha}$$

Resuming, if we suppose that we maintain a certain incidence i (by maneuvering the elevator) and a certain transverse inclination α (by maneuvering the ailerons) the airplane will follow an elicoidal line of path, with speed V_α and inclination to the ground θ_α which are given by the equations:

$$V_\alpha = \frac{V_o}{\sqrt{\cos \alpha}} \quad (13)$$

and

$$\sin \theta_\alpha = \frac{\sin \theta_o}{\cos \alpha} \quad (14)$$

where V_o and θ_o are the speed and the inclination of line of path, corresponding to the rectilinear gliding; it is then easy, from diagram 84, to obtain the couples of values V_α and $\sin \theta_\alpha$ corresponding to each value of α.

In general, equations (13) and (14) tell us that in the spiral descent the angle of incidence being kept the same, an airplane has a speed and an angle of slope of the line of path which are greater than in the rectilinear gliding.

CHAPTER IX

FLYING WITH POWER ON

In the preceding chapter we have studied gliding or flying with the engine off. Let us suppose now, that the pilot, during any course whatever of gliding, starts the engine without maneuvering the elevator. Then a new

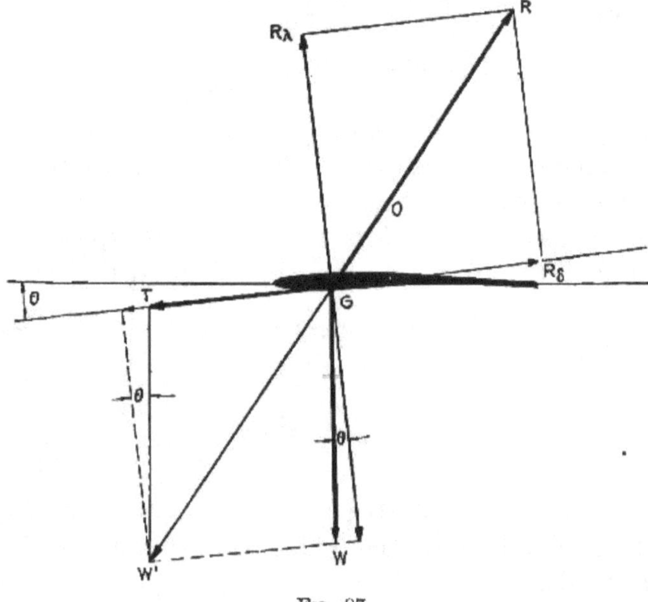

FIG. 87.

force will appear, other than the weight W and air reaction R, namely, the propeller thrust, T.

If, instead of weight W, we consider the fictitious weight W' resulting from W and T (Fig. 87), all the considerations made and notations adopted in the preceding chapter can be applied.

Then

$$R_\delta = T + W \cos (90° - \theta) = T + W \sin \theta$$
$$R_\lambda = W \cos \theta$$

or

$$10^{-4}(\delta A + \sigma)V^2 = T + W \sin\theta \quad (1)$$

$$10^{-4}\lambda A V^2 = W \cos\theta \quad (2)$$

Eliminating $10^{-4} V^2$ from the two equations, we have

$$(\delta A + \sigma) \cdot \frac{W \cos\theta}{\lambda A} = T + W \sin\theta$$

from which

$$T = \left(\frac{\delta}{\lambda} + \frac{\sigma}{\lambda A}\right) W \cos\theta - W \sin\theta \quad (3)$$

Let us suppose that the angle of incidence is fixed, then λ, δ, and σ, will be determined. Equation (3) enables us to find the value of θ for each value of T. For $T = 0$, we return to the case of gliding. As T increases, $\cos\theta$ must increase, and $\sin\theta$ must decrease; that is, the angle θ decreases. Value T_o, for which $\theta = 0$, gives the value of thrust necessary for horizontal flight. For $\theta = 0$, we have $\cos\theta = 1$, and $\sin\theta = 0$; consequently

$$T_o = \left(\frac{\delta}{A} + \frac{\sigma}{\lambda A}\right) \times W \quad (3)$$

for all the values $T < T_o$, the angle θ with the horizontal line is positive; that is, the machine descends. For all the values $T > T_o$, the angle θ with the horizontal line changes sign; that is, the line of path ascends. First of all let us study horizontal flight. Then, as $\theta = 0$ equation (1) and (2) become

$$T = 10^{-4}(\delta A + \sigma) V^2 \quad (4)$$

$$W = 10^{-4}\lambda A V^2 \quad (5)$$

Now the power P_1 in H.P. corresponding to the thrust T in lb. and to the speed V in m.p.h., it is evidently equal to

$$550P_1 = TV \cdot \frac{5280}{3600}$$

$$= 1.47\, TV$$

and because of equation (4)

$$550P_1 = 1.47\ 10^{-4}(\delta A + \sigma) V^3 \quad (6)$$

Equations (5) and (6) enable us to draw a very interesting logarithmic diagram with the method proposed by Eiffel.

Let us have as in the preceding chapter

$$\Lambda = 10^{-4}\,\lambda A$$
$$\Delta = 10^{-4}\,(\delta A + \sigma)$$

Equations (5) and (6) become

$$\frac{W}{V^2} = \Lambda \qquad (7)$$

$$\frac{550P_1}{V^3} = 1.47\,\Delta \qquad (8)$$

Let us consider then the airplane of the example used in the preceding chapter, that is, the airplane having the following characteristics:

$$W = 2700 \text{ lb.}$$
$$A = 270 \text{ sq. ft.}$$
$$\sigma = 160$$

and whose diagrams of λ and δ are those given in Fig. 83. Based upon the table given in the preceding chapter we can compile the following table:

TABLE 5

i	0°	1°	2°	3°	4°	5°	6°	7°	8°	9°
Λ	0.11	0.16	0.22	0.26	0.32	0.35	0.39	0.43	0.48	0.52
$1.47\,\Delta$	0.0398	0.0410	0.0413	0.0437	0.0480	0.0527	0.0787	0.0655	0.0755	0.0853

This table gives a certain number of pairs of values corresponding to Λ and Δ and therefore enables us to draw the diagram of Λ as as function of Δ. Now instead of drawing the diagram on paper graduated with uniform scales, let us draw the same diagram on paper with logarithmic graduation (Fig. 88).

We shall have a logarithmic diagram which gives

$$\Lambda = f\,(1.47\Delta)$$

or

$$\frac{W}{V^2} = f\left(\frac{550P_1}{V^3}\right)$$

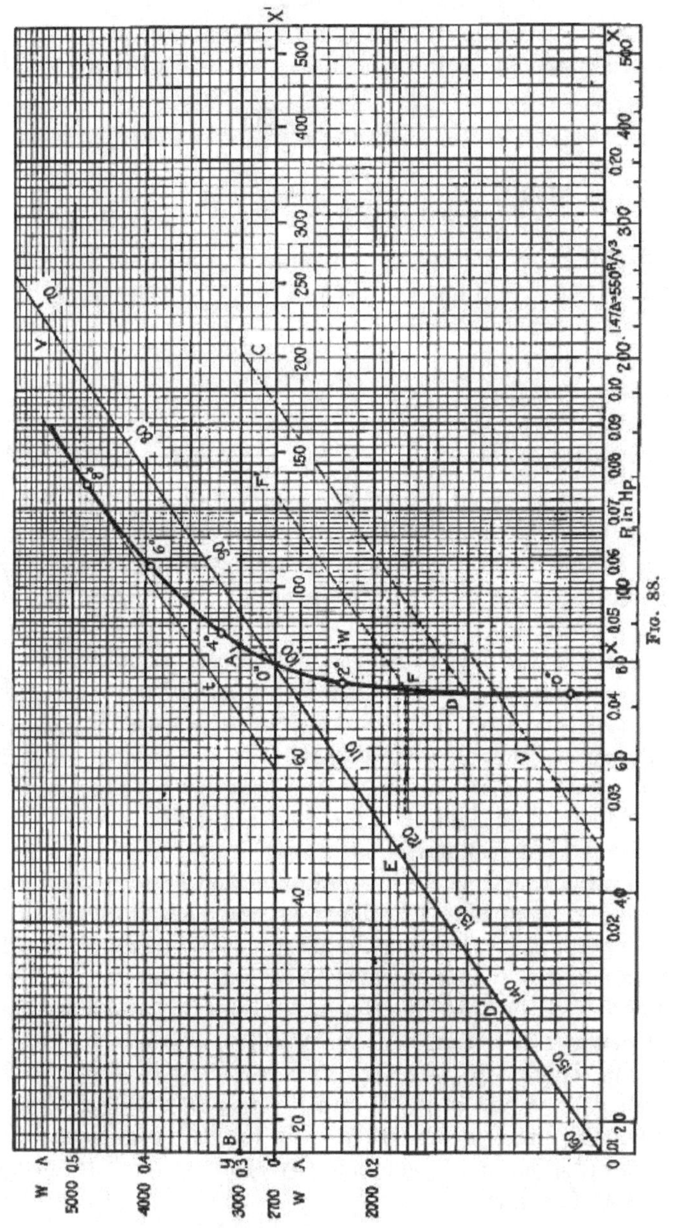

Fig. 88.

Let us consider then any point whatever of this curve for instance the point A; the abscissa OX of this point is

$$OX = \log \frac{550P_1}{V^3}$$

Now $\log \frac{550P_1}{V^3} = \log 550P_1 - 3 \log V$; thus we can consider OX as the algebraic sum of segment $\log 550P_1$, and segment $-3 \log V$. Analogously the ordinate of point A is

$$OY = \log \frac{W}{V^2}$$

and as $\log \frac{W}{V^2} = \log W - 2 \log V$ we can consider OY as the algebraic sum of the two segments $\log W$ and $-2 \log V$. Thus, in order to pass from the origin O to point A of the diagram, it is sufficient to add the segment $\log 550 P_1$ and $-3 \log V$ along the axes OX and $\log W$ and $-2 \log V$ along the axes OY.

Since evidently these segments can be added in any order whatever, we can take first $\log 550P_1$ parallel to the axes of abscissa, then $-3 \log V$ also parallel to the axes of abscissa, then $-2 \log V$ parallel to the axes of ordinates and finally $\log W$ parallel to the axes of ordinates.

Now it is evident that the two segments $-3 \log V$ and $-2 \log V$ corresponding to V, can be replaced by a single oblique segment whose inclination is 2 on 3 and whose length is $-\sqrt{2^2 + 3^2} \,.\, \log V$. Thus we can pass from the origin 0 to point A by drawing three segments, one parallel to the axes OX, the second parallel to an axes of an inclination of 2 on 3 and the third parallel to the axes OY which segments measure in the respective scales P_1, V and W.

The condition necessary and sufficient in order that a system of values of P_1, V and W, may be realized with the given airplane is evidently that the three corresponding segments, summed geometrically starting from the origin, end on the diagram.

The units of measure selected for drawing the diagram of Fig. 88 are the following:

P_1 in H.P.
V in m.p.h.
W in lb.

In order to determine the relation between the scales of Λ and Δ and the scales of P_1, V and W, it is necessary to fix the origin of the scale of V; we shall suppose to assume as origin $V = 100$ m.p.h. Then for $V = 100$ m.p.h., the coördinates Λ and Δ measure also W and P; in fact for the particular value $V = 100$ the segment to be laid off parallel to the scale of V becomes zero and so we go from the origin to the diagram through the sum of the only two segments W and P. Let us consider then the point A whose coördinates are

$$\Lambda = 0.3 \text{ and } \Delta = 0.0463$$

Corresponding to these points we shall have

$$\frac{W}{100^2} = 0.3 \text{ and } \frac{550P_1}{100^3} = 0.0463$$

which gives

$$W = 3000 \text{ lb.} \qquad P_1 = 84.2 \text{ H.P.}$$

Thus the scales of W and P_1 are determined.

In order to determine the scale of V we proceed as follows: Let us give to W and P_1 two values whatever, for instance

$$W = 3000 \text{ and } P_1 = 200 \text{ H.P.}$$

Applying the usual construction we shall lay off $OB = 3000$, $BC = 200$ in the respective scales; from point C we draw a parallel to the scale V to meet the diagram in point D. We shall have in CD the corresponding speed.

Now for $D\Lambda = 0.153$. Consequently, as we have

$$0.153 = \frac{3000}{V^2}$$

we will have

$$V = 140 \text{ m.p.h.}$$

that is, the segment CD laid off in $O''D'$ gives the scale of V. The scales being known it is easy to study the way the airplane acts, that is, it is possible to find for each value of the speed the value of the power necessary to fly.

In Fig. 88 we have disposed the scales so as to facilitate the readings; that is we have made the origin O'' of the scale of V coincide with the intersection of this scale and a line $O'X'$ parallel to the axis OX and passing through the value $W = 2700$ which is the weight of the airplane; and we have furthermore repeated on $O'X'$ the scale of power.

Then, in order to have two corresponding values of P and V we draw from any point whatever E on the scale of the speed, the parallel to OX up to F, point of intersection with the diagram; we draw then FF' parallel to the scale of the speed and we have in F' on $O'X'$ the value of the power P_1 corresponding to a speed E. The examination of the diagram enables us to make some interesting observations.

Let us draw first the tangent t to the diagram which is parallel to scale V; this tangent will cut the axis $O'X'$ in a point corresponding to a power of 58 H.P.; this is the minimum power at which the airplane can sustain itself and the corresponding speed $V_{\min}$ is 72.3 m.p.h.

An airplane having an engine capable of giving no more than this power, could hardly sustain itself; it would be, as one says, *tangent*, and could only fly horizontally or descend, but could by no means follow an ascending line of flight.

For all the values of speed greater or lower than the above value, the necessary power for flying increases. The phenomenon of power increasing for the decreasing speed may seem strange; even more so, if the comparison is made with all other means of locomotion, for which the necessary power for motion is so much greater as the speed of motion increases. But we must reflect that in the airplane, the power necessary for motion is partly absorbed in overcoming the passive resistances, partially in order to insure sustentation; this dynamical sustentation admits a maximum efficiency corresponding to a given value of speed, below which, consequently, the efficiency itself decreases.

Practically, the speed V_{min} corresponds to the minimum value which the speed of the airplane can assume. It is quite true that theoretically the speed of the airplane can still decrease, but the further decrease is of no interest, as it requires increase of power which makes the sustentation more difficult, and therefore the flight more dangerous.

When the speed increases to values greater than V_{min}, the power necessary for sustentation rapidly increases. The maximum value the airplane speed can assume, evidently depends upon the maximum value of useful power the propeller can furnish.

Let P_2 be the power of the engine, and ρ the propeller efficiency; the useful power furnished by the propeller is evidently ρP_2.

To study flying with the engine running, it is necessary to draw the diagram ρP_2 as a function of V, in order to be able to compare for each value of V, the power ρP_2 available for that speed, and the power necessary for flying, also at that speed.

Therefore, it is necessary to know the following diagrams:

(1) $P_2 = f(n)$

(2) $\alpha = f\left(\frac{V}{nD}\right)$, which gives the value of coefficient α

of the formula $P_p = \alpha n^3 D^5$, corresponding to the power absorbed by the propeller, and

(3) $\rho = f\left(\frac{V}{nD}\right)$

The first of the three diagrams must be determined in the engine testing room, and the other two in the aerodynamical laboratory. When they are known, the determination of values ρP_2 as a function of V becomes possible by using a method also proposed by Eifell, and which is interesting to expose diffusely.

Let us consider the equation

$$P_p = \alpha n^3 D^5$$

or

$$\frac{P_p}{n^3 D^5} = \alpha$$

As we have seen in chapter 6, $\alpha = f\left(\frac{V}{nD}\right)$, therefore

$$\frac{P_p}{n^3 D^5} = f\left(\frac{V}{nD}\right)$$

Now, instead of drawing the diagram by taking the values of $\frac{V}{nD}$ as abscissæ, and those of $\frac{P_p}{n^3 D^5}$ as ordinates on uniform scales, let us take these values, respectively, as abscissæ and as ordinates, on paper with logarithmic graduation (Fig. 89).

Let us now consider a point on the curve $\frac{P_p}{n^3 D^5} = f\left(\frac{V}{nD}\right)$; for instance, point A. The abscissa of this point is $OX = \log \frac{V}{nD}$; but $\log \frac{V}{nD} = \log V - \log n - \log D$, consequently we can consider OX as the algebraical sum of the following three, $\log V$, $- \log n$, and $- \log D$. Analogously, the ordinate OY of point A, is $OY = \log \frac{P_p}{n^3 D^5}$, and we can write $OY = \log P_p - 3 \log n - 5 \log D$, considering OY as the algebraic sum of the following, $\log P$, $- 3 \log n$ and $- 5 \log D$. Then, in order to pass from the origin O, to point A of the diagram, it is sufficient to add $\log V$, $- \log n$ and $- \log D$ following axis OX, and $\log P_p$, $- 3 \log n$ and $- 5 \log D$ following axis OY.

Since evidently these segments can be added in any order whatever, we can first take $\log V$, then $- \log n$ parallel to axis OX, and $- 3 \log n$ parallel to the axis of the ordinates, then again $- \log D$ parallel to the axis of the abscissæ, and $- 5 \log D$ parallel to the axis of the ordinates, and finally $\log P_p$. Now it is evident that the two segments $- \log n$ and $- 3 \log n$ corresponding to n, can be replaced by a single oblique segment with an inclination of 3 on 1 and having a length proportional to log n. Analogously, the two segments $-\log D$ and $-5\log D$ corresponding to D,

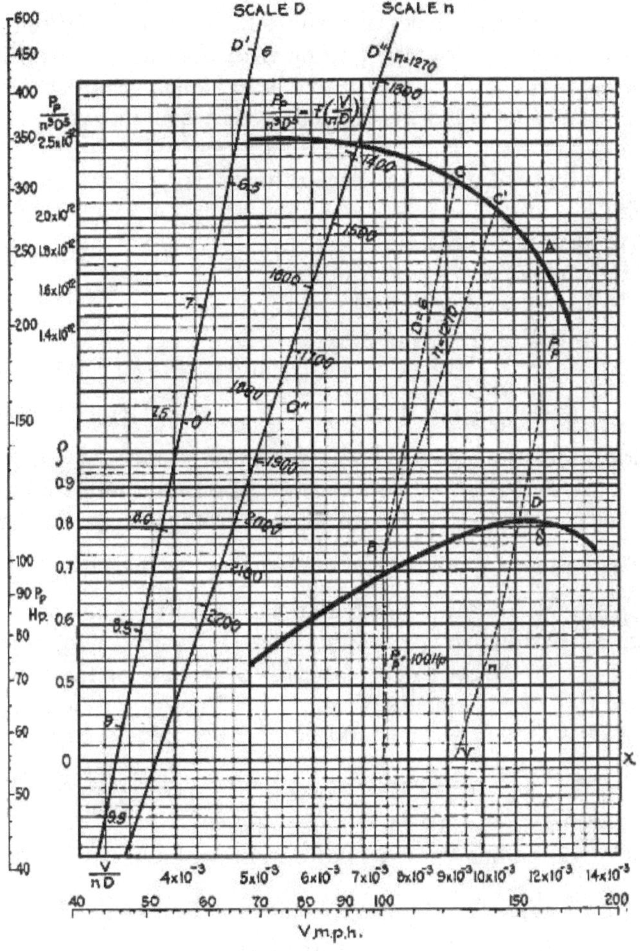

FIG. 89.

can be replaced by a single oblique segment with inclination of 5 on 1 and having a length proportional to log D.

We can definitely pass from origin O to point A of the diagram, by drawing four segments parallel respectively to axis OX, to an axis of inclination 3 on 1, to an axis of inclination 5 on 1, and to axis OY, and which measure V, n, D, and P_p, in their respective scales.

The condition necessary and sufficient for a system of values of V, n, D and P_p to be realizable with a propeller corresponding to the diagram, is evidently that the four corresponding segments (added geometrically starting from the origin) terminate on the diagram.

The units of measure selected for drawing the diagram of Fig. 89 are:

V, in miles per hour
n, in revolutions per minute
D, in feet and
P_p in H.P.

In order to determine the relation between the scales of $\frac{V}{nD}$ and $\frac{P_p}{n^3D^5}$ and those of V, P_p, n, and D, it is necessary to fix the origin of the scales of n and D. Let us suppose that the origin of the scale n be 1800 r.p.m. and that of scale D be 7.5 ft. Then for $n = 1800$ and $D = 7.5$ the coördinates $\frac{V}{nD}$ and $\frac{P_p}{n^3D^5}$ evidently also measure V and P_p; in fact for these particular values, the segments to be laid off parallel to the scales n and D, become zero, and so we go from origin to the diagram by means of the sum of only the two segments V and P_p. Then, considering for instance the speed $V = 100$ m.p.h., it must be marked on the axis OX at the point where $\frac{V}{nD} = \frac{100}{1800 \times 7.5} = 0.0074$. In this way the scale of V is determined.

Corresponding to $V = 100$ m.p.h. we have (see diagram Fig. 89) $\frac{P_p}{n^3D^5} = 2.46 \times 10^{-12}$; thus, making $n = 1800$ and $D = 7.5$ we shall have $P_p = 340$ H.P.; marking the value of $P = 340$ in correspondence to $\frac{P_p}{n^3D^5} = 2.46 \times 10^{-12}$ determines the scale of powers P_p.

In order to find the scale of D, make n equal to 1800, for which the segment n is equal to zero.

Now, by giving V and P_p any two values whatever (for instance $V = 100$ m.p.h. and $P_p = 100$ H.P.) by means of the usual construction a segment BC is determined, which measures the diameter D on the scale of D. The value of D results from the value $\frac{P_p}{n^3D^5}$, which is read on the diagram at point C; in our case, this value is 2.22×10^{-12} and consequently, as $P_p = 100$ and $n = 1800$, we shall have

$$\frac{100}{1800^3 \times D^5} = 2.2 \times 10^{-12}$$

which gives $D = 6$ ft. Thus, by taking to the scale of D, starting from origin O' (which is supposed to correspond to $D = 7.50$ ft.), a segment $O'D' = BC$, and marking the value 6 ft. on the point D', the scale of D is obtained.

Finally, to find the scale of n, it is sufficient to make $D = 7.5$, $V = 100$ m.p.h. and $P_p = 100$, and by repeating analogous construction we find that the segment BC' corresponding to C' is $\frac{P_p}{n^3D^5} = 2.06$; then for $P_p = 100$ and $D = 7.5$ the result is $n = 1270$. Then, by taking to the scale of n, starting from origin O'' (which by hypothesis is equal to $n = 1800$), a segment $O''D'' = BC'$, and marking the value 1270 r.p.m. on the point D'', the scale of n is defined.

Analogously, we can also draw the diagram $\rho = f\left(\frac{V}{nD}\right)$, on the logarithmic paper, by selecting the same units of measure (Fig. 89).

Let us suppose that we know the diagram $P_2 = f\ (n)$, (Fig. 90), which is easily determined in the engine testing room; we can then draw that diagram by means of the scale n, and the scale of the power shown in Fig. 89 (Fig. 91).

Disposing of the three diagrams

$$\frac{P_p}{n^3D^5} = f\left(\frac{V}{nD}\right)$$

$$\rho = f\left(\frac{V}{nD}\right)$$

$$P_2 = f\ (n)$$

drawn on logarithmic paper, it is easy to find the values ρP_2 corresponding to the values of V.

In fact let us draw in Fig. 91, starting from the origin of the scale of n, a segment equal to diameter D of the propeller adopted, measuring D to the logarithmic scale of Fig. 89, in magnitude and direction. We shall have

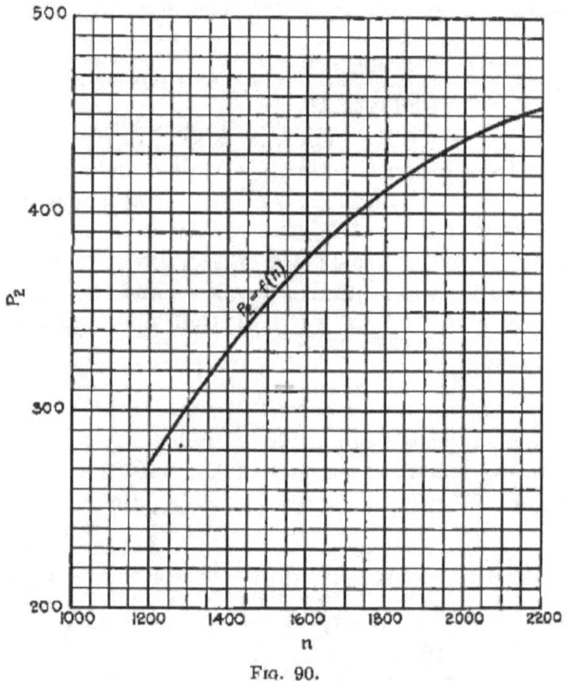

Fig. 90.

point V'; then draw the horizontal line $V'x$. Supposing that Fig. 91 be drawn on transparent paper, let us take it to the diagram of Fig. 89, making $V'x$ coincide with axis OX, and the point V' with any value V whatever, of the speed.

Fig. 92 shows how the operation is accomplished, supposing V' to be made coincident with V = 100 m.p.h. and supposing D = 9.0 feet.

The point of intersection A between the curves P_p and

$P_2 = f(n)$
500
450
400
350
300
250
200
150
100
0
1200
Scale π
1400
V'
1600
1800
0"
2000
2200

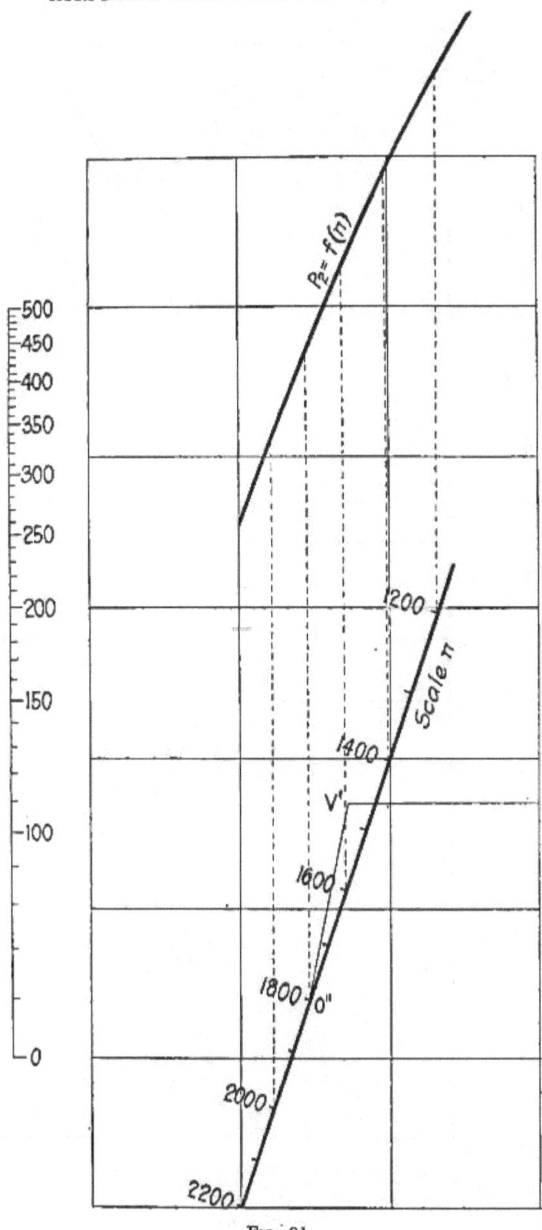

Fig. 91.

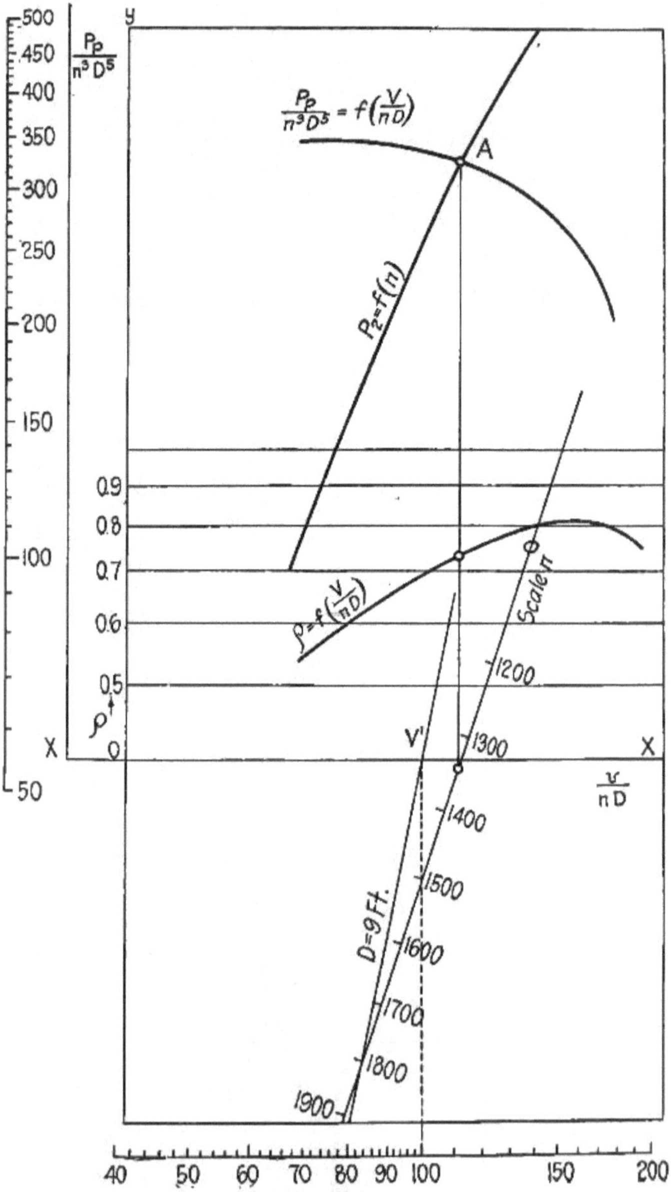
500
450
400
350
300
250
200
150
100
50
$\frac{P_p}{n^3 D^5}$
y
$\frac{P_p}{n^3 D^5} = f\left(\frac{V}{nD}\right)$
A
$P_2 = f(n)$
0.9
0.8
0.7
0.6
0.5
ρ
$\rho = f\left(\frac{V}{nD}\right)$
Scale n
1200
1300
1400
1500
1600
1700
1800
1900
V'
X
0
X
$\frac{V}{nD}$
D=9 Ft.
40
50
60
70
80
90
100
150
200

Fig. 92.

P_2 determines the values of P_2, ρ and n corresponding to an even speed.[1]

We can then determine for each value of V, the corresponding value P_2, and we can obtain the values $\rho \times P_2$ corresponding to those of V in Fig. 88. This has been done in Fig. 93. Comparing, in this figure, the values of ρP_2 and P_1 corresponding to the various speeds, we see that $\rho P_2 = P_1$ for $V = 160$ m.p.h.; this value represents the maximum speed that the airplane under consideration can attain; in fact for higher values of V, a greater power to the one effectively developed by the engine at that speed, would be required.

For all the speed values lower than the maximum value $V = 160$ m.p.h. the disposable power on the propeller shaft is greater than the minimum power necessary for horizontal flight; the excess of power measured by the difference between the values ρP_2 and P_1, as they are read on the logarithmic scales, can be used for climbing. The climbing speed v is easily found when the weight W of the machine is known. In fact in order to raise a weight W at a speed v, a power of $v \times W$ lb. ft. $= \frac{1}{550} \times v \times W$ H.P. is necessary; we now dispose of a power $\rho P_2 - P_1$, consequently the climbing speed is given by

$$\rho P_2 - P_1 = \frac{1}{550} \times v \times W$$

that is,

$$v = \frac{550}{W} \times (\rho P_2 - P_1)$$

The climbing speed is thus proportional to the difference $\rho P_2 - P_1$; it will be maximum corresponding to the maximum value of $\rho P_2 - P_1$; in our example, this maximum is found for $V = 95$ and corresponding to it $v = 33$ ft. per sec.

[1] In fact, point A determines a pair of values of V and n, which are compatible either to the diagram of the power absorbed by the propeller, or to the diagram of the power developed by the engine.

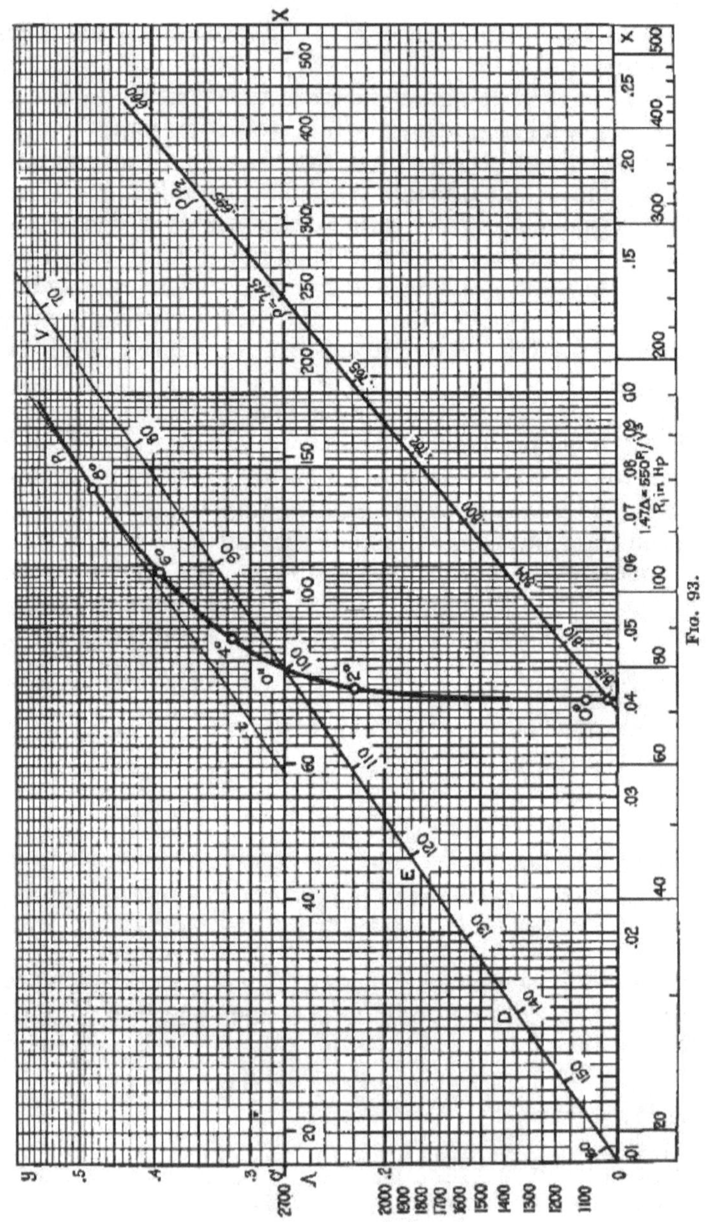

Fig. 93.

The ratio $\frac{v}{V}$ gives the value sin θ which defines the angle θ, as being the angle which the ascending line of path makes with the horizontal line (Fig. 94). We then have

$$v = V \sin \theta$$

This equation shows that the maximum v corresponds to the maximum value of V sin θ, and not to the maximum value of sin θ; that is, it may happen that by increasing the angle θ, the climbing speed will be decreasing instead of increasing.

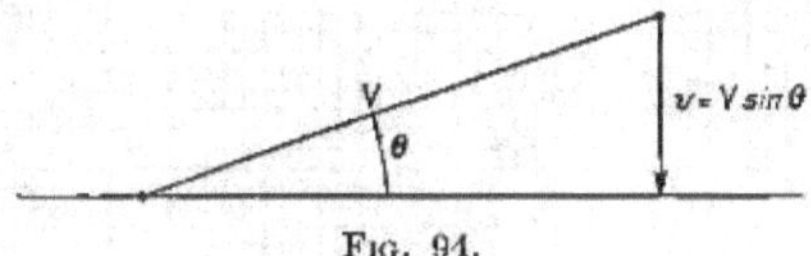

Fig. 94.

In Fig. 95 we have drawn, for the already discussed example, diagrams of v and sin θ as functions of V. We see that v is maximum for sin θ = 0.35; for the value sin θ = 0.425, which represents the maximum of sin θ, we have v = 29, which is less than the preceding value.

We also see that in climbing, the speed of the airplane is less than that of the airplane in horizontal flight, supposing that the engine is run at full power.

The maneuver that must be accomplished by the pilot in order to increase or decrease the climbing speed, consists in the variation of the angle of incidence of the airplane, by moving the elevator.

In fact, as we have already seen,

$$W = 10^{-4} \lambda A V^2$$

Fixing the angle of incidence fixes the value of λ, and consequently that of V necessary for sustentation; the airplane then automatically puts itself in the climbing line of path, to which velocity V corresponds.

But the pilot has another means for maneuvering for height; that is, the variation of the engine power by adjusting the fuel supply. In fact, let us suppose that the pilot reduces the power ρP_2; then the difference $\rho P_2 - P_1$, will decrease, consequently decreasing V and sin θ. If the pilot reduces the engine power to a point where $\rho P_2 - P_1 = 0$; the result will be $v = 0$ and $\sin \theta = 0$. We see then the possibility, by throttling the engine, of flying at a whole

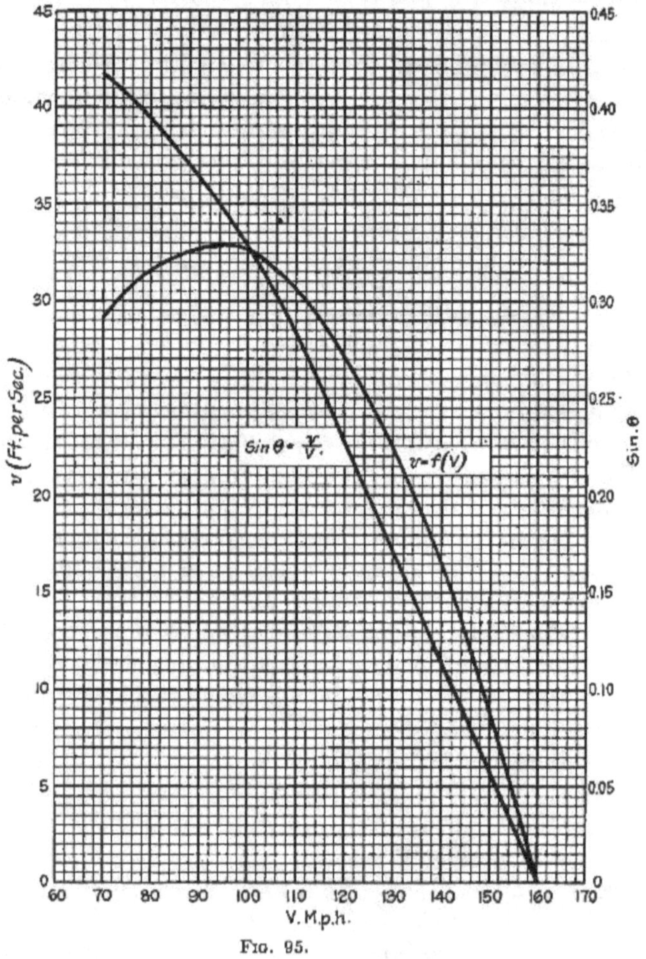

Fig. 95.

series of speeds, varying from a minimum value, which depends essentially upon the characteristics of the airplane, to a maximum value which depends not only upon the airplane, but also upon the engine and propeller.

CHAPTER X

STABILITY AND MANEUVERABILITY

Let us consider a body in equilibrium, either static or dynamic; and let us suppose that we displace it a trifle from the position of equilibrium already mentioned; if the system of forces applied to the body is such as to restore it to the original position of equilibrium, it is said that the body is in a state of stable equilibrium.

In this way we naturally disregard the consideration of forces which have provoked the break of equilibrium. From this analogy, some have defined the *stability of the airplane* as the "*tendency to react on each break of equilibrium without the intervention of the pilot.*" Several constructors have attempted to solve the problem of stability of the airplane by using solely the above criterions as a basis.

In reality in considering the stability of the airplane, the disturbing forces which provoke the break of a state of equilibrium, cannot be disregarded.

These forces are most variable, especially in rough air, and are such as to often substantially modify the resistance of the original acting forces. The knowledge of them and of their laws of variation is practically impossible; therefore there is no solid basis upon which to build a general theory of stability.

Nevertheless, by limiting oneself to the flight in smooth air, it is possible to study the general conditions to which an airplane must accede in order to have a more or less great intrinsic stability.

Let us consider an airplane in normal rectilinear horizontal flight of speed V. The forces to which the airplane is subjected are:

its weight W,

the propeller thrust T, and

the total air reaction R.

These forces are in equilibrium; that is, they meet in one point and their resultant is zero (Fig. 96).

The axis of thrust T generally passes through the center of gravity. Then R also passes through the center of gravity. Supposing now that the orientation of the airplane with respect to its line of path is varied abruptly, leaving all the control surfaces neutral; the air reaction R will change not only in magnitude, but also in position. The variation in magnitude has the only effect of elevating or lowering the line of path of the airplane; instead, the variation in position introduces a couple about the center of gravity, which tends to make the airplane turn. If this turning has the effect of reëstablishing the original position, the airplane is stable. If, however, it has the effect of *increasing* the displacement, the airplane is unstable.

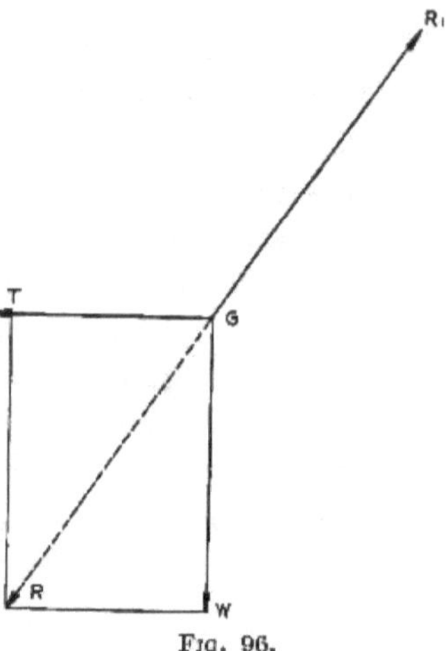

FIG. 96.

For simplicity, the displacements about the three principal axes of inertia, the pitching axis, the rolling axis, and the directional axis (see Chapter II), are usually considered separately.

For the pitching movement, it is interesting only to know the different positions of the total resultant R corresponding to the various values of the angle of incidence. In Fig. 97 a group of straight lines corresponding to the various positions of the resultant R with the variation of the angle of incidence, have been drawn only as a qualitative example. If we suppose that the normal incidence of flight of the airplane is 3°, the center of gravity (because of what has been said before), must be found on the resultant $R_{3°}$. Let us consider the two positions G_1 and G_2. If the center of gravity falls on G_1 the machine is unstable; in fact for angles greater than 3° the resultant is displaced so as to have a tendency to further increase the incidence and *vice versa*. If, instead, the center of gravity falls in G_2, the airplane, as demonstrated in analogous considerations, is stable.

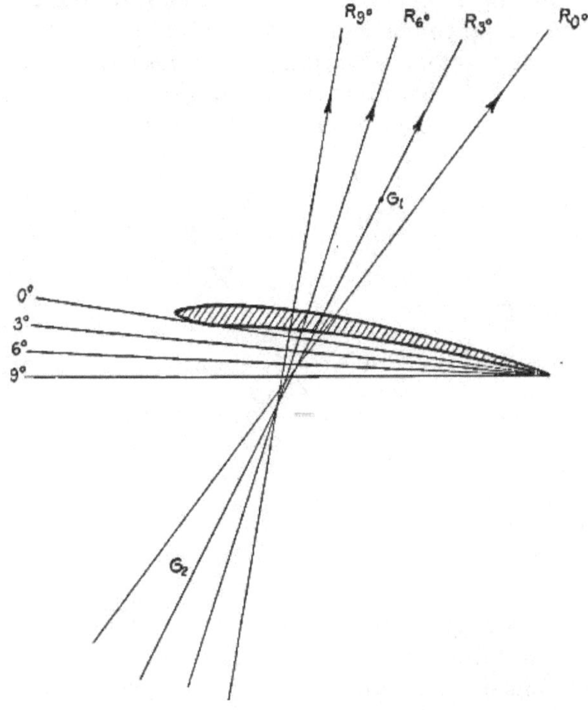

Fig. 97.

In general, the position of the center of gravity can be displaced within very restricted limits, more so if we wish to let the axis of thrust pass near it. On the other hand, it is not possible to raise the wing surfaces much with respect to the center of gravity, because the raising would produce a partial raising of the center of gravity, and also because of constructional restrictions.

Then, in order to obtain a good stability, the adoption of

stabilizers is usually resorted to, which (as we have seen in Chapter II) are supplementary wing surfaces, generally situated behind the principal wing surfaces and making an angle of incidence smaller than that of the principal wing surface. The effect of stabilizers is to raise the zone in which the meeting points of the various resultants are, thus facilitating the placing of the center of gravity within the zone of stability. Naturally it is necessary that the intrinsic stability be not excessive, in order that the maneuvers be not too difficult or even impossible.

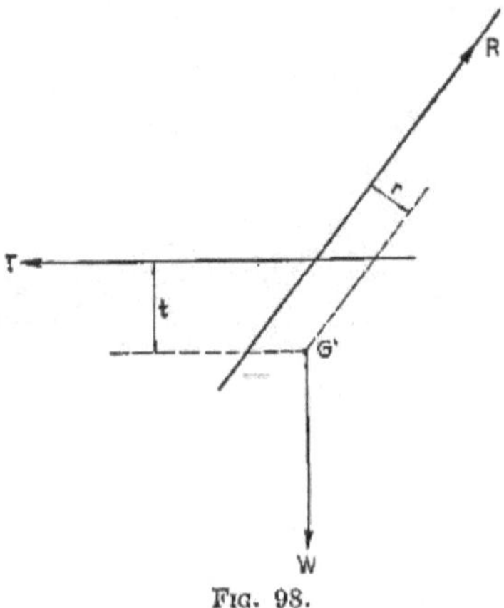

FIG. 98.

The preceding is applied to cases in which the axis of thrust passes through the center of gravity. It is also necessary to consider the case, which may happen in practice, in which the axis of thrust does not pass through the center of gravity. Then, in order to have equilibrium, it is necessary that the moment of the thrust about the center of gravity $T \times t$, be equal and opposite to the moment $R \times r$ of the air reaction (Fig. 98). Let us see which are the conditions for stability.

To examine this, it is necessary to consider the metacentric curve, that is, the *enveloping curve* of all the resultants (Fig. 99). Starting from a point O, let us take a group of segments parallel and equal to the various resultants R_i

corresponding to the normal value of the speed. Let us consider one of the resultants, for instance R_i. At point A, where R_i is tangent to the metacentric curve α, let us draw oa parallel to b, which is tangent to curve β at B the extreme end of R_i.

We wish to demonstrate that the straight line oa is a *locus* of points such that if the center of gravity falls on it, and the equilibrium exists for a value of the angle of incidence, this equilibrium will exist for all the other values of incidence,

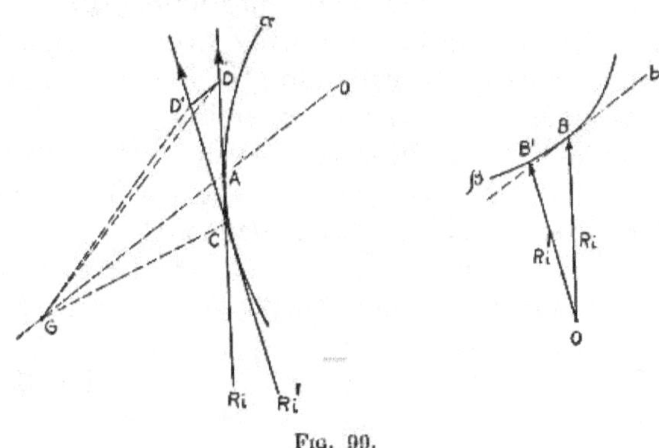

Fig. 99.

(understanding the speed to be constant). In other words, we wish to demonstrate that oa is a *locus* of the points corresponding to the indifferent equilibrium, and consequently it divides the stability zone from the instability zone.

Let us suppose that the center of gravity falls at G on oa, and that the incidence varies from the value i (for which we have the equilibrium) to a value infinitely near i'. If we demonstrate that the moment of R'_i about G is equal to the moment of R_i, the equilibrium will be demonstrated to be indifferent. Starting from C point of the intersection of R_i and R'_i, let us take two segments CD and CD' equal to the value R_i and R'_i respectively. The joining line DD' is parallel to BB'; now when i' differs infinitely little from i, BB' becomes tangent to the curve β at point B; consequently, DD' becomes parallel to tangent b; that is, also to straight line ao. Now point C, if i' differs infinitely little from i, is coincident with A (and consequently the segments GC with GA) then the two triangles GCD' and GCD (which measure the moment of R_i and R'_i with respect to G), become equal, as they have common bases and have vertices situated on a line parallel to the bases: that is, the equilibrium is indifferent.

To find which are the zones of stability and instability, it suffices to suppose for a moment that the center of gravity falls on the intersection of the propeller axis and the resultant R_i, then the center of gravity will be on R_i; and since A is on the line oa, it will be a point of indifferent equilibrium, consequently dividing the line R_i into two half lines corresponding to the zones of stability and instability. From what has already been said, it will be easy to establish the half line which corresponds to the stability, and thus the entire zone of stability will be defined.

The calculation of the magnitude of the moments of stability, is not so difficult when the metacentric curve and the values R_i for a given speed are known.

The foregoing was based upon the supposition that the machine would maintain its speed constant, even though varying its orientation with respect to the line of path. Practically, it happens that the speed varies to a certain extent; then a new unknown factor is introduced, which can alter the values of the restoring couple. Nevertheless, it should be noted that these variations of speed are never instantaneous.

In referring to the elevator, in Chapter II, we have seen that its function is to produce some positive and negative couples capable of opposing the stabilizing couples, and consequently permitting the machine to fly with different values of the angle of incidence. All other conditions being the same (moment of inertia of the machine, braking moments, etc.), the mobility of a machine in the longitudinal sense, depends upon the ratio between the value of the stabilizing moments and that of the moments it is possible to produce by maneuvering the elevator. A machine with great stability is not very maneuverable. On the other hand, a machine of great maneuverability can become dangerous, as it requires the continuous attention of the pilot.

An ideal machine should, at the pilot's will, be able to change the relative values of its stability and maneuverability; this should be easy by adopting a device to vary the ratios of the controlling levers of the elevator. In this way, the other advantage would also be obtained of being able to decrease or increase the sensibility of the controls as the speed increases or decreases. Furthermore, we could resort to having strong stabilizing couples prevail normally in the machine, it being possible at the same time to immediately obtain great maneuverability in cases where it became necessary.

As to lateral stability, it can be defined as the tendency of the machine to deviate so that the resultant of the forces of mass (weight, and forces of inertia) comes into the plane of symmetry of the airplane.

When, for any accidental cause whatever, an airplane inclines itself laterally, the various applied forces are no longer in equilibrium, but have a resultant, which is not contained in the plane of symmetry.

Then the line of path is no longer contained in the plane of symmetry and the airplane *drifts*. On account of this fact, the total air reaction on the airplane is no longer contained in the plane of symmetry, but there is a drift component, the line of action of which can pass through, above or below the center of gravity.

In the first case, the moment due to the drift force about the center of gravity is zero, consequently, if the pilot does not intervene by maneuvering the ailerons, the machine will gradually place itself in the course of drift, in which it will maintain itself. In the other two cases, the drift component will have a moment different from zero, and which will be stabilizing if the axis of the drift force passes above the center of gravity; it will instead, be an overturning moment if this axis passes below the center of gravity. To obtain a good lateral stability, it is necessary that the axis of the drift component meet the plane of symmetry of the machine at a point above the horizontal line contained in the plane of symmetry and passing through the center of gravity; that point is called the center of drift; thus *to obtain a good transversal stability it is necessary that the center of drift fall above the horizontal line drawn through the center of gravity* (Fig. 100). This result can be obtained by lowering the center of gravity, or by adopting a vertical fin situated above the center of gravity, or, as it is generally done, by giving the wings a transversal inclination usually called "dihedral". Naturally what has been said of longitudinal stability, regarding the convenience of not having

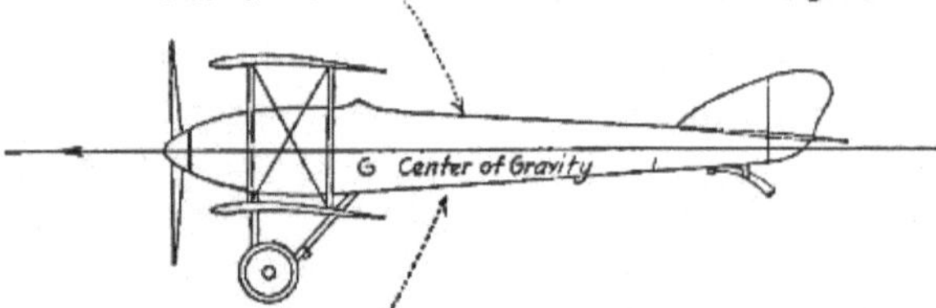

Fig. 100.

it excessive, so as not to decrease the maneuverability too much, can be applied to lateral stability.

Let us finally consider the problems pertaining to directional stability. The condition necessary for an airplane to have good stability of direction is, by a series of considerations analogous to the preceding one, that the center of drift fall behind the vertical line drawn through the center of gravity (Fig. 101). This is obtained by adopting a rear fins.

By adding Figs. 100 and 101, we have Fig. 102 which shows that the center of drift must fall in the upper right quadrant.

Summarizing, we may say that it is possible to build machines which, in calm air, are provided with a great intrinsic stability; that is, having a tendency to react every time the line of path tends to change its orientation relatively to the machine. It is necessary, however, that this tendency be not excessive, in order not to decrease the maneuverability which becomes an essential quality in rough air, or when acrobatics are being accomplished.

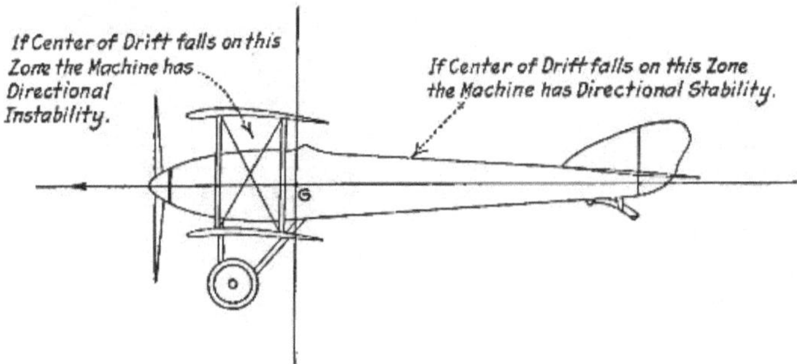

FIG. 101.

Thus far we have considered the flight with the engine running. Let us now suppose that the engine is shut off. Then the propeller thrust becomes equal to zero. Let us

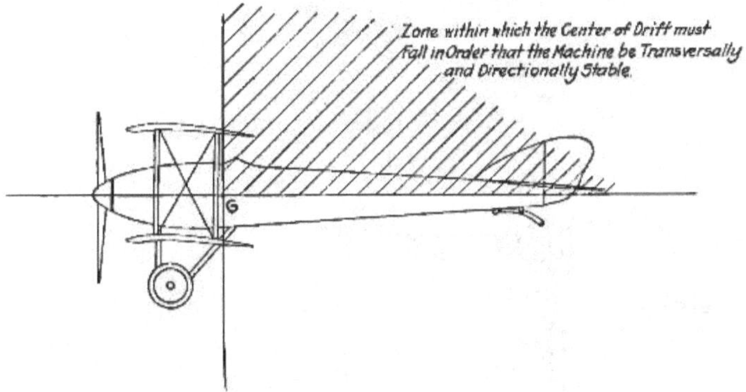

FIG. 102.

first consider the case in which the axis of thrust passes through the center of gravity.

In this case, the disappearance of the thrust will not bring any immediate disturbance in the longitudinal equilibrium of the airplane. But the equilibrium between

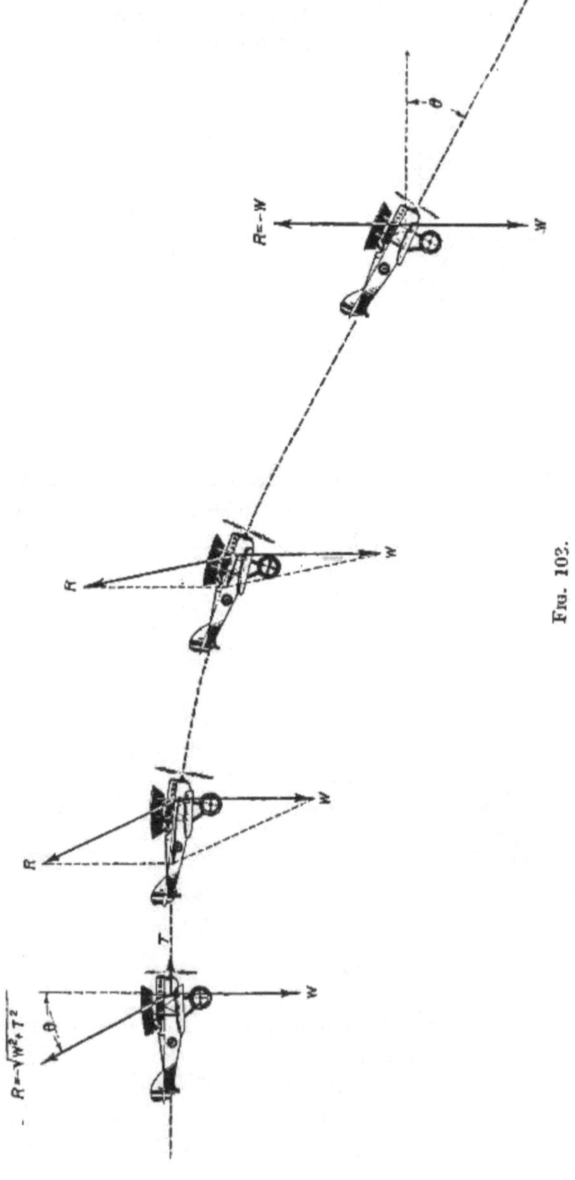

Fig. 102.

weight, thrust, and air reaction, will be broken, and the component of head resistance, being no longer balanced by the propeller thrust, will act as a brake, thereby reducing the speed of the airplane; as a consequence, the reduction of speed brings a decrease in the sustaining force; thus equilibrium between the component of sustentation of the air reaction and the weight is broken, and the line of path becomes descendent; that is, an increase of the angle of incidence is caused; a stabilizing couple is then produced, tending to restore the angle of incidence to its normal value; that is, tending to adjust the machine for the descent.

The normal speed of the airplane then tends to restore itself; the inclination of the line of path and the speed will increase until they reach such values that the air reaction becomes equal and of opposite direction to the weight of the airplane (Fig. 103). Practically, it will happen that this position (due to the fact that the impulse impressed on the airplane by the stabilizing couple makes it go beyond the new position of equilibrium) is not reached until after a certain number of oscillations. Let us note that the gliding speed in this case is smaller than the speed in normal flight; in fact in normal flight, the air reaction must balance W and T, and is consequently equal to $\sqrt{W^2 + T^2}$; in gliding instead, it is equal to W; that is, calling R' and R'' respectively, the air reaction in normal flight and in gliding flight,

$$\frac{R'}{R''} = \frac{\sqrt{W^2 + T^2}}{W} = \sqrt{1 + \frac{T^2}{W^2}}$$

and calling V' and V'' the respective speeds, we will have

$$\frac{V'}{V''} = \sqrt{\frac{R'}{R''}} = \sqrt[4]{1 + \frac{T^2}{W^2}}$$

When the axis of thrust does not pass through the center of gravity, as the engine is shut off a moment is produced equal and of opposite direction to the moment of the thrust with respect to the center of gravity. Thus if the axis of thrust passes above the center of gravity, the moment developed will tend to make the airplane nose up. If instead, it passes below the center of gravity, the moment developed will tend to make the airplane nose down. If the airplane is provided with intrinsic stability, a gliding course will be established, with an angle of incidence different from that in normal flight, and which will be greater in case the axis of thrust passes above the center of gravity, and smaller in the opposite case. The speed of gliding in the first case, will be smaller, and greater in the second case than the speed obtainable when the axis of thrust passes through the center of gravity.

Naturally, the pilot intervening by maneuvering the control surfaces can provoke a complete series of *equilibrium*, and thus, of paths of *descent*.

We have seen that when a stabilizing couple is introduced, the airplane does not immediately regain its original equilibrium, but attains it by going through a certain number of oscillations of which the magnitude is directly proportional to the stabilizing couple; in calm air, the oscillations diminish by degrees, more or less rapidly according to the importance of the dampening couples of the machine.

In rough air, instead, sudden gusts of wind may be encountered which tend to increase the amplitude of the oscillations, thus putting the machine in a position to provoke a definite brake of the equilibrium, and consequently to fall. That is why the pilot must have complete control of the machine; that is, machines must be provided with great maneuverability in order that it may be possible, at the pilot's will, to counteract the disturbing couple, as well as to dampen the oscillations. In other words, if the controls are energetic enough, the maneuvers accomplished by the pilot can counteract the periodic movements, thereby greatly decreasing the pitching and rolling movements.

In order to accomplish acrobatic maneuvers such as turning on the wing, looping, spinning, etc., it is necessary to dispose of the very energetic controls, not so much to start the maneuvers themselves, as to rapidly regain the normal position of equilibrium if for any reason whatever the necessity arises.

Let us consider an airplane provided with intrinsic automatic stability, as being left in the air with a dead engine and insufficient speed for its sustentation. The airplane will be subjected to two forces, weight and air reaction, which do not balance each other, as the air reaction can have any direction whatever according to the orientation of the airplane and the relative direction of the line of path.

Let us consider two components of the air reaction, the vertical component and the horizontal component. The vertical component partly balances the weight; the difference between the weight and this component measures the forces of vertical acceleration to which the airplane is subjected. The horizontal component, instead, can only be balanced by a horizontal component of acceleration; in other words, it acts as a centripetal force, and tends to make the airplane follow a circular line of path of such radius that the centrifugal force which is thereby developed, may establish the equilibrium. Thus, an airplane left to itself, falls in a spiral line of path, which is called spinning. Let us suppose, now, that the pilot does not maneuver the controls; then, if the machine is provided with intrinsic stability, it will tend to orient itself in such a way as to have the line of path situated in its plane of symmetry and making an angle of incidence with the wing surface equal to the angle for which the longitudinal equilibrium is obtained. That is, the machine will tend to leave the spiral fall, and put itself in the normal gliding line of path. Naturally in order that this may happen, a certain time, and, what is more important, a certain vertical space, are necessary. The disposable vertical space may happen to be insufficient to enable the machine to come out of its course in falling; in that case a crash will result.

We see then what a great convenience the pilot has in being able to dispose of the energetic controls which can be properly used to decrease the space necessary for restoring the normal equilibrium.

Summarizing, we can mention the following general criterions regarding the intrinsic stability of a machine:

1. It is necessary that the airplane be provided with intrinsic stability in calm air, in order that it react automatically to small normal breaks in equilibrium, without requiring an excessive nervous strain from the pilot;

2. This stability must not be excessive in order that the maneuvers be not too slow or impossible; and

3. It is necessary that the maneuvering devices be such as to give the pilot control of the machine at all times.

Before concluding the chapter it may not be amiss to say a few words about mechanical stabilizers. Their scope is to take the place of the pilot by operating the ordinary maneuvering devices through the medium of proper servo-motors. Naturally, apparatuses of this kind, cannot replace the pilot in all maneuvers; it is sufficient only to mention the landing maneuver to be convinced of the enormous difficulty offered by a mechanical apparatus intended to guide such a maneuver. Essentially, their use should be limited to that of replacing the pilot in normal flight, thereby decreasing his nervous fatigue, especially during adverse atmospheric conditions.

We can then say at once that a mechanical stabilizer is but an apparatus sensible to the changes in equilibrium which is desired to be avoided, or sensible to the causes which produce them, and capable of operating, as a consequence of its sensibility, a servo-motor, which in turn maneuvers the controls. We can group the various types of mechanical stabilizers, up to date, into three categories:

1. Anemometric,
2. Clinometric, and
3. Inertia stabilizer.

There are also apparatus of compound type, but their parts can always be referred to one of the three preceding categories.

1. The anemometric stabilizers are, principally, speed stabilizers. They are, in fact, sensible to the variations of the relative speed of the airplane with respect to the air, and consequently tend to keep that speed constant.

Schematically an anemometric stabilizer consists of a small surface *A* (Fig. 104), which can go forward or backward under the action of the air thrust *R*, and under the reaction of a spring *S*. The air thrust *R*, is proportional to the square of the speed. When the relative speed is equal to the normal one, a certain position of equilibrium is obtained; if the speed increases, *R* increases and the small disk goes backward so as to further compress the spring. If, instead, the speed decreases, *R* will decrease, and the

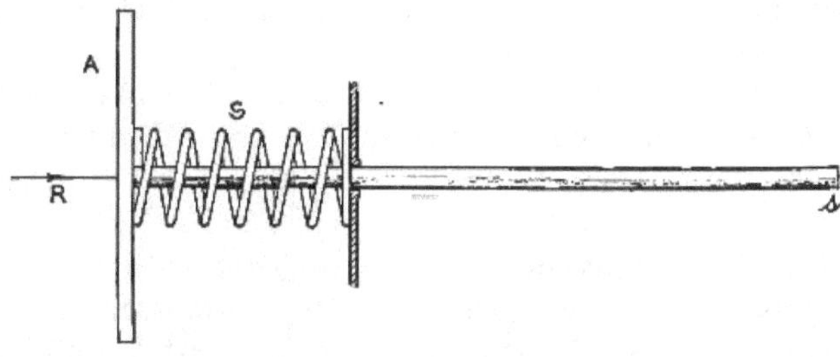

Fig. 104.

small disk will go forward under the spring reaction. Through rod *S*, these movements control a proper servomotor which maneuvers the elevator so as to put the airplane into a climbing path when the speed increases, and into a descending path when the speed decreases.

Such functioning is logical when the increase or decrease of the relative speed depends upon the airplane, for instance, because of an increase or decrease of the motive power. The maneuver, however, is no longer logical if the increase of relative speed depends upon an impetuous gust of wind which strikes the airplane from the bow; in fact, this maneuver would aggravate the effect of the gust, as it would cause the airplane to offer it a greater hold.

Thus we see that an anemometric stabilizer, used by itself, can give, as it is usually said, counter-indications, which lead to false maneuvers.

In consideration of this, the Doutre stabilizer, which is until now, one of the most successful of its kind ever built, is provided with certain small masses sensible to the inertia forces, and of which the scope is to block the small anemometric blade when the increase of relative speed is due to a gust of wind.

2. Several types of clinometric stabilizers have been proposed; the mercury level, the pendulum, the gyroscope, etc.

The common fault of these stabilizers is that they are sensible to the forces of inertia.

The best clinometric stabilizer that has been built, and which is to-day considered the best in existence, is the Sperry stabilizer.

It consists of four gyroscopes, coupled so as to insure the perfect conservation of a horizontal plane, and to eliminate the effect of forces of inertia, including the centrifugal force.

The relative movements of the airplane with respect to the gyroscope system, control the servo-motor, which in turn actions the elevator and the horizontal stabilizing surfaces. A special lever, inserted between the servo-motor and the gyroscope, enables the pilot to fix his machine for climbing or descending; then the gyroscope insures the wanted inclination of the line of path.

There is a small anemometric blade which fixes the airplane for the descent when the relative speed decreases. A special pedal enables the detachment of the stabilizer and the control of the airplane in a normal way.

3. The inertia stabilizers are, in general, made of small masses which are utilized for the control of servo-motors; and which, under the action of the inertia forces and reacting springs, undergo relative displacement.

In general, the disturbing cause, whatever it may be, can be reduced, with respect to the effects produced by it, to a force applied at the center of gravity, and to a couple.

The force admits three components parallel to three principal axes, and consequently originates three accelerations (longitudinal, transversal, and vertical). The couple can be resolved into three component couples, which originate three angular accelerations, having as axis the same principal axis of inertia. A complete inertia stabilizer should be provided with three linear accelerometers and three angular accelerometers, which would measure the six aforesaid components.

CHAPTER XI

FLYING IN THE WIND

Let us first of all consider the case of a wind which is constant in direction as well as in speed.

Such wind has no influence upon the stability of the airplane, but influences solely its speed relative to the ground.

Let V be the speed proper of the airplane, and W the speed of the wind; in flight the airplane can be considered as a body suspended in a current of water, of which the

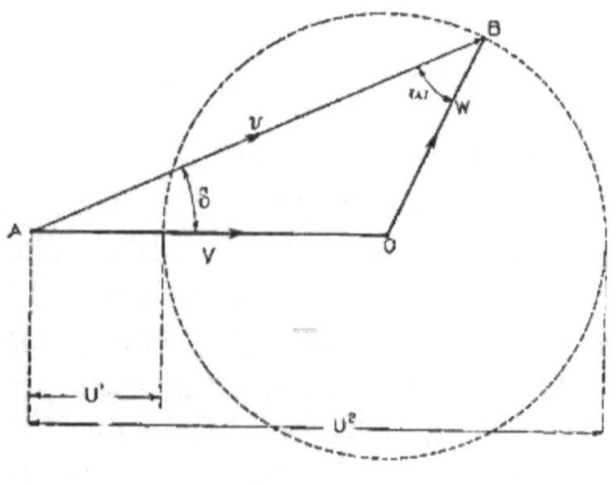

Fig. 105.

speed U, with respect to the ground, becomes equal to the resultant of the two speeds V and W; we can then write (Fig. 105)

$$U = V + W$$

We see then, that the existence of a wind W changes speed V not only in dimension but also in direction.

Furthermore, if from a point A we wish to reach another point B, and ω is the angle which the wind direction makes with the line of path AB, it is necessary to make the airplane fly not in direction AB, but in a direction AO making an angle δ with AB such that the resulting speed U is in the direction AB. By a known geometrical theorem, we have

$$U = \sqrt{V^2 + W^2 - 2UW \cos (180° - \delta - \omega)}$$

and

$$\sin \delta = \frac{W}{V} \sin \omega$$

A simple diagram is given in Fig. 106, which enables the calculation of angle δ, when the speeds V and W, and the angle ω which the wind makes with the line of flight to be covered, is known.

This diagram is constituted of concentric circles, whose radius represents the speed of the wind, and of a series of radii, of which the angles with respect to the line OA give the angles ω between the line of path and the wind. Let us find the angle δ of drift, at which the airplane must fly, for example, with a 30 m.p.h. wind making 90° with the line of path (*the drift angle of the trajectory* must not be confused with the angle of drift of the airplane with respect to the trajectory, of which we have discussed in the chapter on stability). Let us take point B the intersection of the circle of radius 30 with the line BC which makes 90° with OA; making B the center, and speed V of the airplane the radius, which we shall suppose equal to 100 m.p.h., we shall have point C which determines U and δ; in fact OC equals U, and angle BCO equals δ. In our case $U =$ 95.5 m.p.h., and $\sin \delta = 0.3$.

The speed of the wind varies within wide limits, and can rise to 110 miles per hour, or more; naturally it then becomes a violent storm.

A wind of from 7 to 8 miles an hour is scarcely perceptible by a person standing still. A wind of from 13 to 14 miles, moves the leaves on the trees; at 20 miles it moves the small branches on the trees and is strong enough to cause a flag to wave. At 35 miles the wind already gathers strength and moves the large branches; at 80 miles, light obstacles such as tiles, slate, etc., are carried away; the big storms, as we have already mentioned, even reach a speed of

110 miles an hour. As airplanes have actually reached speeds greater than 110 m.p.h. (even 160 m.p.h.), it would be possible to fly and even choose direction from point to point in violent wind storms. But the landing maneuver,

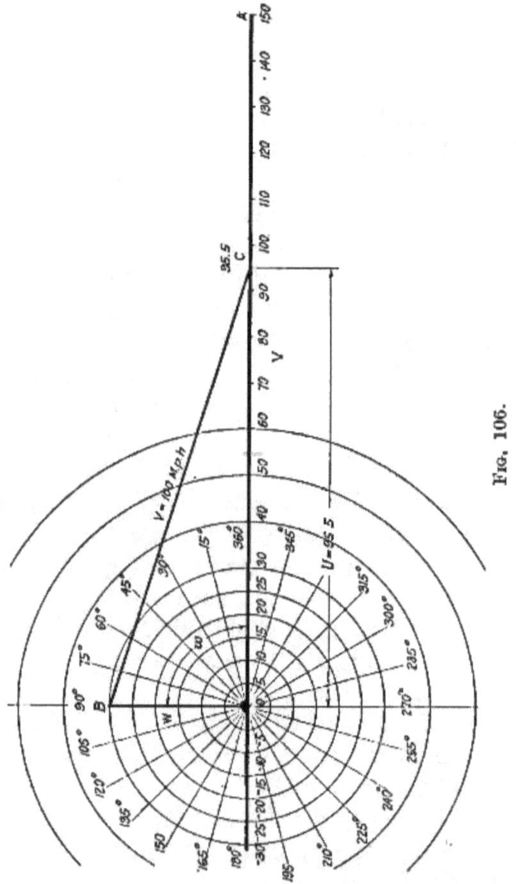

Fig. 106.

consequently, becomes very dangerous. At least during the present stage of constructive technique, it is wise not to fly in a wind exceeding 50 to 60 m.p.h. After all, such winds are the highest that are normally had, the stronger ones being exceptional and localized. On the contrary, for the aims of an organization, for instance, for aerial mail service, it would be useless to take winds higher than 30 to 40 m.p.h. into consideration.

If we call M the distance to be covered in miles, V the speed of the airplane in m.p.h., and W the maximum speed in m.p.h., of the wind to be expected, the travelling time in hours, when the wind is contrary, will be

$$l_w = \frac{M}{V - W} = \frac{M}{V} \times \frac{1}{1 - \frac{W}{V}}$$

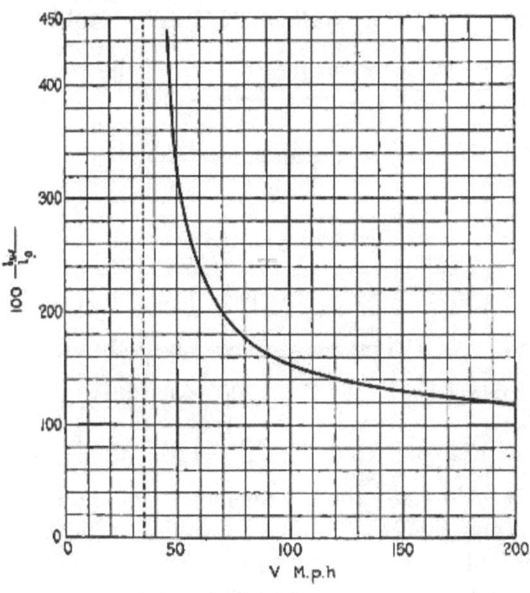

Fig. 107.

When the wind is zero the travelling time will be

$$l_o = \frac{M}{V}$$

consequently

$$l_w = l_o \times \frac{1}{1 - \frac{W}{V}}$$

Supposing that we admit, for instance in mail service, a maximum wind of 35 m.p.h., a diagram can easily be drawn which for every value of speed V, will give the value $100 \frac{l_w}{l_o}$ which measures the percent increase in the travelling time (Fig. 107).

This diagram shows that the travelling time tends to become infinite when V approaches the value of 35 m.p.h. For each value of V lower than 35 m.p.h. the value $100 \frac{l_w}{l_o}$ is negative; that is, the airplane having such a speed, and flying against a wind of 35 m.p.h. would, of course,

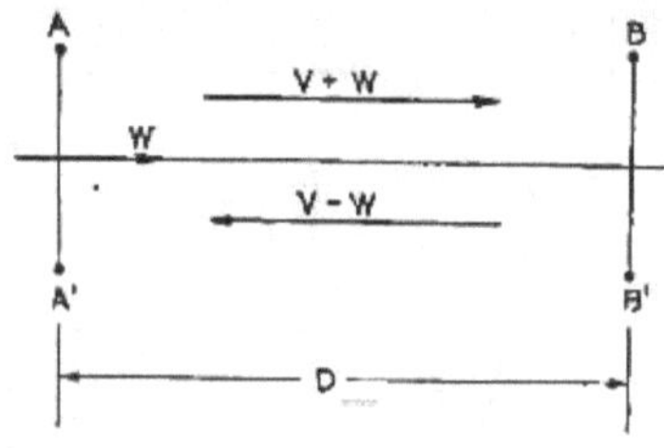

Fig. 108.

retrocede. As V increases above the value 35, the term $100 \frac{l_w}{l_o}$ decreases; for $V = 100$ we have for instance $100 \frac{l_w}{l_o} = 154$ per cent.; for $V = 130$, $100 \frac{l_w}{l_o} = 137$ per cent., etc. We see then, because of contrary wind, that the per cent increase in the travelling time, is inversely proportional to the speed.

Before beginning a discussion on the effect of the wind upon the stability of the airplane, it is well to guard against an error which may be made when the speed of an airplane is measured by the method of crossing back and forth between two parallel sights. Let AA' and BB' be the two parallel sights (Fig. 108). Let us suppose that a wind of speed W is blowing parallel to the line joining the parallel sights. Let t_1 be the time spent by the airplane in covering the distance D in the direction of AA' to BB', and t_2 the time spent to cover the distance in the opposite direction. It would be an error to calculate the speed of the airplane by dividing the space $2D$ by the sum $t_1 + t_2$. In fact the speed in going from AA' to BB' is equal to

$$V + W = \frac{D}{t_1}$$

and in going the other way

$$V - W = \frac{D}{t_2}$$

By adding the two above equations: member to member, we have

$$2V = \frac{D}{t_1} + \frac{D}{t_2}$$

that is

$$V = \frac{1}{2}\left(\frac{D}{t_1} + \frac{D}{t_2}\right)$$

Now this expression has a value absolutely different from the other $\frac{2D}{t_1 + t_2}$. For example: supposing $D = 2$ miles, $t_1 = 0.015$ hours, and $t_2 = 0.023$ hours, we will have

$$\frac{1}{2}\left(\frac{D}{t_1} + \frac{D}{t_2}\right) = \frac{1}{2}\left(\frac{2}{0.015} + \frac{2}{0.023}\right) = 110 \text{ m.p.h.}$$

while

$$\frac{2D}{t_1 + t_2} = \frac{4}{0.015 + 0.023} = 105 \text{ m.p.h.}$$

When the speed of the wind is constant in magnitude and direction, the airplane in flight does not resent any effect as to its stability. But the case of uniform wind is rare, especially when its speed is high. The amplitude of the variation of normal winds can be considered proportionally to their average speed. Some observations made in England have given either above or below 23 per cent. as the average oscillations; and either more or less than 33 per cent. as the maximum oscillation. In certain cases, however, there can be brusque or sudden variations of even greater amplitude.

Furthermore, the wind can vary from instant to instant also in direction, especially when close to broken ground. In fact, near broken ground, the agitated atmosphere produces the same phenomena of waves, suctions, and vortices, which are produced when sea waves break on the rocks.

If the airplane should have a mass equal to zero, it would instantaneously follow the speed variations of the air in which it is located; that is, there would be a complete *dragging* effect. As airplanes have a considerable mass they consequently follow the disturbance only partially.

It is then necessary to consider beside the partial *dragging* effect, also the *relative action* of the wind on the airplane, action which depends upon the temporary variation of the relative speed in magnitude as well as in direction. The reaction of the air upon the airplane takes a different value than the normal reaction, and the effect is that at the center of gravity of the airplane a force and a couple (and consequently a movement of translation and of rotation), are produced.

We have seen that in normal flight the sustaining component L of the air reaction, balances the weight. That is, we have

$$10^{-4}\lambda A V^2$$

If the relative speed V varies in magnitude and direction, the second term of the preceding equation will become $10^{-4}\lambda^1 A V'^2$, and in general we will have

$$10^{-4}\lambda' \times A \times V'^2 A \times V^2 \gtrless 10^{-4}\lambda A V^2$$

Consequently we shall have first of all, an excess or deficiency in sustentation and then the airplane will take either a climbing or descending curvilinear path, and will undergo such an acceleration that the corresponding forces of inertia will balance the variation of sustentation.

If, for instance, the sustentation suddenly decreases, the line of path will bend downward. In such a case, all the masses composing the airplane, including the pilot, will undergo an acceleration g' contrary to the acceleration due to gravity g.

If m is the mass of the pilot, his apparent weight will no longer be mg but $m(g-g')$; if it were that $g'>g$, the relative weight of the pilot with respect to the airplane would become negative, and tend to throw the pilot out of the airplane. Thence comes the necessity of pilots and passengers strapping themselves to their seats.

Let us suppose that an airplane having a speed V undergoes a frontal shock of a gust increasing in intensity from W to $W + \Delta W$; if the mass of the airplane is big enough, the relative speed (at least at the first instant), will pass from the value V to that of $V + \Delta W$; the value of the air reaction which was proportional to V^2 will become proportional to $(V + \Delta W)^2$; the percentual variation of reaction on the wing surface will then be

$$\frac{(V + \Delta W)^2 - V^2}{V^2} = \frac{2 \times V \times \Delta W + (\Delta W^2)}{V^2}$$

$$= 2 \frac{\Delta W}{V} + \left(\frac{\Delta W}{V}\right)^2$$

that is, it will be inversely proportional to the speed of the airplane. Great speeds consequently are convenient not only for reducing the influence of the wind on the length of time for a given space to be covered, but also in order to become more independent of the influence of the wind gusts.

Let us now consider a variation in the direction of the wind. Let us first suppose that this variation modifies only the angle of incidence i; then the value λ will change. For a given variation Δi of i, the percent variation of λ will be inversely proportional to the angle i of normal flight. From this point of view, it would be convenient to fly with high angles of incidence; this, however, is not possible, for reasons which shall be presented later.

Let us now suppose that the gust be such as to make the direction of the relative wind depart from the plane of symmetry; there will then be an angle of drift. A force of drift will be produced, and if the airplane is stable in calm air, a couple will be produced tending to put the airplane against the wind and to bank it on the side opposite to that from which the gust comes. Naturally it is necessary that these phenomena be not too accentuated in order not to make the flight difficult and dangerous with the wind across. We find here the confirmation of the statement that stabilizing couples be not excessive.

PART III

CHAPTER XII

PROBLEMS OF EFFICIENCY

Factors of Efficiency and Total Efficiency

The efficiency of a machine is measured by the ratio between the work expended in making it function and the useful work it is capable of furnishing. For a series of machines and mechanisms which successively transform work, the whole efficiency (that is, the ratio between the energy furnished to the first machine or mechanism and the useful energy given by the last machine or mechanism), is equal to the product of the partial efficiencies of the successive transformations.

To be able to effect the calculation of efficiency in an airplane, it is necessary to consider two principal groups of apparatus: the engine-propeller group and the sustentation group. There is no doubt of the significance of the engine-propeller group efficiency; it is the ratio between the useful power given by the propeller and the total power supplied to it by the engine. The sustentation group comprises the wings, the controlling surfaces, the fuselage, the landing gear, etc.; that is, the mass of apparatus which forms the actual airplane.

For the sustentation group, the efficiency, as it was previously defined, has no significance, because neither supplied energy nor returned energy is found in it. The function of the sustentation group is to insure the lifting of the airplane weight, with a head resistance notably less than the weight itself. The ratio between the lifted weight and the head resistance is usually taken as the measure of the efficiency of the sustentation group.

The lifted load of an airplane is given by the expression

$$L = 10^{-4}\ \lambda A V^2$$

and the head resistance is equal to the sum of two terms; one referring to the wing surface, the other to the parasite resistances:

$$D = 10^{-4}\ (\delta A + \sigma) V^2$$

Thus the efficiency of the sustaining surface can be measured by

$$\epsilon = \frac{L}{D} = \frac{\lambda A}{\delta A + \sigma}$$

If ρ is the propeller efficiency, the product $r = \rho \times \epsilon$ can serve well enough to characterize the total efficiency of the airplane. Naturally the number r cannot be considered as a ratio between two works; and it differs from a true and proper efficiency (which is always smaller than unity) because it is in general greater than unity, as it contains the factor ϵ which is always greater than 1. Let us immediately note that the value of r is not constant, because the values of ϵ and ρ are not constant. In fact ϵ is a function of λ and δ, which vary with the variation of the angle of incidence i, and ρ is a function of the speed V and of the number of revolutions n of the engine. Practically, it is interesting to know the value of r as a function of the speed, which is possible by remembering the equation

$$W = L = 10^{-4}\ \lambda A V^2$$

In fact W being constant, this equation permits the determining of a corresponding value V for each value of i, and therefore the making of a diagram of efficiency ϵ as a function of speed V. Moreover, by what has already been mentioned in Chapter IX, when the engine propeller group is fixed, the value of ρ as a function of V can be found and then it is easy to draw the diagram of r as a function of V.

It is possible to give r a much simpler expression than the preceding one; thus

$$r = \rho \times \frac{\lambda A}{\delta A + \sigma} \qquad (1)$$

obtaining λA and $(\delta A + \sigma)$ from the equations

$$W = 10^{-4}\ \lambda A V^2$$
$$550 P_1 = 1.47\ 10^{-4}\ (\delta A + \sigma) V^3$$

and substituting in (1) we have

$$r = 0.00267 \rho \frac{WV}{P_1} \qquad (2)$$

Knowing W, the diagrams $\rho = f(V)$ and $P_1 = f(V)$, we can draw the diagram $r = f(V)$.

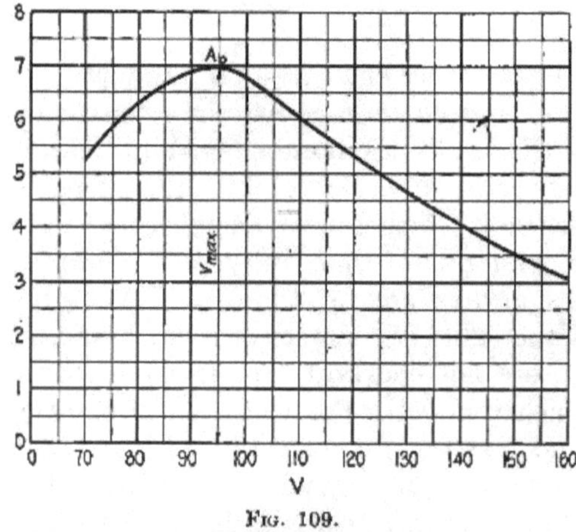

FIG. 109.

Let us draw, for instance, this diagram for the airplane of the example of Chapter IX. For this airplane we have $W = 2700$ lb., consequently

$$r = 7.2 \rho \frac{W}{P_1}$$

Fig. 93 gives the values of P_1 and ρ corresponding to the various speeds for the propeller which has already been considered in Chapter IX. We can then obtain the value of r corresponding to each value of V and draw the diagram of Fig. 109.

This diagram shows that r is maximum and equal to 6.9 for a value of speed, $V = 95$ m.p.h., after which it decreases; for $V = 160$ m.p.h. for instance (which represents the maximum speed of the airplane under consideration) $r = 3.12$; that is r is equal to 45 per cent. of the maximum value. In other words our airplane running at its maximum speed, has an efficiency equal to less than one-half the efficiency it has at the speed of 95 m.p.h. to which corresponds to the maximum climbing speed.

Let us consider again formula (2); since $P_1 = \rho P_2$ when the airplane flies horizontally at its maximum speed, equation (2) can also be written

$$r = 0.00267 \times \frac{W \times V}{P_2}$$

Practically then when we know the maximum speed of the airplane and the corresponding maximum power of the engine, it is possible to have the value of r corresponding to the maximum speed.

This value is much lower than the maximum which the airplane can give; thus calculating r based on the maximum speed of the airplane and on the maximum power of its engine, we would have an imperfect idea of the real total efficiency.

Now we intend to show that to measure the efficiency corresponding to the maximum climbing speed is not a difficult matter.

Let us suppose in fact that the airplane makes a climbing test and let n be the number of revolutions of the engine while climbing. Let V be the speed of translation measured by one of the usual speedometers. Knowing n, we know the value P'_2 corresponding to the power developed by the engine.

Such power is absorbed partly by the airplane, and partly by the work necessary to do the lifting. Let $v_{max.}$ be the maximum climbing speed, which can be measured by ordinary barographs. The power absorbed by flying will be

$$P'_2 = \frac{Wv_{max.}}{550\rho'}$$

where ρ' is the propeller efficiency which can be estimated with sufficient approximation knowing $\frac{V'}{nD}$ (V' is the horizontal speed corresponding to $v_{max.}$).

We then have

$$r_{max.} = \frac{V'W}{P'_2 - \frac{v_{max.} W}{550\rho'}}$$

that is, by measuring V', v and n, and by estimating ρ', it is possible to have a value approximate enough to the maximum value of the total efficiency. Bréguet has proposed an expression which he calls *motive quality,* whose magnitude can be used to give an idea of the efficiency of the airplane. Let us remember the two equations:

$$W = 10^{-4}\lambda A V^2$$
$$\rho P_2 = 0.267\ 10^{-6}\ (\delta A + \sigma) V^3$$

By eliminating V from the two preceding equations, we have

$$P_2 = 0.267\ W^{3/2} \times \frac{1}{\sqrt{A}} \times \frac{1}{\rho} \times \frac{\delta + \frac{\sigma}{A}}{\lambda^{3/2}} \qquad (3)$$

The motive quality q is the expression

$$q = \rho \frac{\lambda^{3/2}}{\delta + \frac{\sigma}{A}}$$

Let us remember that

$$r = \rho \frac{\lambda A}{\delta A + \sigma} = \rho \frac{\lambda}{\delta + \frac{\sigma}{A}}$$

We see that

$$q = r\sqrt{\lambda}$$

That is, q is proportional to r and therefore it measures the efficiency of the airplane.

Equation (3) can be written

$$P_2 = 0.267 \frac{W^{3/2}}{\sqrt{A} \times q}$$

from which we have

$$q = \frac{0.267\ W^{3/4}}{P_2 \sqrt{A}}$$

Also q assumes various values, and its maximum value corresponds to the maximum of ascending speed $v_{max.}$ That is, we have by expressing $v_{max.}$ in ft. per second that

$$q_{max} = \frac{0.267\ W^{3/4}}{\sqrt{A}\left(P'_2 - \frac{W \times v_{max.}}{550\rho'}\right)}$$

which can also be written

$$q_{max.} = \frac{147\sqrt{\frac{W}{A}}}{\frac{550\ P'_2}{W} - \frac{v_{max.}}{\rho'}}$$

Since $\frac{W}{A}$ is the load per sq. ft. of the wing surface, and $\frac{P_2}{W}$ is the weight per horsepower of the airplane, $v_{max.}$ and ρ' being known, $q_{max.}$ is easily calculated. In the preceding example we have for instance

$$\frac{W}{A} = 10;\ \frac{P'_2}{W} = 7.3;\ v' = 33;\ \rho' = 0.695$$

consequently

$$q_{max.} = 0.177$$

CHAPTER XIII

THE SPEED

In ordinary means of locomotion, speed is usually considered as a luxury, but in the airplane, it represents an essential necessity, for the whole phenomenon of sustentation is based upon the relative speed of the wing surfaces with respect to the surrounding air.

The future of the airplane, as to its application in everyday life, stands essentially upon its possibility of reaching average commercial speeds far superior to those of the most rapid means of transportation.

When the airplane is in flight, high speeds present dangers incommensurably smaller than those which threaten a train or a motor car running at high speed. On the contrary we have seen that the faster an airplane is, the better it fights against the wind. It is quite true that high speeds present real dangers when landing, but modern speedy airplanes are designed so as to permit a strong reduction in speed when they must return to earth.

Let us remember that the two general equations of the flight of an airplane are:

$$W = 10^{-4}\,\lambda A V^2 \tag{1}$$

$$550\rho \times P_2 = 1.47\,10^{-4}\,(\delta A + \sigma)V^3 \tag{2}$$

by expressing P_2 in H.P. and V in m.p.h. Equation (2) gives,

$$V = 155\,\frac{\rho^{1/3}\,P_2^{1/3}}{(\delta A + \sigma)^{1/3}} \tag{3}$$

We see then, that if we wish to increase V we must increase ρ and P_2, decrease δ, A and σ.

The improvement of ρ is of the greatest importance not only in order to obtain a higher speed but also in order to improve the total efficiency. In regard to propellers, we have discussed their efficiency and the factors which have influence upon it, and we have seen that ρ is a function of the ratio $\frac{V}{\pi n D}$.

By drawing the diagram ρ as a function of $\frac{V}{\pi n D}$, we see that ρ passes through a maximum value $\rho_{max.}$ after which it decreases.

The value $\frac{V}{\pi n D}$ (to which the value $\rho_{max.}$ corresponds) is directly proportional to the ratio $\frac{\text{Pitch}}{\text{Diameter}}$. Let us consider, for instance, three propellers of diameter D', D'', D''' and of pitch p', p'', p''' such that $p'/D' < p''/D'' < p'''/D'''$; the curves of the efficiencies ρ', ρ'', ρ''' will be such that $\rho'_{max.}$, $\rho''_{max.}$, and $\rho'''_{max.}$ correspond to the three values $\frac{V'}{\pi n D} < \frac{V''}{\pi n D} < \frac{V'''}{\pi n D}$ (Fig. 110).

Now, if with a given machine we wish to have the maximum horizontal speed, it is convenient to select the propeller of such pitch and diameter so as to give the maximum efficiency at that speed. In formula (3), the propeller efficiency is seen to be to the $\frac{1}{3}$ power; this means that for each 1 per cent. of increase of the efficiency, the speed increases only by $\frac{1}{3}$ per cent.

The increase of the motive power P_2 is another means of increasing the speed; also P_2 is seen to be to the $\frac{1}{3}$ power and consequently at first glance, we may think that for a percentual increase of P_2 the same may be applied as that which has been said for a per cent. increase of ρ. Practically though, to increase P_2 means adopting an engine of higher power, consequently of greater weight and different incumbrance. Thus the change of P_2 is reflected upon the terms δ, A and σ. It is not possible to translate into a formula the relation which exists between P_2, δ, A and σ. It is necessary then to make proper verifications for each successive case.

The value of δ depends upon the form and profile of the wing surface; it is smaller for the wings with very flat

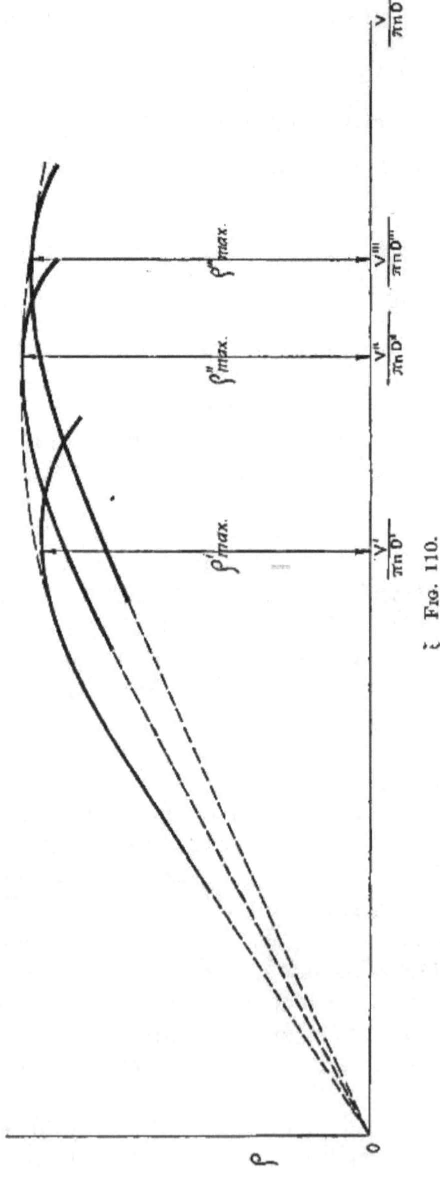

Fig. 110.

aerofoil, and which for this reason are usually called "*wings for speed.*" For very fast machines, some designers have even adopted wings with convex instead of concave bottoms. Naturally this convexity is smaller than that of the wing back (Fig. 111). We then also have a negative pressure below the bottom, and the sustentation is then due to the excess of negative pressure on the back with respect to that on the bottom.

The decrease of sustaining surface A also has influence upon the increase of speed.

Fig. 111.

From this point of view it would then be convenient to greatly increase the load per unit of the wing surface $\frac{W}{A}$. But remembering equation (1) we have that

$$V = 100 \sqrt{\frac{W}{A}} \frac{1}{\sqrt{\lambda}}$$

This expression states that when $\frac{W}{A}$ is given, the value of V is inversely proportional to $\sqrt{\lambda}$.

Let us give λ the maximum value $\lambda_{max.}$ which it is practically possible to give (the one corresponding to $i = 8°$ to $10°$). Then the preceding formula gives the minimum value of the speed it is possible to attain.

$$V_{min.} = 100 \frac{1}{\sqrt{\lambda_{max.}}} \sqrt{\frac{W}{A}} = \alpha \sqrt{\frac{W}{A}}$$

that is, the minimum speed at which the airplane can sustain itself is directly proportional to $\sqrt{\frac{W}{A}}$. Consequently if we wish to keep the value of $V_{min.}$ within reasonable limits of safety, it is necessary not to excessively increase the value of $\frac{W}{A}$; that is, not to excessively reduce the value of A. Practically the value of $\frac{W}{A}$ is kept between 6 and 10 lb. per sq. ft.

For the sake of interest we shall recall that in the Gordon Bennett race of 1913, machines participated with a unit load up to 13 lb. per sq. ft. Such machines are difficult to maneuver; are the worst *gliders*, and naturally require a great mastery in landing; their practical use would have been excessively dangerous. For sport and touring machines, the value of $\frac{W}{A}$ must be lowered to values of 6 to 4 and even 3 lb. per sq. ft.

The decrease of σ, analogous to the increase of ρ, constitutes one of the most interesting means of increasing speed. Let us remember that

$$\sigma = \Sigma KA$$

that is, it is equal to the sum of all the passive resistances due to the various parts of the airplane. For decreasing σ it is then necessary:

1. To reduce the coefficients of head resistance of the various parts to a minimum,

2. To reduce the corresponding major sections to a minimum.

In order that the reader may have an idea of the influence of the five factors ρ, P_2, δ, A, and σ upon the speed, let us suppose that for a given airplane any four of the above terms are known, and let us see how V varies with a variation of the 5th element.

Suppose for instance that

$\rho = 0.7; P_2 = 350$ H.P.; $\delta = 0.6; A = 340$ sq. ft.; $\sigma = 200.$ Then, giving P_2, δ, A, and σ the preceding values, let us draw the diagram of the equation

$$V = 155 \frac{\rho^{1/3} \sqrt[3]{350}}{\sqrt[3]{0.6 \times 340 + 200}} \text{ (Fig. 112).}$$

By making δ vary from the value 0.7 to the value 0.8, we see that while for $\rho = 0.7$, the speed is about 130 m.p.h.; for $\rho = 0.8$ it is above 136 m.p.h.; that is, while the efficiency increases by 14.3 per cent., the speed increases by 4.6 per cent.

Analogously the diagram $V = f(P_2)$, $V = f(\delta)$, $V = f(A)$, and $V = f(\sigma)$, have been drawn respectively in Figs.

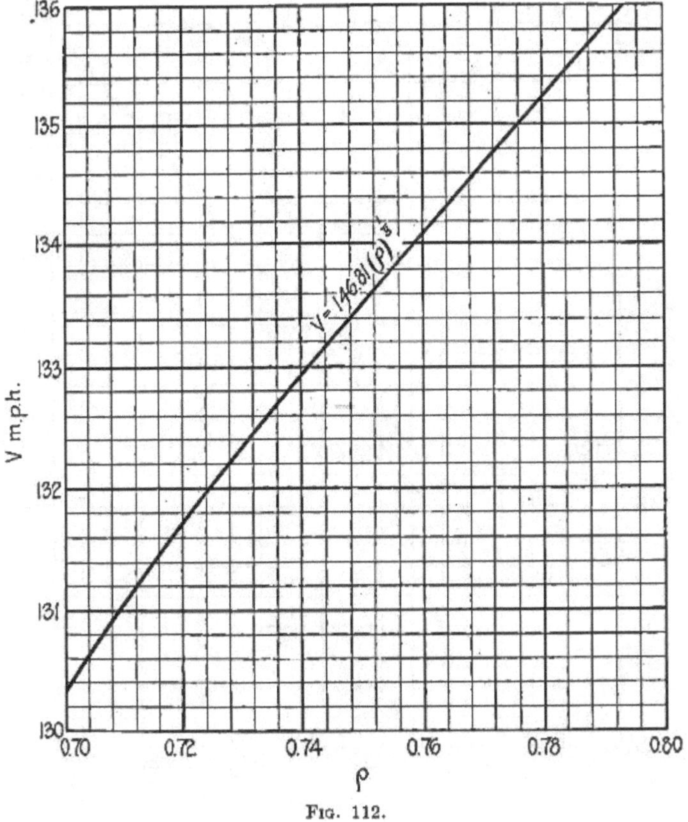

Fig. 112.

113, 114, 115, and 116, always adopting the preceding values for the constant terms.

All the foregoing presupposes the air density constant and equal to the normal density; that is, to the one corresponding to the pressure of 33.9 ft. of water and to the temperature of 59°F.

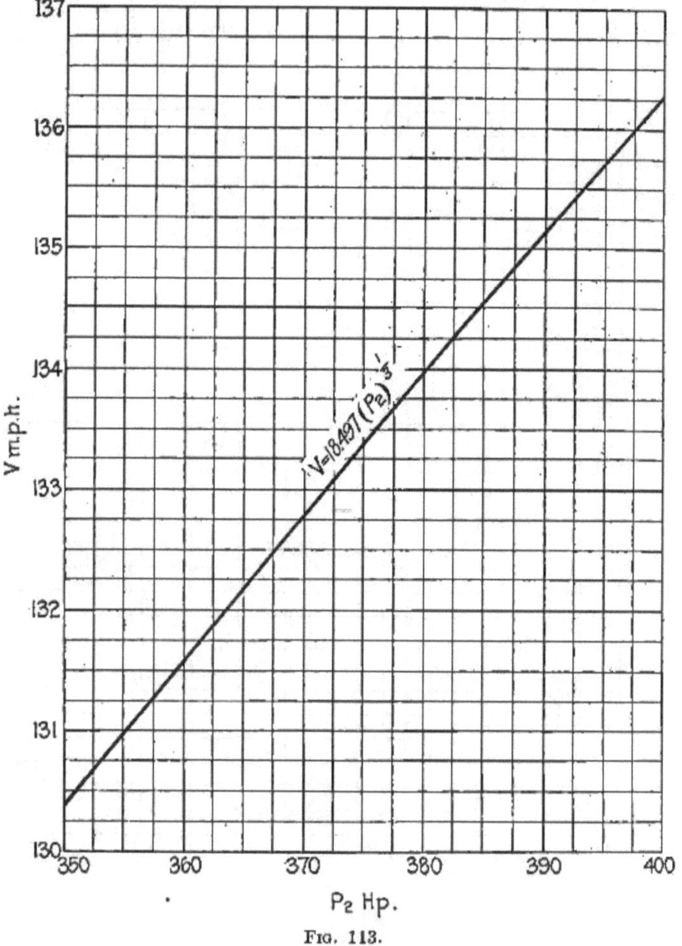

Fig. 113.

Now as it is known the density of the air decreases as we rise in the atmosphere (see Chapter V), following a logarithmic law given by the equation

$$H = 60720 \log \frac{P_o}{P} \times \frac{460 + t^\circ}{519} = 60720 \log \frac{1}{\mu} \quad (1)$$

Where H is the height in feet,

$\frac{P_o}{P}$ is the ratio between the pressure at sea level and the pressure at height H;

t^o is the Fahrenheit temperature at sea level, and

μ is the ratio between the density at height H and the normal density defined above.

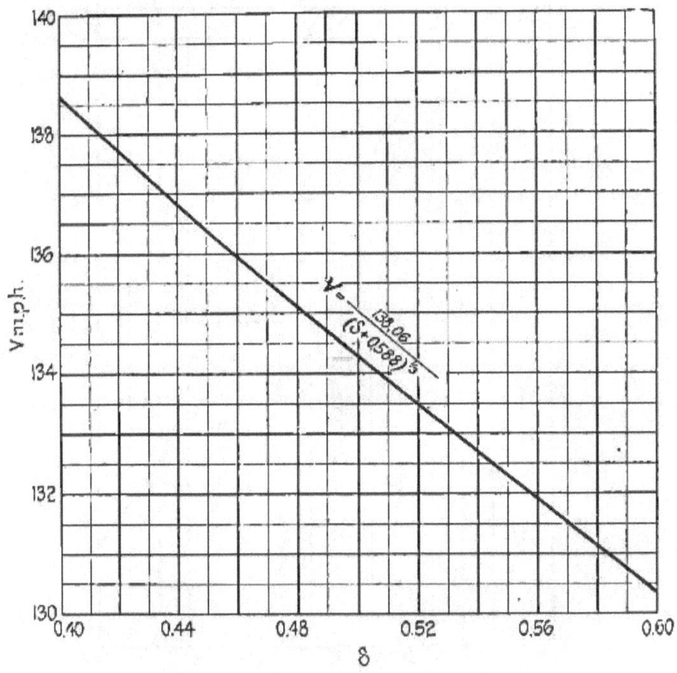

FIG. 114.

Equation (1) can be translated into linear diagrams by using a paper graduated with a logarithmic scale on the ordinates, and with a uniform scale on the abscissæ, giving to t^o successively various values. In Fig. 117 these lines are drawn for $t^o = 0°, 20°, 40°, 59°$ and $80°F$.

By using these diagrams, the density corresponding to a given height for a given value of the temperature at ground level, is easily found.

Then let us again take up the examination of the formula for speed

$$V = 155 \times \frac{\rho^{1/3} P_2^{1/3}}{(\delta A + \sigma)^{1/3}}$$

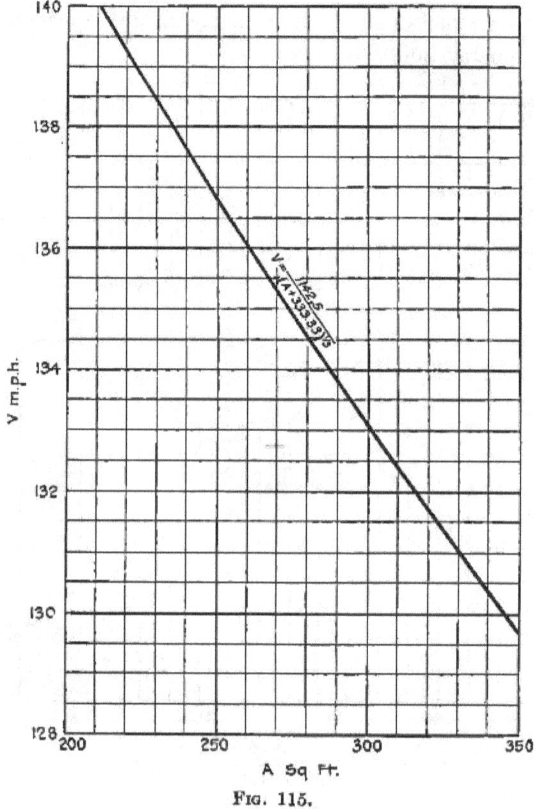

Fig. 115.

and let us place in evidence the influence of the variation of the density on various parameters which appear in it.

The efficiency ρ is a function of $\frac{V}{nD}$; now this ratio is influenced by the variation of the density, since V and n vary; then also ρ varies with a variation of μ.

We have already spoken of the influence of the density

on the motive power in Chapter V, where we saw that the ratio between the power at height H and that at ground level is equal to μ.

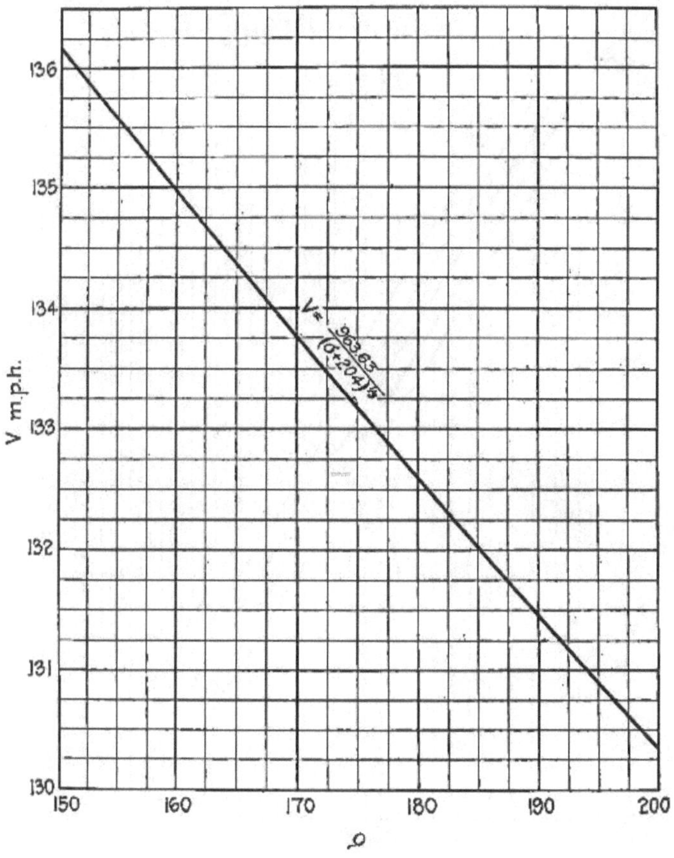

FIG. 116.

The useful power $_\rho P_2$ given by the engine propeller group is thus a function of the air density; therefore the diagram $_\rho P_2 = f(V)$ changes completely with a variation of μ. In Chapter IX we saw how to draw that diagram when the density is normal; that is, $\mu = 1$. Let us now consider the case of $\mu < 1$. The ratio $\frac{P_p}{n^3D^5} = \alpha$ is not only a function of $\frac{V}{nD}$, but also of μ; and precisely that ratio is proportional

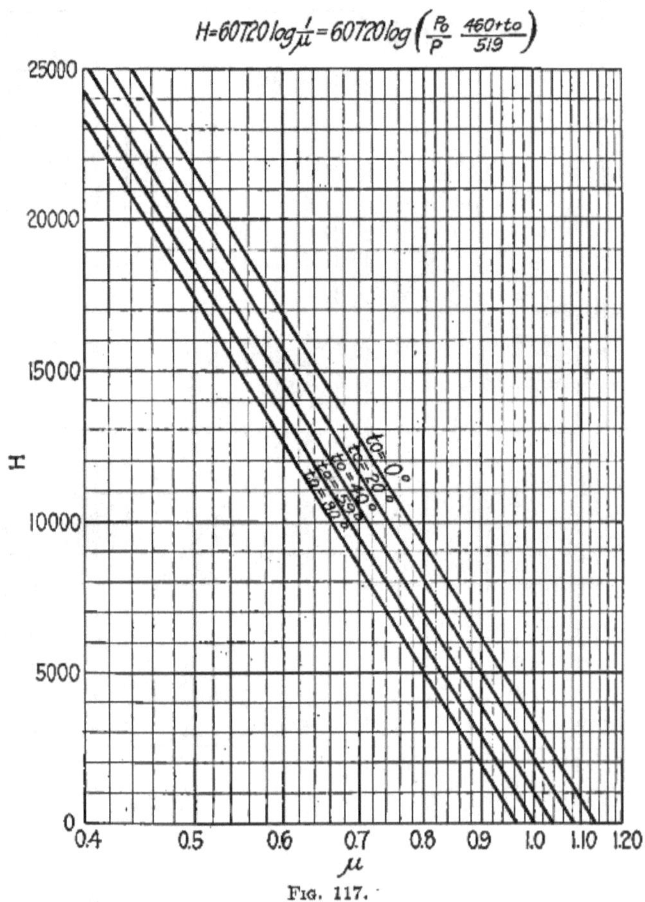

Fig. 117.

to μ. Consequently for each value of μ a diagram $\frac{\mu P_p}{n^3D^5}$ needs to be drawn. In Fig. 118 such diagrams have been drawn ona logarithmic scale for the propeller family to which Fig. 89 of Chapter IX refers, and for the values μ =

1.0, 0.55, 0.41, 0.25, corresponding, for a temperature of 59° F., to the heights of 0, 16,000, 24,000 and 28,000 ft.

The diagram which gives the motive power P_2 as function of the number of revolutions is also to be decreased propor-

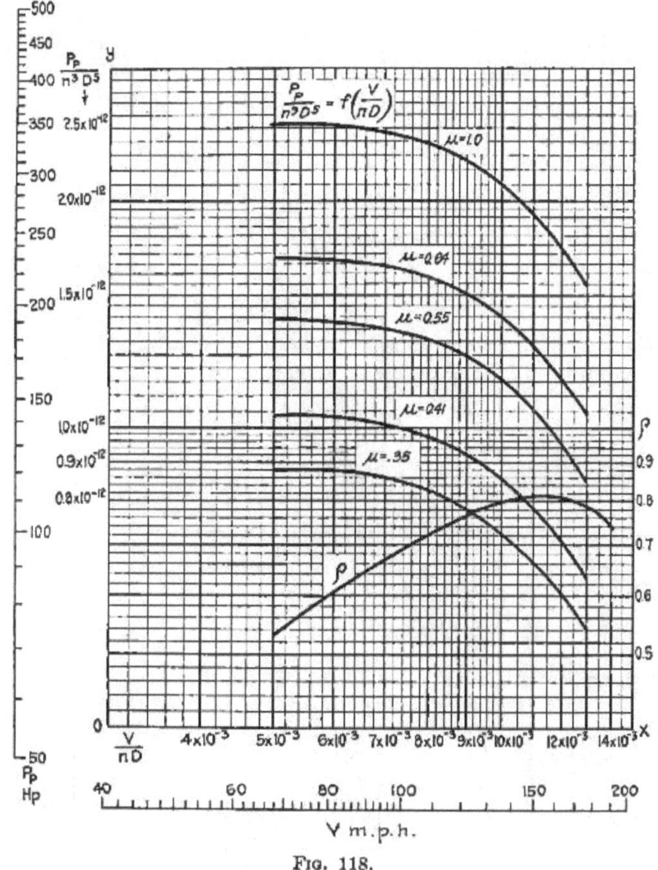

Fig. 118.

tional to μ. In Fig. 119 we have taken up again the diagram of Fig. 91 of Chapter IX, drawing it for the preceding values μ.

Then by the known construction, we can draw the diagrams $\rho P_2 = f(V)$ for the preceding values of μ (Fig. 120).

Horsepower
550
500
450
400
350
300
250
200
150
100
0
50
μ=1.0
P2=f(n)
μ=0.55
μ=0.41
μ=0.35
Y
X
1200
1400
1600
1800
Scale n

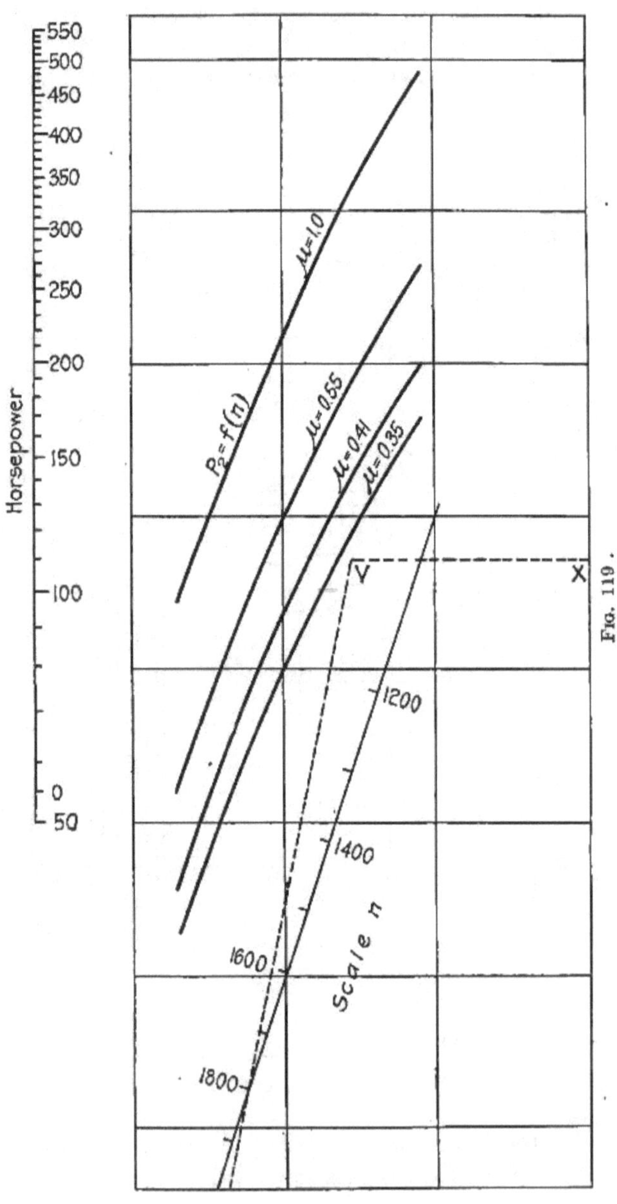

Fig. 119.

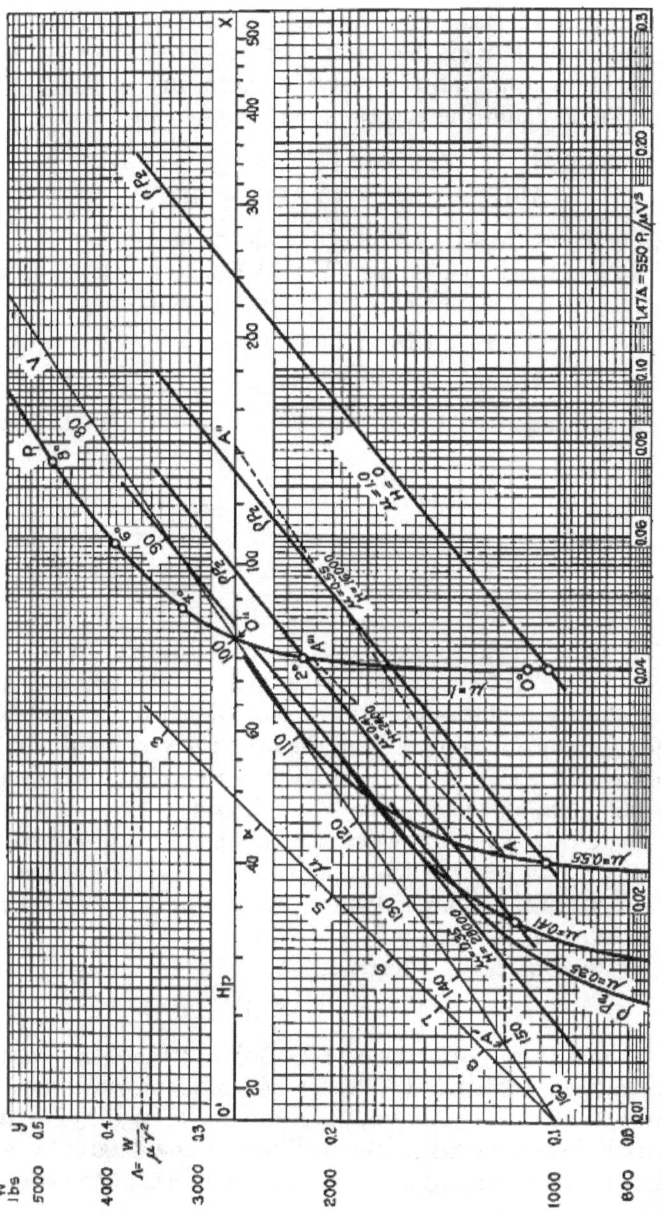

FIG. 120.

In order to make evident the influence of the decrease of the air density on the parameter proper of the airplane, or in other words on the power P_1 necessary to flying, let us take up again the general equation of flight

$$W = 10^{-4}\,\lambda A V^2$$
$$550P_1 = 1.47 \times 10^{-4}\,(\delta A + \sigma)\,V^3$$

and make evident the influence of the air density.

We have seen in Chapter VII that λ, δ, and σ vary proportionally to μ; consequently the preceding equations become

$$W = 10^{-4}\,\mu\lambda A V^2$$
$$550P_1 = 1.47 \times 10^{-4}\,\mu\,(\delta A + \sigma)\,V^3$$

that is, remembering what has been said in Chapters VIII and IX

$$\frac{W}{\mu V^2} = \Lambda$$

and

$$\frac{550P_1}{\mu V^3} = 1.47\Delta$$

Then considerations analogous to those developed in the preceding chapters enable us to take μ into account by introducing a new scale with a slope of 1/1 on the axis of the abscissæ and to pass from the origin to any point whatsoever of the diagram by summing geometrically four segments equal and parallel to W, P, V and μ.

As the weight of the airplane is constant and equal to 2700 lb., it is possible according to what has been said also in Chapter IX, to simplify the interpretation of the diagram, proceeding as follows:

Let us consider the diagram

$$\Lambda = f(1.47 \times \Delta)$$

for $\mu = 1$ (Fig. 120). From each point of this diagram let us draw segments parallel to the scale of μ and which measures to this scale, the value $\mu = 0.55$. Let us join the ends of these segments. We shall have a new diagram $\Lambda = f$ (1.47Δ) corresponding to $\mu = 0.55$. We intend to demonstrate that if from any point whatsoever A of this diagram we draw a parallel to the scales of V and P, we shall have in A' and A'' respectively a pair of corresponding values of speed V of power P_1 for $\mu = 0.55$, that is at the height of 16,000 ft. In fact let us call A''' the meeting point of the straight line AA''' drawn parallel to the scale of μ, on the original diagram. By construction AA''' is equal to 0.55. Let us suppose now that we wish to find the corresponding pairs of values V and P_1 for $W =$ 2700 and $\mu = 0.55$. Then it will be sufficient to draw from O' corresponding to 2700 lb. a parallel to the scale of power and from A, extreme point of the segment AA''' corresponding to the value $\mu = 0.55$ a parallel to the scale of speed. These two straight lines will meet in A'' and will individuate two segments $O' A''$ and AA'' as measure of the corresponding power and speed.

Thus, as $AA'' = O''A'$, if we wish to study the flight at a height of 16,000 ft., it is possible to use the diagram $\Lambda = f$ (1.47Δ) drawn, by adopting the same scales as said above.

Based upon analogous considerations the diagrams $\Lambda =$ $f(1.47\Delta)$ for $\mu = 0.41$ and $\mu = 0.35$, have been drawn.

We then dispose, in Fig. 120 of four pairs of diagrams, which give the values of P_1 and ρP_2 corresponding to $\mu = 1$; 0.55; 0.41 and 0.35, that is, for the heights of 0, 16,000, 24,000 and 28,000 ft. The meeting points of these diagrams define the maximum value of the speed which the airplane can reach with that given engine-propeller group at the various heights. The diagrams corresponding to the height of 28,000 ft. do not intersect. This means that for the airplane of our case the flight would not be possible at this height.

For the lower altitudes it is possible to draw the diagrams of the corresponding maximum and minimum speeds (Fig. 121). Let us note immediately that while the maximum speeds depend essentially upon the engine-propeller group and consequently can be varied with a variation of the characteristic of this group the minimum speeds depend exclusively upon the airplane. From the examination of the diagrams of Fig. 120 we see that as we raise in the atmosphere the maximum speed which the airplane can reach diminishes gradually while the minimum flying speed increases accordingly.

It is interesting to study the case (merely theoretical at the present stage of the technique of the engines) in which

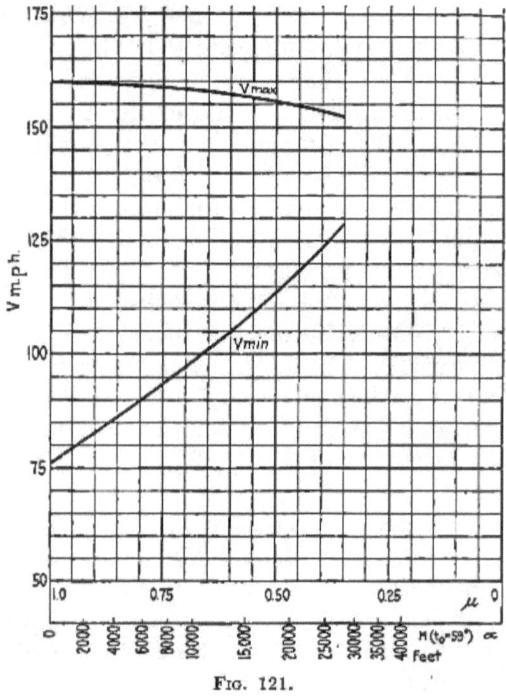

Fig. 121.

the motive power is not effected by the variation of the air density but keeps constant at the various heights. We shall see immediately that in this case the propeller will greatly increase the number of revolutions; it is then necessary to extend the characteristics of the engine above 2200 revolutions per minute.

Let us suppose that this characteristic be the one of Fig. 122. We can then draw by the usual construction the

550
500
450
400
350
300
250
200
150
Hp.
100
50
μ=1.0
$P_2=f(n)$
1200
Scale=n
1400
V
X
1600
1800
2000
2200
2400
2600

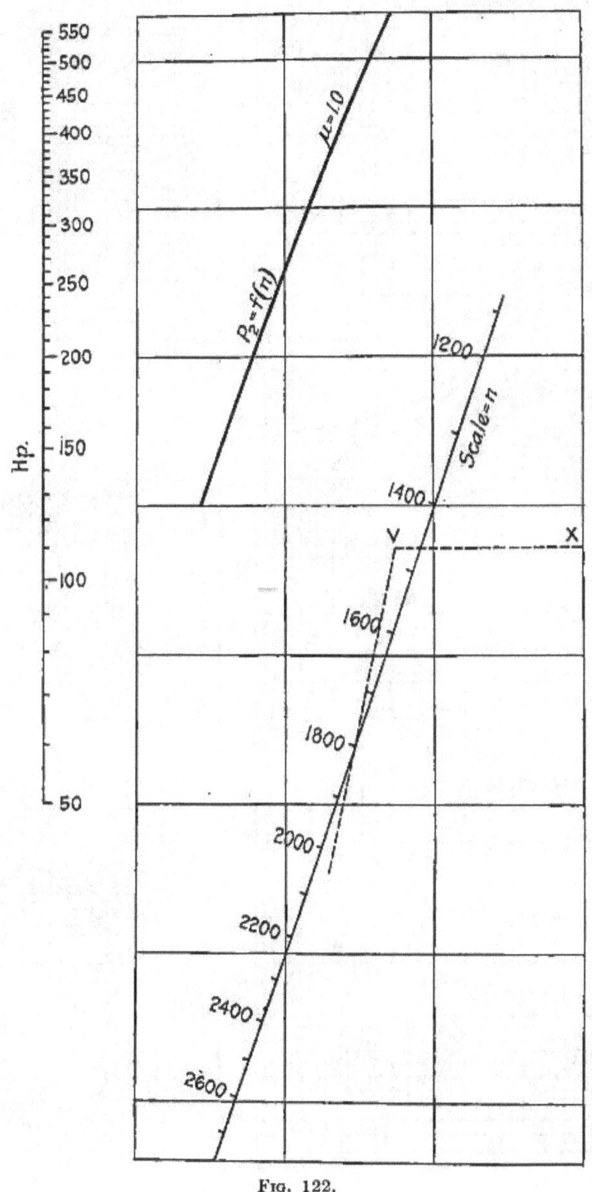

FIG. 122.

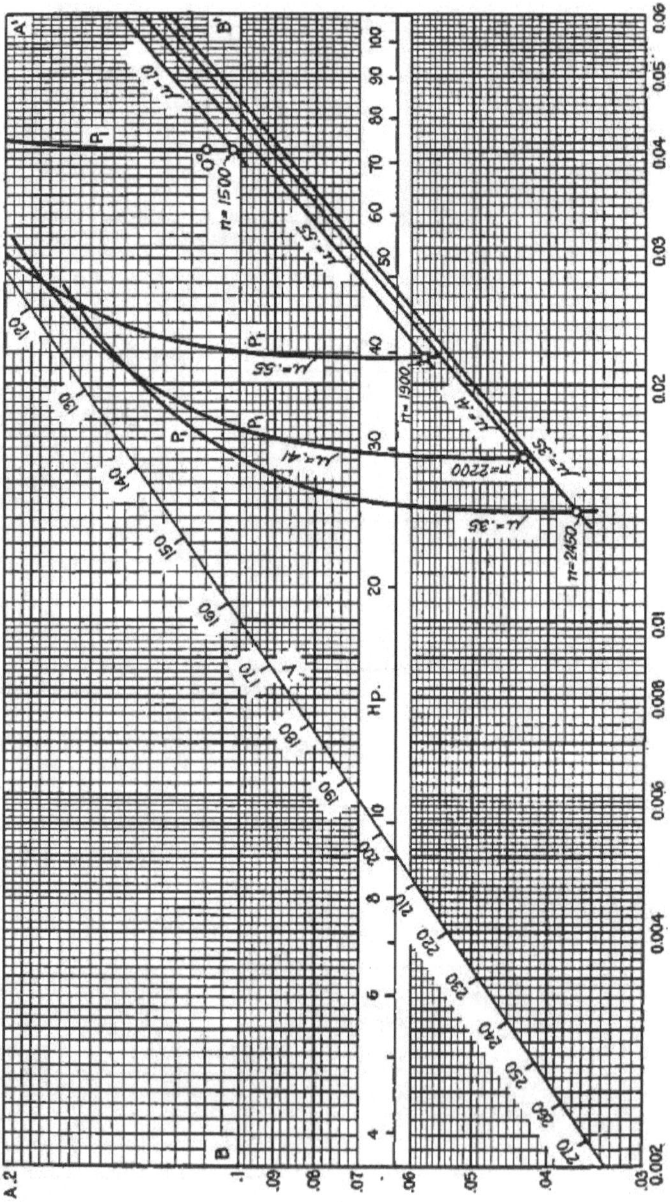

FIG. 123.

pairs of corresponding diagram, which give P_1 and ${}_\rho P_2$. This has been done in Fig. 123, in which has been drawn

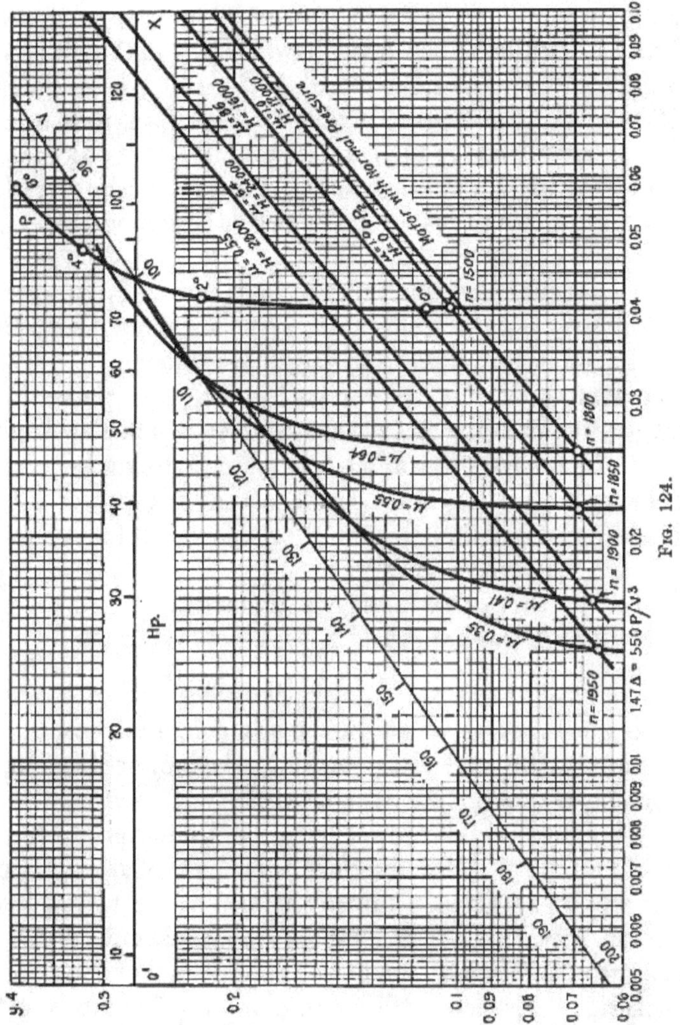

Fig. 124.

only part of the diagrams containing the intersections which define the maximum speeds. We see how these speeds vary, as they increase and how flight becomes possible even at 28,000 ft. and for greater altitudes. For our example we find that the speed at 28,000 ft. is equal to 265 m.p.h., while at sea level it was 160 m.p.h. Thus we also find that the number of revolutions of the propeller at 28,000 ft. is of 2450 r.p.m. against 1500 r.p.m. at sea level.

Let us note first of all that in practice it would not be possible to run the engine at 2450 r.p.m. without risking or breaking it to pieces, if the engine is designed for a maximum speed of say 1800 r.p.m.

In second place we shall note that it would be practically impossible to build an engine or a special device such as to keep the same power at any height whatsoever.

The utmost we can suppose is that the power is kept constant for instance up to 12,000 ft., after which it will naturally begin to decrease again. In order to make a more likely hypothesis, we shall suppose that the power is kept constant up to 12,000 ft. and then decreases following the usual law of proportionality.

Based on this hypothesis we have drawn the diagram of Fig. 124 for the values

$$\mu = 1.00;\ 0.64;\ 0.55;\ 0.41;\ 0.35$$

We see then that as we raise, the speed increases but much less than in the preceding case; furthermore after 12,000 ft. the speed remains about constant.

If we could build propellers with diameter and pitch variable in flight, the operation of the engine-propeller group would be greatly improved and a great step would be made toward the solution of the aviation engine for high altitudes, because the problem of propeller is one of the most serious obstacles to be overcome for the study of the devices which make it possible to feed the engine with air at normal pressure at least up to a certain altitude.

CHAPTER XIV

THE CLIMBING

In Chapter IX we have seen that the climbing speed can be easily calculated as a function of V, when the power $\rho \times P_2$ furnished by the propeller and the power P_1 necessary for the sustentation of the airplane at that speed, are known; and we have seen that the climbing speed v (expressed in feet per second), is given by

$$v = 550 \frac{\rho P_2 - P_1}{W}$$

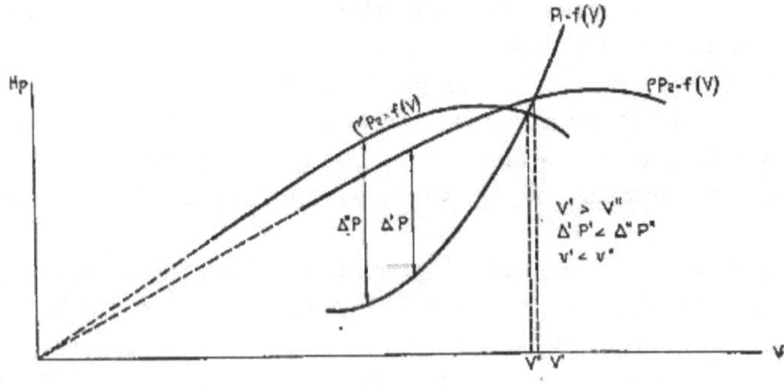

Fig. 125.

Practically, the maximum value $v_{max.}$ of the climbing speed, obtained when the difference $\rho P_2 - P_1$ is maximum, is of interest to us

$$v_{max.} = 550 \frac{(\rho P_2 - P_1)_{max.}}{W}$$

Thus if we wish to increase the climbing speed it is necessary to make the value $(\rho P_2 - P_1)_{max.}$ the maximum possible.

Let us suppose that the power P_2 be given; then first of all it is necessary that the airplane be built so that the minimum value of P_1 be the lowest possible; in the second place it is necessary that the propeller be selected so as to give the maximum efficiency, not at the maximum speed of the airplane, but at lower speeds, in order to increase the difference $\rho P_2 - P_1$.

Fig. 125 shows how this can be accomplished; the diagrams ρ' and ρ'' correspond to two propellers having different ratio p/D. While the propeller ρ' is better for speed than ρ'', the propeller which corresponds to the lower value of p/P is decidedly better for climbing.

Thus, practically, it is possible to adopt an entire series of propellers on a machine, to each one of which corresponds two special values for the maximum horizontal and climbing speeds. Naturally the selection of the propeller will be made according to whether preference is given to the horizontal speed or to the climbing speed.

In order to study in full details, the climbing of an airplane in the atmosphere, it is necessary to study the influence the decrease of the air density has upon the climbing speed.

Let us, as before, call μ the ratio between the air density at height H, and at sea level. At sea level $\mu = 1$ and the maximum climbing speed is the one given by formula (1).

As the airplane rises, the value μ decreases and then formula (1) should be written

$$v_{\text{max.}} = f(\mu)$$

Referring to what has been said in the preceding Chapter when the characteristics of the airplane for $\mu = 1$ are known, it is easy to draw for different values of μ, the curves

$$\rho P_2 = f(V) \text{ and } P_1 = f(V)$$

In Fig. 120 of the preceding chapter we have drawn these curves for the example of Chapter IX, and for values of

$$\mu = 1.0, 0.55, 0.41$$

For convenience, these curves are reproduced in Fig. 126.

Comparing the pairs of curves corresponding to the same value of μ, it is easy to plot the diagram which gives

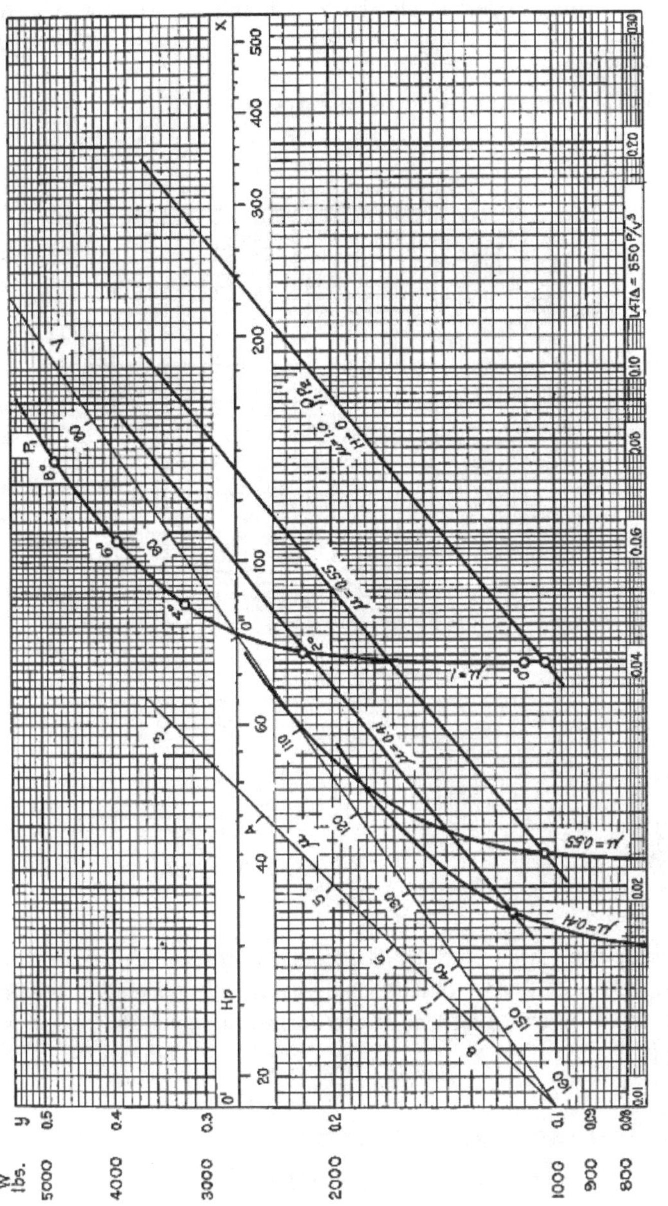

Fig. 126.

the climbing speed at the various heights. In Fig. 127 we have drawn this diagram, taking $v_{max.}$ as abscissæ and H as ordinates.

It is interesting to draw the diagram

$$t = f(H)$$

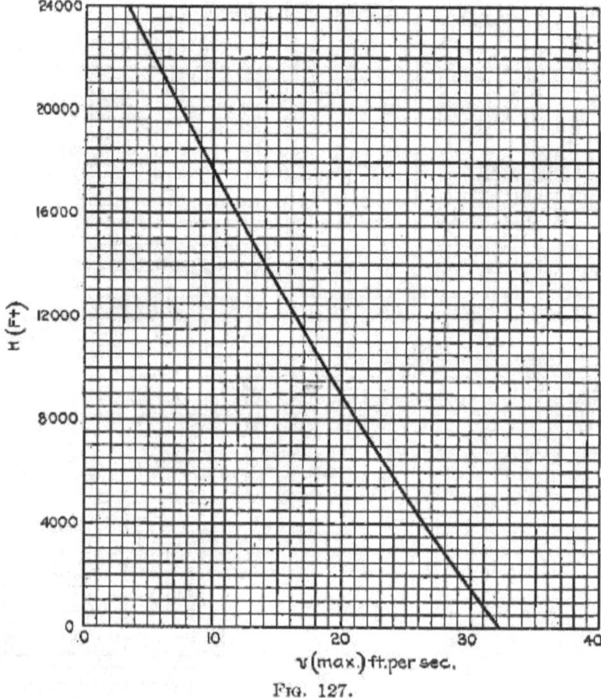

Fig. 127.

giving the time spent by the airplane in reaching a certain height H. To construct this diagram it is necessary first of all to draw the diagram of the equation

$$\frac{1}{v} = f(H) \quad \text{Fig. 128}a,$$

which is easily obtained, from

$$v = f(H)$$

0.40
0.30
0.20
0.10
0
$t_o = 59°F.$
$\frac{1}{V}$
$\frac{1}{V}$
ΔH
0
6000
12000
18000
24000
H (Ft.)
a

2000
1500
1000
500
0
t (Sec)
$t_o = 59°F$
$\int \frac{dH}{v}$
0
6000
12000
18000
24000
H (Ft.)
b

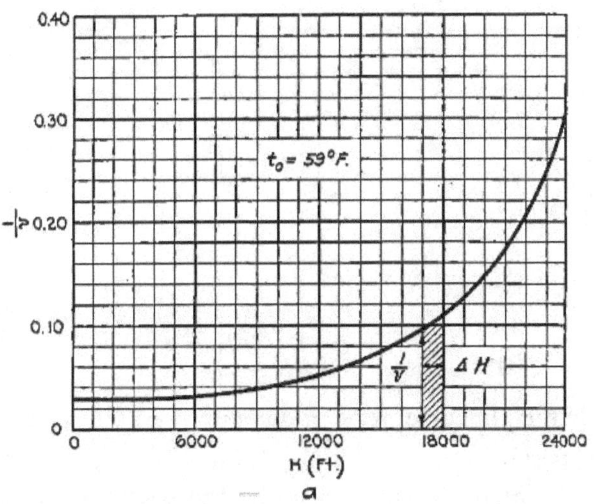

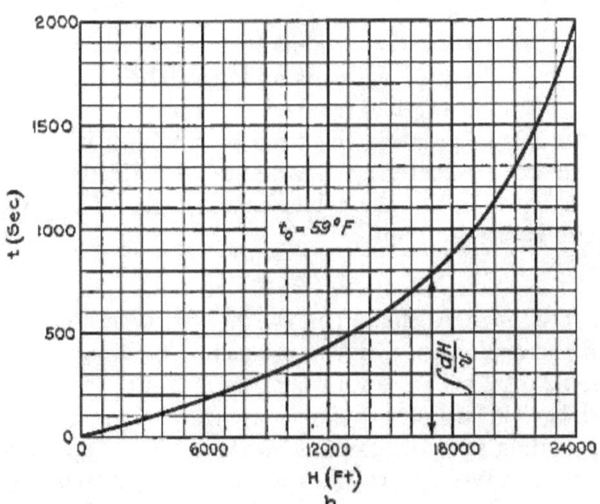

Fig. 128.

By integrating $\frac{1}{v} = f(H)$ we have $t = f(H)$, (Fig. 128 *b*). In fact the elementary area of the diagram $\frac{1}{v} = f(H)$ is equal to

$$\frac{1}{v} \times dH$$

but

$$v = \frac{dH}{dt}$$

consequently

$$\frac{1}{v} \times dH = dt$$

and

$$\int \frac{1}{v} \times dH = t$$

that is, the integration of diagram $\frac{1}{v} = f(H)$ gives t.

In Fig. 128 *a*, *b*, we have drawn the scales of H for $t =$ 59°. Since by increasing H the value v tends toward zero, that of $\frac{1}{v}$ tends toward ∞, and consequently that of t also tends toward ∞. That is to say, when the airplane reaches a certain height, it no longer rises. It is said then, that the airplane has reached its *ceiling*.

In actual practice the time of climbing is measured by means of a registering barograph. In Fig. 129 an example of a barographic chart has been given. This chart gives directly the diagram

$$H = f(t)$$

that is, it gives the times on the abscissæ and the heights on the ordinates. Since to reach its *ceiling*, the airplane would take an infinitely long time, practically the *ceiling* is usually defined as the height at which the ascending speed becomes less than 100 ft. per minute.

It is advisable to stop a little longer in studying the influence the various elements of the airplane have upon the *ceiling*.

HEIGHT
6000 M. 19680 Ft.
5000 M. 16400 Ft.
4000 M 13120 Ft.
3000 M 9840 Ft.
2000 M. 6560 Ft.
1000 M. 3280 Ft.
0
5'
5
5'
5'
5'
5'
ATMOSPHERIC PRESSURE
400 mm 15.75 Ft.
450 mm 17.72 Ft.
500 mm 19.69 Ft.
550 mm 21.65 Ft.
600 mm 23.62 Ft.
650 mm 25.59 Ft.
700 mm 27.56 Ft.
760 mm 29.92 Ft.

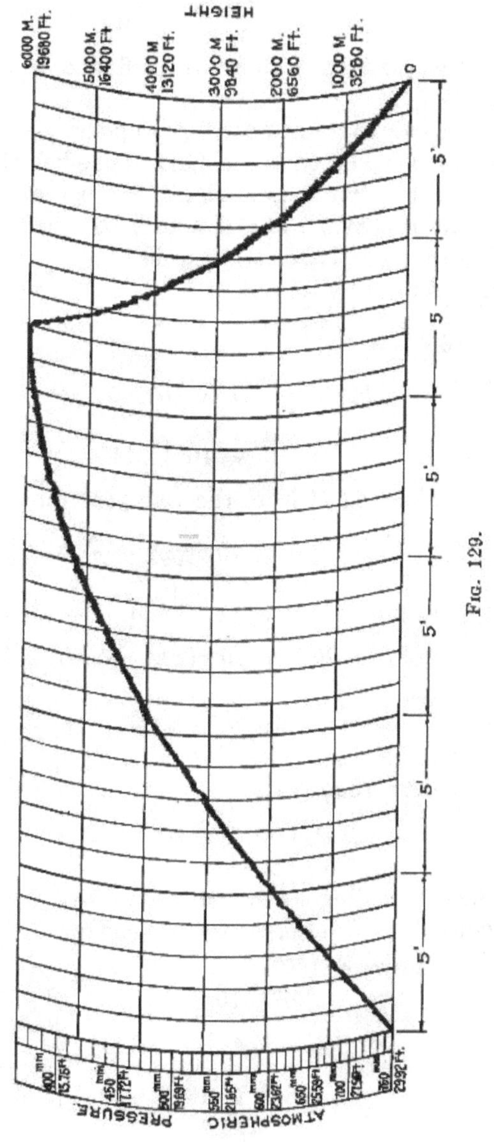

Fig. 129.

Let us again consider the formula

$$v = 550 \times \frac{\rho P_2 - P_1}{W}$$

and let us place in evidence the influence of μ on the difference $\rho P_2 - P_1$.

Supposing that we adopt a propeller best for climbing; that is, one which gives the maximum efficiency corresponding to the maximum ascending speed, we can, with sufficient practical approximation, assume ρ constant; then, since P_2 varies proportionally to μ, the useful power available, can be represented by

$$\mu\rho P_2$$

As for P_1,

$$550P_1 = 1.47 \times 10^{-4} (\delta A + \sigma) V^3$$

but

$$W = 10^{-4} \lambda A V^2$$

thus eliminating V from the two preceding equations

$$P_1 = 267 \times 10^{-3} (\delta A + \sigma) \times \frac{W^{3/2}}{\lambda^{3/2} A^{3/2}}$$

Now δ, σ, and λ are proportional to μ, therefore

$$P_1 = 267 \times 10^{-3}(\mu\delta A + \mu\sigma) \frac{W^{3/2}}{\mu^{3/2}\lambda^{3/2}A^{3/2}}$$

$$= 267 \times 10^{-3} \frac{1}{\sqrt{\mu}} (\delta A + \sigma) \frac{W^{3/2}}{\lambda^{3/2}A^{3/2}}$$

and we can then write

$$v = \frac{550}{W}\left[\mu\rho P_2 - 267 \times 10^{-3} \frac{1}{\sqrt{\mu}} (\delta A + \sigma) \frac{W^{3/2}}{\lambda^{3/2}A^{3/2}}\right]$$

Since the ceiling is reached when $v = 0$, it will correspond to value μ', which makes the second term of the preceding equation equal to zero.

$$\mu'\rho P_2 - \frac{267 \times 10^{-3}}{\sqrt{\mu'}} (\delta A + \sigma) \frac{W^{3/2}}{\lambda^{3/2}A^{3/2}} = 0$$

That is

$$\mu' = \frac{267^{2/3} \times 10^{-2}}{\rho^{2/3}P_2^{2/3}} (\delta A + \sigma)^{2/3} \frac{W}{\lambda A}$$

Remembering that

$$H = 60{,}720 \log \frac{1}{\mu}$$

the maximum value $H_{max.}$ of ceiling will be

$$H_{max.} = 60{,}720 \times \log \frac{1}{\mu'}$$

that is

$$H_{max.} = 60{,}720 \log \frac{\rho^{2/3} P_2^{2/3} \lambda A}{(\delta A + \sigma)^{2/3}\, W\, 267^{2/3} \times 10^{-2}} \qquad (1)$$

We can then enunciate the following general principles:

1. Every increase of ρ, P_2, and λA increases the *ceiling* of the airplane and *vice versa.*

2. Every decrease of δA, σ, and W similarly increases the *ceiling* and *vice versa.*

Equation (1) can also be put into the following form:

$$H_{max.} = 60{,}720 \log \frac{\rho^{2/3} \lambda}{\left(\frac{W}{P_2}\right)^{2/3} \left(\frac{\delta A + \sigma}{A}\right)^{2/3} \left(\frac{W}{A}\right)^{1/3} 267^{2/3} \times 10^{-2}} \qquad (2)$$

where

ρ = propeller efficiency

λ = lift coefficient of wing surface

$\frac{W}{P_2}$ = weight lifted per horsepower

$\frac{\delta A + \sigma}{A}$ = total resistance per square foot of wing surface.

$\frac{W}{A}$ = load per square foot of wing surface.

We then have five well-determined physical quantities which influence the value $H_{max.}$. As an example, and with a proceeding analogous to that adopted for the study of horizontal speed, we shall give to these parameters a series of values, and then, making them variable one by one, we shall study the influence of this variation upon $H_{max.}$.

Let us suppose for instance that

$$\rho = 0.8;\ \lambda = 22$$
$$\frac{W}{P_2} = 6 \text{ lb. per H.P.}$$
$$\frac{\delta A + \sigma}{A} = 1.2$$
$$\frac{W}{A} = 6 \text{ lb. per sq. ft.}$$

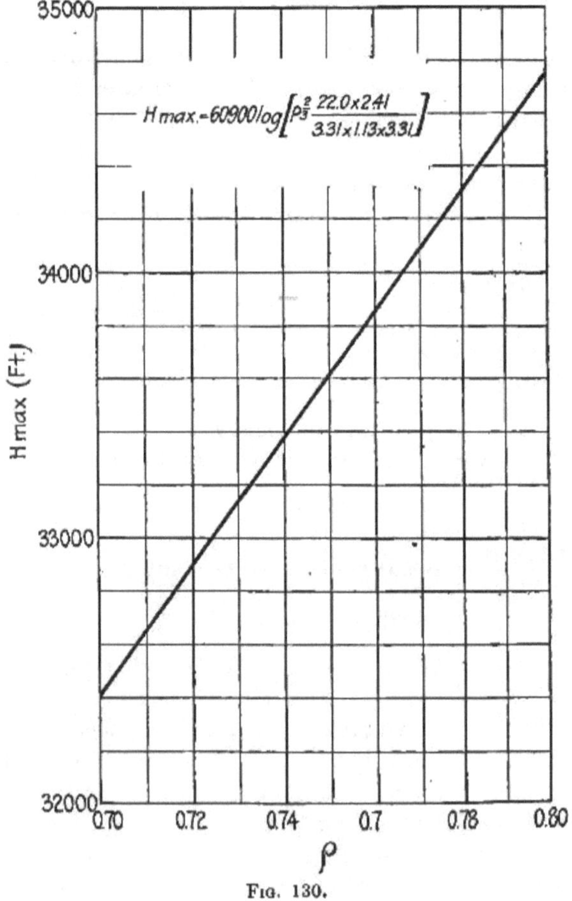

Fig. 130.

Then it is easy to draw the following diagrams on a paper having the logarithmic graduation on the axis of the abscissæ OX, and the normal graduation on the axis OY:

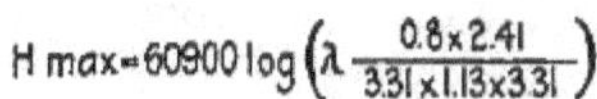

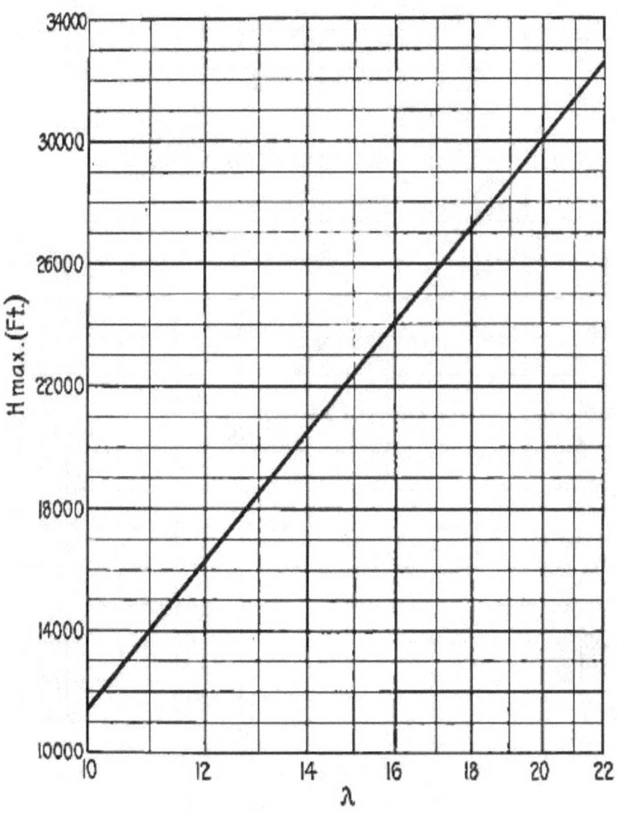

Fig. 131.

$H_{max.} = f(\rho)$ for ρ variable from 0.7 to 0.8 (Fig. 130)
$H_{max.} = f(\lambda)$ for λ variable from 10 to 22 (Fig. 131)
$H_{max.} = f\left(\frac{W}{P_2}\right)$ for $\frac{W}{P_2}$ variable from 6 to 14 lb. per H.P. (Fig. 132)

$H_{max.} = f\left(\frac{\delta \times A + \sigma}{A}\right)$ for $\frac{\delta \times A + \sigma}{A}$ variable from 1.2 to 1.8 (Fig. 133)

$H_{max.} = f\left(\frac{W}{A}\right)$ for $\frac{W}{A}$ variable from 6 to 9 lb. per sq. ft. (Fig. 134)

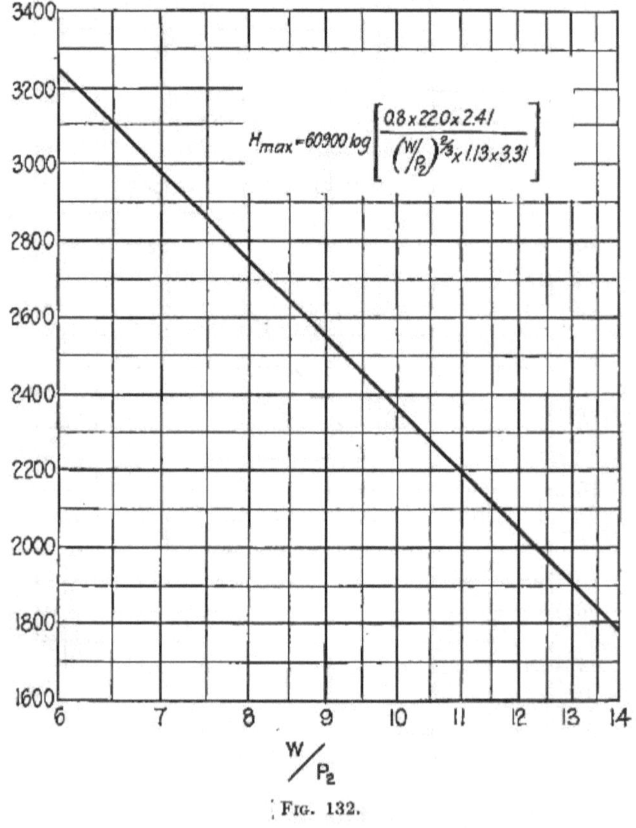

FIG. 132.

We wish to show now, how, with sufficient practical approximation, it is possible to reduce the formula which gives $H_{max.}$, to become solely a function of W, P_2 and A; that is, of the three elements which are always known in an airplane. In fact the values of ρ and $\lambda_{max.}$ for the greatest parts of the airplanes are values differing but little from each other and which can be considered with sufficient approximation equal to

$$\rho = 0.75 \qquad \lambda = 16$$

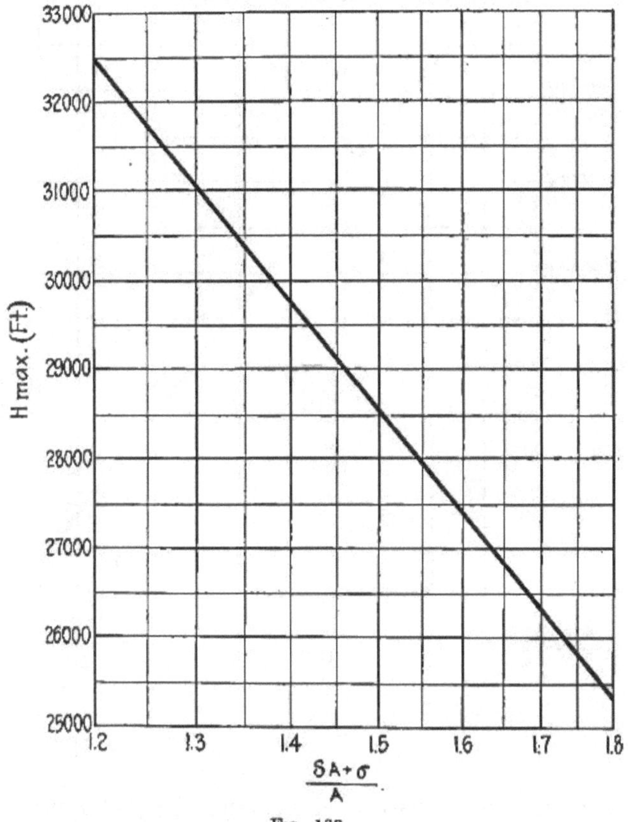

Fig. 133.

Let us furthermore remember that the head resistance R_δ and sustaining force R_λ are expressed by

$$R_\delta = 10^{-4}(\delta A + \sigma)V^2$$
$$R_\lambda = 10^{-4}\lambda A V^2$$

and consequently

$$\frac{R_\delta}{R_\lambda} = \frac{\delta A + \sigma}{\lambda A}$$

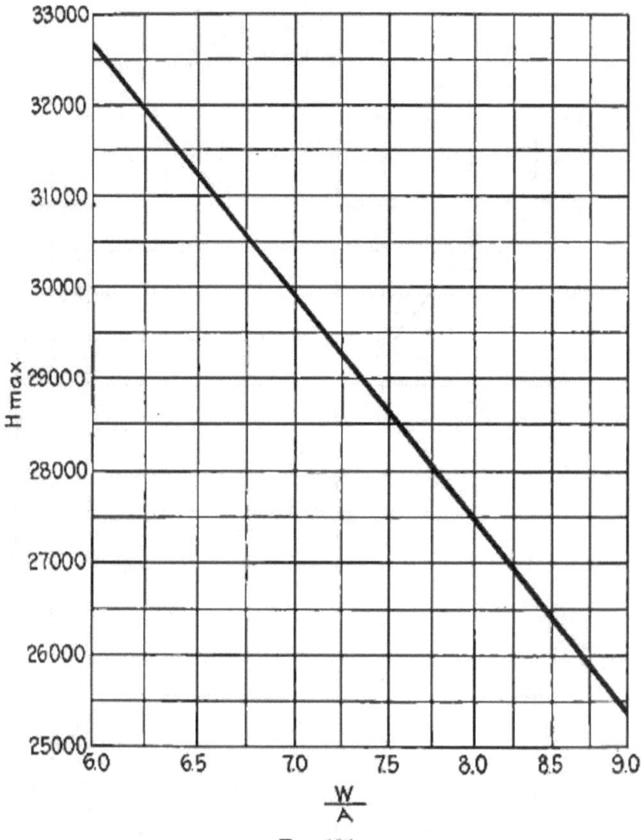

Fig. 134.

Now, in a well-constructed airplane, the minimum value of $\frac{R_\delta}{R_\lambda}$ is between 0.15 and 0.18. Assuming 0.15, we shall have

$$\frac{\delta A + \sigma}{\lambda A} = 0.15$$

and for $\lambda = 16$

$$\frac{\delta A + \sigma}{A} = 2.4$$

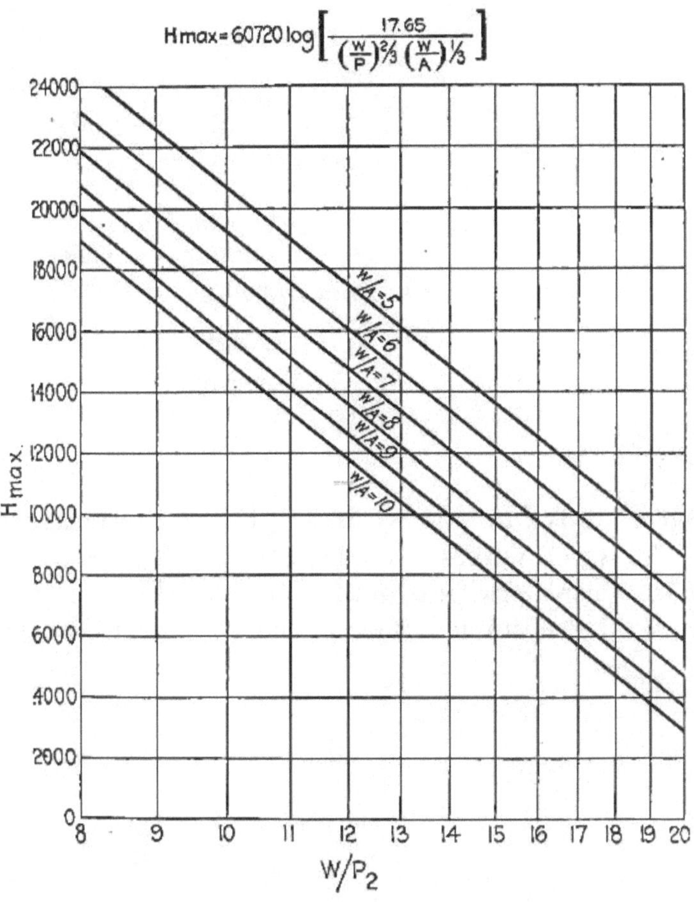

Fig. 135.

Then formula (2) becomes

$$H_{max.} = 60{,}720, \log \frac{0.75^{2/3} \times 16 \times 10^{+2}}{\left(\frac{W}{P_2}\right)^{2/3} 2.4^{1/3} \left(\frac{W}{A}\right)^{1/3} 267^{2/3}}$$

that is

$$H_{\text{max.}} = 60{,}720 \log \frac{17.65}{\left(\frac{W}{P_2}\right)^{3/4}\left(\frac{W}{A}\right)^{1/3}}$$

Based on this formula, we have plotted the diagrams of Fig. 135 which makes it possible to find $H_{\text{max.}}$ rapidly and with sufficient practical approximation when the weight, power and sustaining surface of the airplane are known.

CHAPTER XV

GREAT LOADS AND LONG FLIGHTS

In studying the history of aviation, the continuous increase of the dimensions of airplanes and of the power of engines, is decidedly marked. From the small units of 30 to 40 H.P. with which aviation started, we have to-day attained engines which develop 600 H.P. and more.

It is interesting to transfer to a diagram the history of the increase of the power of the engines from 1909 to the end of 1918, that is the progress of aviation engines in 9 years (Fig. 136).

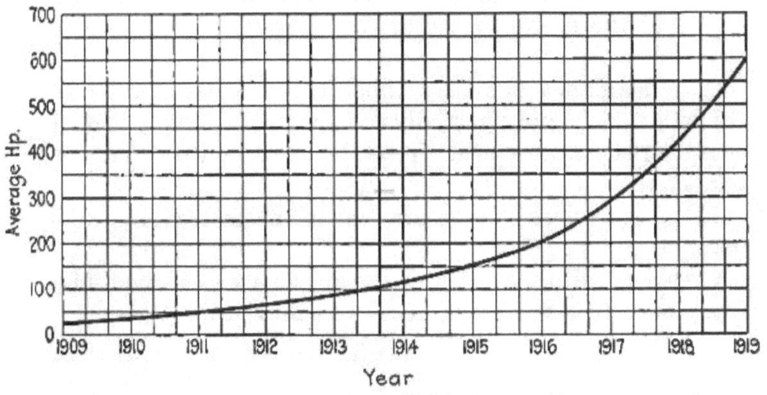

Fig. 136.

The great war which has just ended, while it gave a great impulse to many problems of aviation, has demanded that the high power available should be almost exclusively employed in raising the horizontal and ascending speeds under the urgency of military needs, leaving as secondary the research of great loads and great cruising radii, incompatible with too high horizontal and climbing speeds. We then find military machines, single seater scout planes, that with 300 H.P. can barely carry a total load of 600 lb.

(including pilot, gasoline and armament), and two seater machines that with 400 H.P. and more can barely carry a total useful load of 1300 lb.

Now certainly it is not by carrying some hundred pounds of useful load and by having the possibility of covering two or three hundred miles without stopping, that the airplane will be able to make its entrance among the practical means of locomotion. It is necessary that the hundreds of pounds and miles, become respectively thousands. To be able to traverse great distances of land and sea with safety, carrying a load such as to make these crossings commercial, is the great future of mercantile aviation.

To-day then, the vital problems of aviation are: the increase of the useful load and the increase of the cruising radius.

At first glance one may think that the two problems coincide; this is only partially true, each one having proper characteristics, as it will better be seen in the following part of this chapter.

Let us start with the examination of the problem of useful load.

Let us call W the weight of the airplane and U the useful load; since U is a fraction of W we can write

$$U = uW$$

where u is naturally less than 1.

Remembering the expression of total efficiency of the airplane

$$r = 0.00267 \frac{WV}{P_2}$$

we can also write

$$U = 375u \frac{r}{V} P_2$$

That equation shows that in order to increase the useful load it is necessary to increase u, the ratio $\frac{r}{V}$, and P_2.

(*a*) The coefficient $u = \frac{U}{W}$ gives the per cent. which is represented by the useful load with respect to the total weight of the airplane. Let us consider two airplanes having equal dimensions and forms; let us suppose that the weight be W for both, and the useful loads instead be different and equal to U_1 and U_2. Then we shall have respectively

$$U_1 = u_1 W$$
$$U_2 = u_2 W$$

Let us further suppose that the engine be the same for both airplanes, and that its weight be equal to $e \times W$; then, calling $a_1 \times W$ and $a_2 \times W$, the weights of the structure, that is, the weights of the airplanes properly speaking considered without engine and without useful load, we will have

$$W = u_1 W + eW + a_1 W$$
$$W = u_2 W + eW + a_2 W$$

and subtracting member from member

$$u_1 - u_2 = a_2 - a_1$$

That is to say if $u_1 > u_2$, W_1 shall have $a_2 > a_1$, and *vice versa;* that is, if the useful load of the first machine is greater than that of the second, the weight of its structure will instead be less. Now the weight of the structures, if the airplanes are studied with the same criterions and calculated with the same method, evidently characterize the solidity of the machine; and in that case the airplane having a lesser weight of structure, also has a smaller factor of safety, and if this is under the given limits, it may become dangerous to use it. Therefore, it is undesirable to increase the value of $u = \frac{U}{W}$ by diminishing the solidity of a machine.

It may also happen that two machines having different weights of structure, can have the same factor of safety, and in that case, the machine having less weight of structure is better calculated and designed than the other. The effort of the designer must therefore be to find the maximum possible value of coefficient u, assigning a given value to the factor of safety and seeking the materials, the forms and the dispositions of various parts which permit obtaining this coefficient with the minimum quantity of material, that is, with minimum weight. In modern airplanes, the coefficient u varies from the minimum value 0.3 (which we have for the fastest machines, as for instance the military scouts), to the value of 0.45 for slow machines.

The low value of u for the fastest machines depends upon two causes:

1. The factor of safety, necessary for very fast machines, must be greater than that necessary for the slow ones, therefore the value of coefficient a in the fast machines is greater than in the slow ones, with a consequent reduction of the value u.

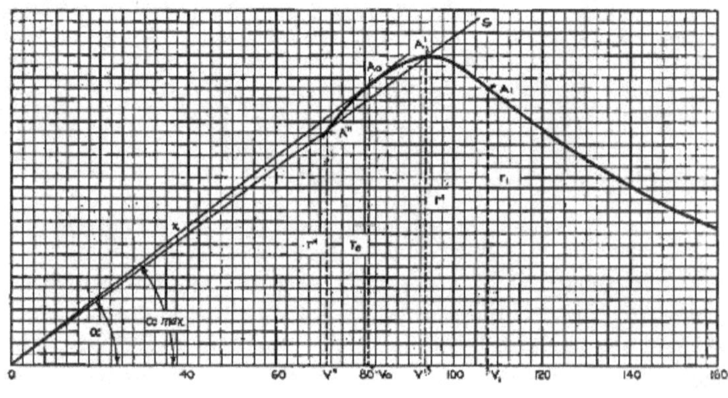

Fig. 137.

2. A fast machine having the same power, must be lighter than a slow machine (see the formula of total efficiency). That is to say, the importance of coefficient e increases, and therefore u diminishes.

(b) In Chapter XII, we studied coefficient r and saw that it was a function of V. Let us now study ratio $\frac{r}{V}$ and find in it the maximum value to be put in the formula of useful load.

Fig. 137 shows the diagram $r = f(V)$ already given in Fig. 109 of that chapter. The diagram refers to a particular example; its development, however, enables making some considerations of general character. From origin O let us draw any secant whatever to the diagram. This, in general, will be cut in two points A' and A''; let us call r' and r'' the values of efficiency and V' and V'' the values of speeds corresponding to these points.

Then evidently

$$\frac{r'}{V'} = \frac{r''}{V''} = \tan \alpha$$

Since we seek the maximum value of $\frac{r}{V}$, in order to have two values r_o and V_o such that their ratios will be the maximum possible, it will suffice to draw tangent t from origin O to point A_o of the diagram,

$$\frac{r_o}{V_o} = \tan \alpha_{max.}$$

Therefore infinite pairs of speeds V' and V'' exist, respectively greater and smaller than V_o, which individualize equal values of ratio $\frac{r}{V}$; naturally one would choose only the values of speed V', which are greater.

Practically it is not possible to adopt the maximum value $\frac{r_o}{V_o}$, as the airplane would be tangent, and could therefore scarcely sustain itself; it is then necessary to choose a lower value of $\frac{r_1}{V_1}$ and corresponding to a speed $V_1 > V_o$. The value $\frac{r_1}{V_1}$ must be inversely proportional to the height to be reached. In fact the equation

$$r = 0.00267 \frac{WV}{P_2}$$

states that $\frac{r}{V}$ is proportional to $\frac{W}{P_2}$. Now as the maximum height $H_{max.}$ is a function of $\frac{W}{P_2}$, consequently it is also a function of $\frac{r}{V}$.

(*c*) We treat finally the problems which relate to the increase of power P_2.

The increase of motive power has the natural consequence of immediately increasing the dimensions of the airplane.

The question naturally arises, "up to what limit is it possible to increase the dimensions of the airplane?"

First of all it is necessary to confute a reasoning false in its premises and therefore in its conclusions, sustained by some technical men, to demonstrate the impossibility of an indefinite increase in the dimensions of the airplane.

The reasoning is the following:

Consider a family of airplanes geometrically similar, having the same coefficient of safety.

In order that this be so, it is necessary that they have a similar value for the unit load of the sustaining surface $\frac{W}{A}$, and for the speed, as it can be easily demonstrated by virtue of noted principles in the science of constructions. Let us furthermore suppose that the airplanes have the same total efficiency r.

Then, as

$$r = 0.00267 \frac{WV}{P_2}$$

and as r and V are constant, W will be proportional to P_2; that is the total weight of the airplane with a full load will be proportional to the power of the engine

$$W = pP_2$$

The weight of structure $a \times W$ of airplanes geometrically similar, is proportional to the cube of the linear dimensions, which is equivalent to the cube of the square root of the sustaining surface; then

$$aW = a'A^{3/2}$$

but $\frac{W}{A}$ = constant, therefore A is proportional to W and consequently we may write

$$aW = a''W^{3/2}$$

that is

$$a = a''W^{1/2}$$

Since the weight of the motor group $e \times W$ is proportional to the power P_2,

$$e \times W = e' \times P_2$$

but

$$P_2 = \frac{W}{p}$$

so

$$e \times W = \frac{e'}{p} W$$

that is

$$e = \text{constant}$$

Then as

$$u + a + e = 1$$

we will have

$$u = 1 - e - a''\sqrt{W}$$

and this formula states that the value of coefficient u diminishes step by step as W increases, that is, as the dimensions of the machine increase step by step, until coefficient u becomes zero for that value of W which satisfies the equation

$$1 - e - a''\sqrt{W} = 0$$

that is

$$W = \left(\frac{1 - e}{a''}\right)^2$$

Thus the useful load becomes zero and the airplane would barely be capable of raising its own dead weight and the engine. So for example supposing

$$e = 0.25 \qquad a'' = 0.004$$

we shall have

$$W = \left(\frac{1 - 0.25}{0.004}\right)^2 = 35{,}000 \text{ lb.}$$

Now all the preceding reasoning has no practical foundation, because it is based on a false premise, that is, that the airplanes be geometrically similar. In fact, it is not at all necessary that it be so; on the contrary, the preceding reasoning demonstrates that to enlarge an airplane in geometrical ratio would be an error.

Nature has solved the problem of flying in various ways. For example, from the bee to the dragon fly, from the fly to the butterfly, from the sparrow to the eagle, we find wing structures entirely different in order to obtain the maximum strength and elasticity with the minimum weight.

It may be protested that flying animals have weights far lower than those of airplanes; but if we recall, that alongside of insects weighing one ten thousandth of a pound, there are birds weighing 15 lb., we will understand that if nature has been able to solve the problem of flying within such vast limits, it should not be difficult for man, owing to his means of actual technical knowledge, to create new structures and new dispositions of masses such as to make possible the construction of airplanes with dimensions far greater than the present average machines.

For example, one of the criterions which should be followed in large aeronautical constructions is that of distributing the masses. The wing surface of an airplane in flight must be considered as a beam subject to stresses uniformly distributed represented by the air reaction, and to concentrated forces represented by the various weights. Now by distributing the masses respectively on the wing surface, we obtain the same effect as for instance in a girder or bridge when we increase the supports; that is, there will be the possibility of obtaining the same factor of safety by greatly diminishing the dead weight of the structure.

Another criterion which will probably prevail in large aeronautical constructions, is the disposition of the wing surfaces in tandem, in such a way as to avert the excessive wing spans.

The multiplane dispositions also offer another very vast field of research.

As we see, the scientist has numerous openings for the solution; so it is permissible to assume that with the increase of the airplane dimensions not only may it be possible to maintain constant the coefficient of proportionality u but even to make it smaller. Thus with the increase of power we shall be able to notably increase the useful load.

Concluding, we may say that the increase of useful load can be obtained in three ways:

(a) Perfecting the constructive technique of the airplane and of the engine, that is reducing the percentage of dead weights in order to increase u,

(b) Perfecting the aerodynamical technique of the machine, reducing the percentage of passive resistance and increasing the wing efficiency and the propeller efficiency, so as to increase the value of ratio $\frac{r}{V}$ corresponding to the normal speed V, and

(c) Finally, increasing the motive power.

Let us now pass to the problem of increasing the cruising radius. Let us call $S_{max.}$ the maximum distance an airplane can cover, and let us propose to find a formula which shows the elements having influence upon $S_{max.}$

The total weight W of the airplane is not maintained constant during the flight because of the gasoline and oil consumption; it varies from its maximum initial value W_i to a final value W_f, which is equal to the difference between W_i and the total quantity of gasoline and oil consumed.

Let us consider the variable weight W at the instant t, and let us call dW its variation in time dt.

If P is the power of the engine and c its specific consumption (pounds of gasoline and oil per horsepower), the consumption in time dt will be

$$cPdt$$

and since that consumption is exactly equal to the decrease of weight in the time dt, we shall have

$$dW = -cPdt \quad (1)$$

From the formula of total efficiency we have

$$P = 0.00267\,\frac{WV}{r}$$

then substituting that value in (1)

$$dW = -\ 0.00267cW\ \frac{V}{r}\ dt$$

and since

$$V = \frac{dS}{dt}$$

$$\frac{dW}{W} = -\ 0.00267c\ \frac{dS}{r}$$

and integrating

$$\int \frac{dW}{W} = -\ 0.00267 \int \frac{cdS}{r}$$

The value of c, specific consumption of the engine, can, with sufficient approximation, be considered constant for the entire duration of the voyage.

Regarding r, we have already seen that it is a function of V; we shall now see that it is also a function of W. In fact, let us suppose that we have assigned a certain value V_1 to V; then the total efficiency will be

$$r = 0.00267 V_1 \frac{W}{P} = \text{const} \times \frac{W}{P}$$

Supposing now that W is made variable; it would also vary P, following a law which cannot be expressed by a certain simple mathematical equation; it will then also vary ratio $\frac{W}{P}$ and consequently r.

Practically, however, it is convenient, by regulating the motive power and therefore the speed, to make value r about constant and equal to the maximum possible value. We can also consider an average constant value for r. Thus the preceding integration becomes very simple. In fact, as $W = W_i$ for $S = 0$, and $W = W_f$ for $S = S_{max.}$, we shall have,

$$\log_e W_f = -\ 0.00267 \frac{c}{r} S_{max.} + \log_e W_i$$

that is

$$S_{max.} = 375 \times \frac{r}{c} \times \log_e \frac{W_i}{W_f}$$

and introducing the decimal logarithm instead of the Napierian

$$S_{max.} = 865 \times \frac{r}{c} \times \log \frac{W_i}{W_f} \tag{1}$$

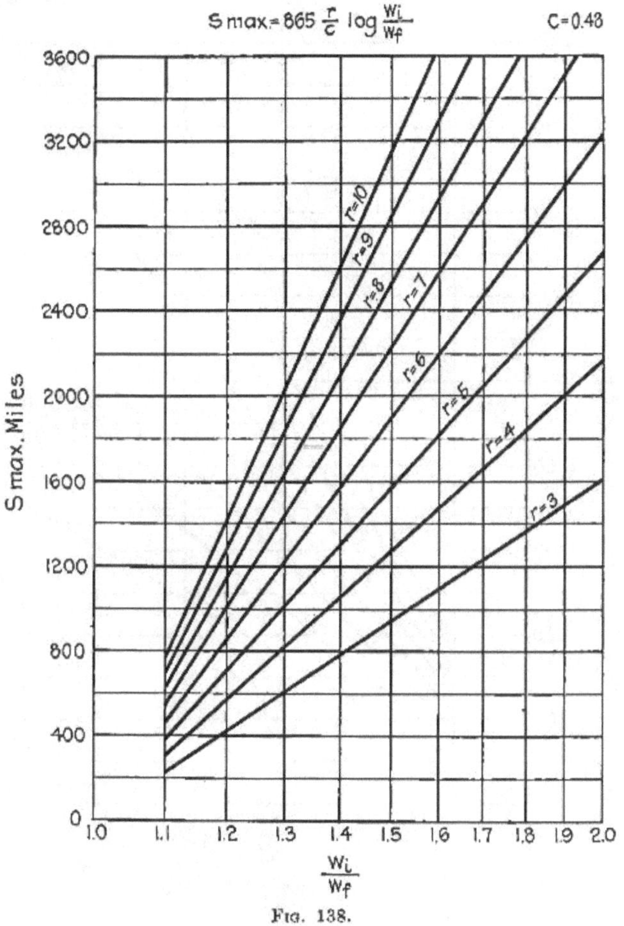

Fig. 138.

The cruising radius therefore depends upon three factors:

1. Upon the total aerodynamical efficiency. This dependency is linear; that is to say, an increase of say 10 per

cent. of aerodynamical efficiency, equally increases the maximum distance which can be covered by 10 per cent.

2. Upon the specific consumption of the engine. That dependency is inverse; thus, for example, if for we could

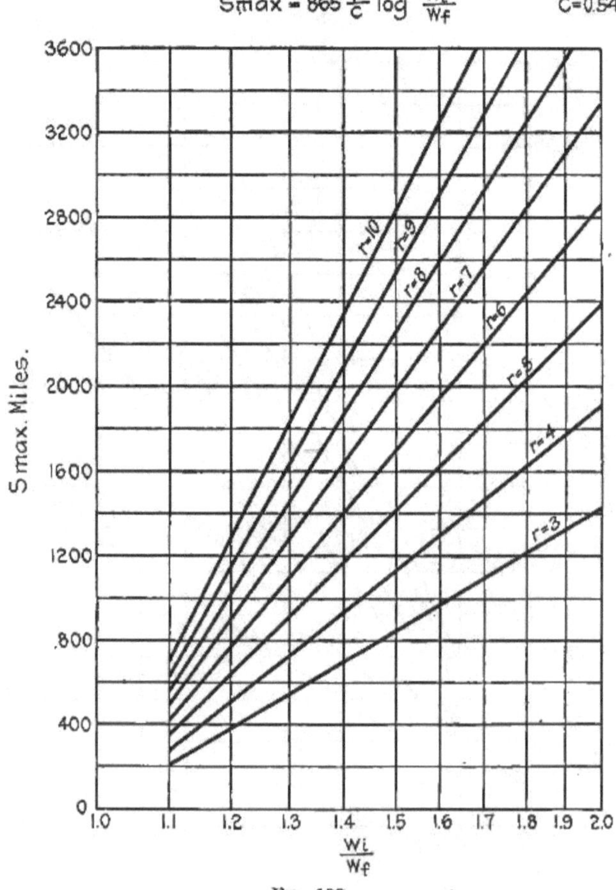

FIG. 139.

reduce the specific consumption to half, the radius of action would be doubled.

3. Upon the ratio between the total weight of the airplane and this weight diminished by the quantity of gasoline and

oil the airplane can carry. That ratio depends essentially upon the construction of the airplane; that is, upon the ratio between the dead weights and the useful load.

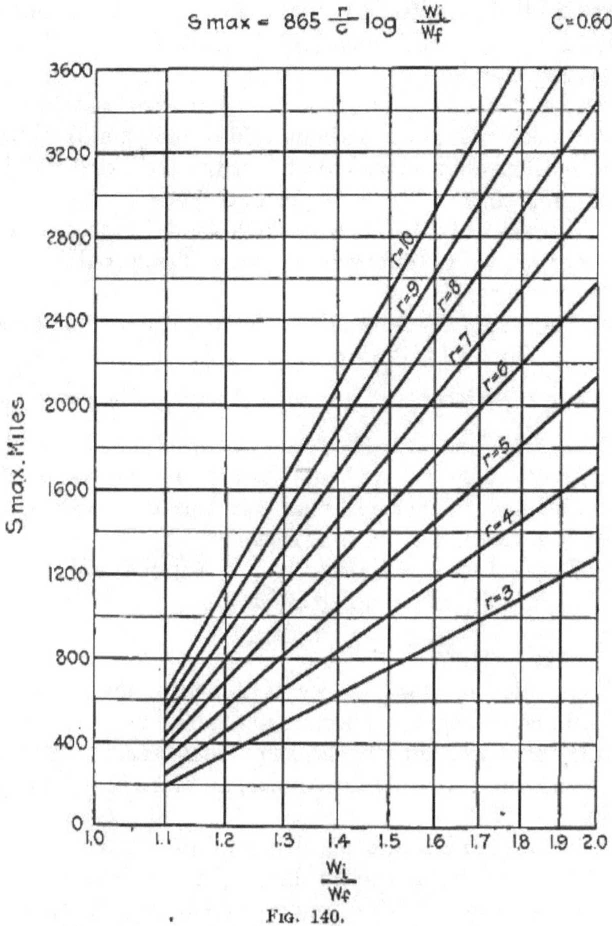

Fig. 140.

We see, consequently, that the essential difference between the formula of the useful load and that of the cruising radius is in the fact that in the latter the total specific consumption of the engine, an element which did not even appear in the other formula, intervenes and has a great importance. From that point of view, almost all modern aviation engines leave much to be desired; their low weight per horsepower (2 lb. per H.P. and even less), is obtained at a loss of efficiency; in fact they are enormously strained in their functioning and consequently their thermal efficiency is lowered.

The total consumption per horsepower in gasoline and oil, for modern engines is about 0.56 to 0.60 per H.P. hour; while gasoline engines have been constructed (for dirigibles), which only consume 0.47 lb. per H.P. hour.

A decrease from 0.60 to 0.48 would lead, by what we have seen above, to an increase in the cruising radius of 25 per cent.

Starting from formula (1) we have constructed the diagrams of Figs. 138, 139 and 140 which give the values of $S_{max.}$ as a function of $\frac{W_i}{W_f}$ for the different values of r and c. In Fig. 138 it has been supposed that $c = 0.48$ lb. per H.P. hour, in Fig. 139 $c = 0.54$ and in Fig. 140 $c = 0.60$. The diagrams have a normal scale on the ordinates and a logarithmic scale on the abscissæ.

The use of the diagrams is most simple, and permits rapidly of finding the radius of action of an airplane when r, c and $\frac{W_i}{W_f}$, are known.

Before closing this chapter, it is interesting to examine as table resuming the characteristics of the best types of military airplanes adopted in the recent war, for scouting, reconnaissance, day bombardment, and for night bombardment.

In Table 6 the following elements are found:

W_i = weight of the airplane with full load.

W_f = weight of the empty machine with crew and instruments necessary for navigation.

$\frac{W_i}{W_f}$ = ratio between initial weight and final weight.

We shall suppose therefore that all the useful load, comprising military loads, consists of gasoline and oil.

P = maximum power of the engine.

$\frac{W_i}{P}$ = weight per horsepower.

$\frac{W_i}{A}$ = load per unit of the wing surface.

$V_{max.}$ = the maximum horizontal speed of the airplane.

$v_{max.}$ = the maximum ascending speed averaged from ground level to 10,000 ft.

$P' = \frac{W_i \times v_{max.}}{0.75 \times 550}$ is the power absorbed in horsepower to obtain the ascending speed $v_{max.}$, supposing a propeller efficiency equal to 0.75.

$V' = V_{max.}\sqrt[3]{\frac{P - P'}{P}}$ is the horizontal speed of the airplane for which we have the maximum ascending speed $V_{max.}$.

$r = 0.00267\frac{W_i \times V_{max.}}{P}$ is the total efficiency corresponding to $V_{max.}$.

$r' = 0.00267\ \frac{W_i \times V'}{P - P'}$ is the total efficiency corresponding to V'.

S and S' = the maximum distances covered in miles corresponding to $V_{max.}\ r, \frac{W_i}{W_f}$ and $V', r', \frac{W_i}{W_f}$ respectively, supposing $c = 0.60$.

$\frac{S'}{S}$ = the gain in distance covered, flying at speed V' instead of V.

$W'_i = 375 \times r' \times \frac{0.85\,P}{V'}$ is the total weight the airplane can lift at speed V', supposing an allowance of excess power of 15 per cent.

$W'_f = W_f + \frac{1}{3}\ (W'_i - W_i)$ is the new value of the final weight, supposing that $\frac{1}{3}$ of the gain in weight is necessary to reinforce the airplane so as to have the same factor of safety.

$\frac{W'_i}{W'_f}$ = the new ratio between the new initial weight

TABLE 6

Type of machine	Wi, lb.	Wf lb.	$\frac{Wi}{Wf}$	P, h.p.	$\frac{Wi}{P}$ lb. per h.p.	$\frac{Wi}{A}$ lb. per sq. ft.	Vmax. m.p.h.	s max., ft. per sec.	P', h.p.	V' m.p.h.	r	r'	S, miles	S', miles	$\frac{S'}{S}$	Wi' lb.	Wf lb.	$\frac{Wi'}{Wf'}$	$\frac{Wi'}{A}$ lb. per sq. ft.	$\frac{Wi'}{P}$ lb. per sq. ft.	S'', miles	$\frac{S''}{S'}$
Single-seater fighter.	2100	1840	1.14	275	7.6	8.2	140	21	107	119	2.85	3.98	236	326	1.38	2930	2117	1.38	11.44	10.65	802	2.46
Two-seater reconnaissance..........	3300	2660	1.24	300	11.0	7.2	120	11	88	107	3.52	4.45	476	599	1.26	3980	2887	1.38	8.69	13.27	896	1.50
Two-seater day bomber...........	4200	2600	1.62	420	10.0	7.8	120	10	102	109	3.20	3.84	967	1160	1.20	4720	2773	1.70	8.77	11.24	12.76	1.10
Three-seater night bomber...........	12000	8000	1.50	1100	10.9	8.6	100	6	175	94	2.91	3.26	739	826	1.12	12160	8053	1.51	8.68	11.06	840	1.02

and the new final weight.

$\frac{W'_i}{A}$ = the new load per unit of wing surface.

$\frac{W'_i}{P}$ = the new load per horsepower.

S'' = the maximum distance covered corresponding to $\frac{W'_i}{W'_f}$ and to r'.

The examination of Table 7 enables making the following deductions:

1. Whatever be the type of machine it is convenient to fly at a reduced speed V', because in that way the cruising radius increases.

2. All war airplanes are utilized very little as to useful load and consequently as to cruising radius. As column $\frac{S''}{S'}$ shows, they could have a radius of action far superior if their enormous excess of power could be renounced. The gain is naturally stronger for the more light, quick airplanes, as for instance the scout machines, than for the heavier types.

PART FOUR

DESIGN OF THE AIRPLANE

CHAPTER XVI

MATERIALS

The materials used in the construction of an airplane are most varied. The more or less suitable quality of material for aviation can be estimated by the knowledge of three elements: specific weight, ultimate strength and modulus of elasticity.

Knowing these elements it is possible to calculate the coefficients

$$A_1 = \frac{\text{ultimate strength in pounds per square inch}}{\text{specific weight in pounds per cubic inch}}$$

and

$$A_2 = \frac{\text{modulus of elasticity in pounds per square inch}}{\text{specific weight in pounds per cubic inch}}$$

The coefficients A_1 and A_2 are not plain numbers, but have a linear dimension, and a very simple physical interpretation can be given to them; that is, A_1 measures the length in inches which, for instance, a wire of constant section of a certain material should have in order to break under the action of its own weight; A_2 instead, measures the length in inches which a wire (also of constant section) of the material should have in order that its weight be capable of producing an elongation of 100 per cent.

The higher the coefficients A_1 and A_2, the more suitable is a material for aviation.

It may be that two materials have equal coefficients A_1 and A_2, but different specific weights. In that case the material of lower specific weight is preferable when there are no restrictions as to space; instead, preference will be given to the material of higher specific weight when space is limited. This because of structural reasons, or in order to decrease head resistance.

In all of the following tables whenever possible, we shall give the values of specific weight and coefficients A_1 and A_2.

We shall briefly review the principal materials, grouping them into the following broad categories:

A. Iron, steel and their manufactured products.

B. Various metals.

C. Wood and veneers.

D. Various materials (fabrics, rubbers, glues, varnishes, etc.).

A. IRON, STEEL AND THEIR COMMON FORM AS USED IN AVIATION

Iron and steel are employed in various forms and for various uses; for forged or stamped pieces, in rolled form for bolts, in sheets for fittings, plates, joints, in tinned or leaded sheets for tanks, etc.

Fig. 141.

In Table 7 are shown the best characteristics required of a given steel according to the use for which it is intended.

Steel wires and cables are of enormous use in the construction of the airplane. Tables 8 and 9 give respectively tables of standardized wires and cables.

TABLE 7

Uses	Kind of steel	Tensile strength minimum		Yield point minimum		Elongation minimum per cent.	Reduction of area minimum per cent.	Carbon	Manganese	Phosphorus (max.)	Sulphur (max.)	Nickel	Chromium	Vanadium (min.)
		Lb. per sq. in.	Kg. per sq. mm.	Lb. per sq. in.	Kg. per sq. mm.									
For Forgings	Annealed carbon steel......	65000	45.70	36000	25.31	in 2″ = 25	50							
	Heat-treated carbon steel...	80000	56.24	60000	42.18	= 22	45	0.30–0.40	0.50–0.80	0.045	0.050			
	Heat-treated carbon steel...	95000	66.79	70000	49.21	= 18	45	0.40–0.50	0.50–0.80	0.045	0.050			
	Alloy steel..............	100000	70.30	80000	56.24	= 20	50	0.15–0.25	0.30–0.60	0.040	0.045	3.25–3.75 1.00–1.50	 0.45–0.75 0.60–0.90	0.15
	Alloy steel..............	120000	84.37	95000	66.79	= 18	50	0.25–0.35	0.50–0.80	0.040	0.045	3.25–3.75 1.00–1.50	 0.45–0.75 0.80–1.10	0.15
	Alloy steel..............	130000	91.40	115000	80.85	= 17	50							
For Sheets	Extra soft carbon.........	38000	26.70	20000	14.60	in 4″ = 30	..	0.05–0.15	0.30–0.60	0.045	0.050			
	Soft carbon..............	45000	31.64	25000	17.58	= 25								
	Tinned steel..............	45000	31.64	30000	21.09	= 20	..	0.05–0.15	0.30–0.60	0.045	0.050			
	Terne steel..............	45000	31.64	30000	21.09	= 20	..	0.05–0.15	0.30–0.60	0.045	0.050			
	Mild carbon..............	55000	38.67	36000	25.31	= 20	..	0.15–0.25	0.30–0.60	0.045	0.050			
	Mild carbon—vanadium...	60000	42.18	40000	28.12	= 20	..	0.20–0.30	0.50–0.80	0.045	0.050	....	....	0.15
	Alloy steel	100000	70.30	75000	52.70	= 15	..	0.15–0.25	0.30–0.60	0.040	0.045	3.25–3.75 1.00–1.50	 0.45–0.75 0.60–0.90	0.15
For Bolts	Cold drawn or cold rolled:													
	Alloy steel—Ni. Cr.......	100000	70.30	80000	56.24	in 2″ = 10	40	0.15–0.25	0.30–0.60	0.040	0.045	1.00–1.50	0.45–0.75	
	Alloy steel—Ni. Cr.......	125000	87.88	95000	66.79	= 17	50	0.25–0.35	0.50–0.80	0.040	0.045	1.00–1.50	0.45–0.75	

TABLE 8.—SIZES, WEIGHTS AND PHYSICAL PROPERTIES OF STEEL WIRE

English Units

American wire gage	Diameter, in.	Weight per 100 ft.-lb.	Torsion test*	Bend test	Breaking strength	Tensile strength
			Number of turns (minimum)	Number of bends (minimum)	Pounds (minimum)	Lb. per sq. in. (minimum)
6	0.162	7.010	16	5	4500	219,000
7	0.144	5.560	19	6	3700	229,000
8	0.129	4.400	21	8	3000	233,000
9	0.114	3.500	23	9	2500	244,000
10	0.102	2.770	26	11	2000	245,000
11	0.091	2.200	30	14	1620	249,000
12	0.081	1.744	33	17	1300	252,000
13	0.072	1.383	37	21	1040	255,000
14	0.064	1.097	42	25	830	258,000
15	0.057	0.870	47	29	660	259,000
16	0.051	0.690	53	34	540	264,000
17	0.045	0.547	60	42	425	267,000
18	0.040	0.434	67	52	340	270,000
19	0.036	0.344	75	70	280	275,000
20	0.032	0.273	84	85	225	280,000
21	0.028	0.216	96	105	175	284,000

METRIC UNITS

American wire gage	Diameter, mm.	Weight per 100 m., kg.	Torsion test*	Bend test	Breaking strength	Tensile strength
			Number of turns (minimum)	Number of bends (minimum)	Kilograms (minimum)	Kg. per sq. mm. (minimum)
6	4.115	10.440	16	5	2041.0	154.0
7	3.665	8.280	19	6	1678.0	161.1
8	3.264	6.550	21	8	1361.0	163.8
9	2.906	5.210	23	9	1134.0	171.6
10	2.588	4.120	26	11	907.0	172.2
11	2.305	3.280	30	14	735.0	175.0
12	2.053	2.597	33	17	590.0	177.2
13	1.828	2.060	37	21	472.0	179.4
14	1.628	1.635	42	25	376.5	181.5
15	1.450	1.295	47	29	299.4	182.1
16	1.291	1.028	53	34	244.9	185.6
17	1.150	0.814	60	42	192.8	187.7
18	1.024	0.646	67	52	154.2	189.8
19	0.912	0.512	75	70	127.0	193.4
20	0.813	0.406	84	85	102.1	196.8
21	0.724	0.322	96	105	79.4	199.6

*The minimum number of complete turns which a wire must withstand may be computed from the formula:

$$\text{Number of turns} = \frac{2.7}{\text{diameter in inches}} = \frac{68.6}{\text{dia. in millimeters}}$$

TABLE 9.—WEIGHTS, SIZES AND STRENGTH OF 7 × 19 FLEXIBLE CABLE

English units			Metric units		
Diameter, in.	Approximate weight, lb. per 100 ft.	Breaking strength, lb. (minimum)	Diameter, mm.	Approximate weight, kg. per 100 m.	Breaking strength kg. (minimum)
0.375 (3/8)	26.45	14,400	9.525	39.36	6,532
0.344 (11/32)	22.53	12,500	8.731	33.53	5,670
0.312 (5/16)	17.71	9,800	7.938	26.35	4,445
0.281 (9/32)	14.56	8,000	7.144	21.67	3,629
0.250 (1/4)	12.00	7,000	6.350	17.86	3,175
0.218 (7/32)	9.50	5,600	5.556	14.14	2,540
0.187 (3/16)	6.47	4,200	4.763	9.63	1,905
0.156 (5/32)	4.44	2,800	3.969	6.61	1,270
0.125 (1/8)	2.88	2,000	3.175	4.29	907

The formation of cables is shown in Fig. 141. The cable is made of 7 *strands of* 19 *wires each;* the figure shows how these strands are formed. The smaller diameters are extra-flexible so that they can be used as control wires as they well adapt themselves in pulleys.

Recently, steel *streamline* wires have been introduced to replace cables, in order to obtain a better air penetration. Fig. 142 shows the section of one of such wires. Their use

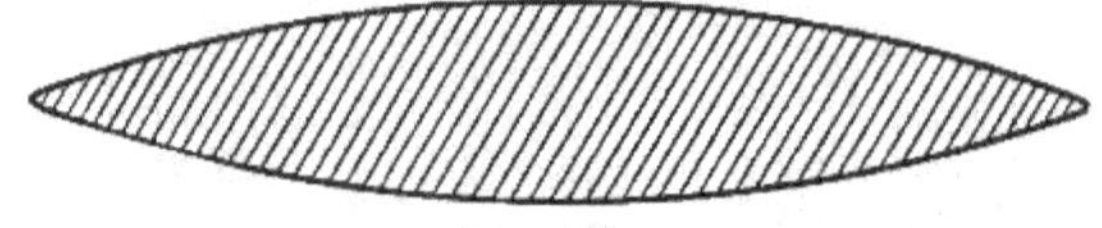

FIG. 142.

has not yet greatly broadened, especially because their manufacture has until now not become generalized. It is foreseen though, that the system will rapidly become popular.

We shall now take up the attachments of wires and cables. The attachment most commonly used for wires, is the so-called "*eye*" (Fig. 143). It is an easy attachment to make, but it reduces, however, the total resistance of the wire by 20 to 30 per cent. depending on the diameter of the wire.

Wires with larger threaded ends (called "tie rods") (Fig. 144), are becoming of general use. A very good attachment can be obtained by covering the bent wire with brass wire and soldering the whole with tin (Fig. 145); in this way an attachment is obtained which gives

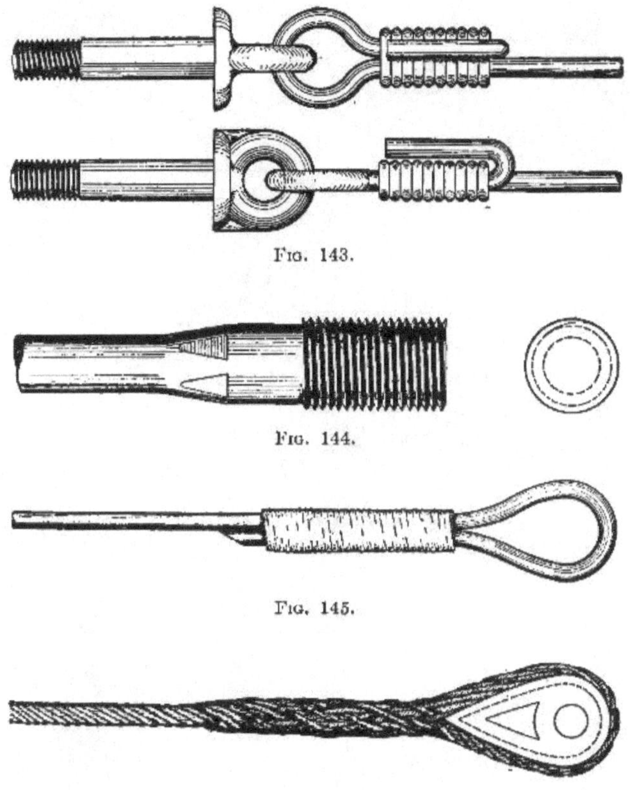

Fig. 143.

Fig. 144.

Fig. 145.

Fig. 146.

100 per cent. of the wire resistance. The soldering is made with tin in order to avoid the annealing of the wire.

The best attachment of cables is made by so-called splicing after bending it around a thimble (Fig. 146), which is made either of stamped sheets or of aluminum (Fig. 147).

Steel is also much used in tube form, either seamless, cold rolled, or welded. Table 10 gives the characteristics of the steel of various tube types.

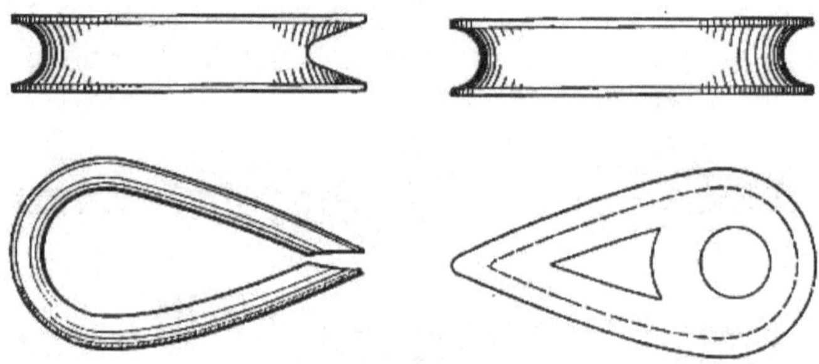

Fig. 147.

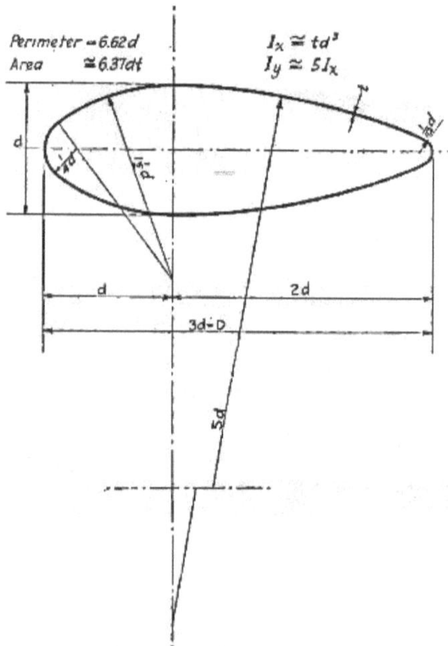

Fig. 148.

Tables 11 and 12 give the standard measurements of round tubes with the values of weight in pounds per foot and the values of the polar moment of inertia in in.[4]

Steel tubing having a special profile formed so as to give a minimum head resistance is also greatly used for interplane struts as well as for all other parts which must necessarily be exposed to the relative wind.

The best profile (that is, the profile which unites the best requisites of mechanical resistance, lightness and air penetration) is given in Fig. 148 which also shows how it is drawn, and gives the formulæ for obtaining the perimeter, the area, and the moments of inertia I_x and I_y about the two principal axes as function of the smaller diameter d and thickness t.

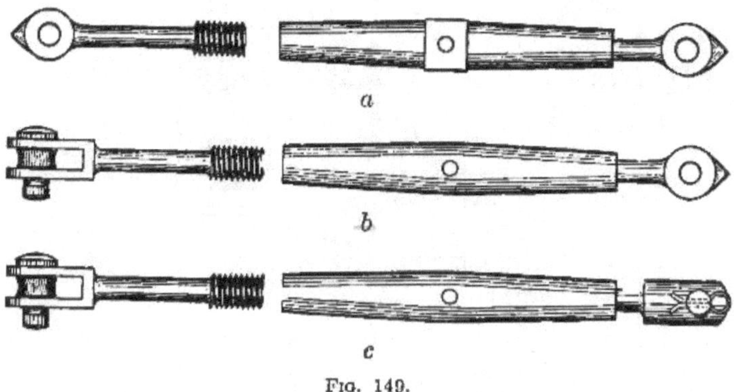

Fig. 149.

Tables 13 and 14 give all the above mentioned values, and furthermore the weight per linear foot for the more commonly used dimensions.

A greatly used fitting in aeronautical construction is the *turnbuckle*, which is designed to give the necessary tension to strengthening or stiffening wires and cables.

A turnbuckle is made of a central *barrel* into which two *shanks* are screwed with inverse thread; the shanks have either eye or fork ends; thus we have three classes of turnbuckles:

Double eye end turnbuckle (Fig. 149*a*)
Eye and fork end turnbuckle (Fig. 149*b*)
Double fork end turnbuckle (Fig. 149*c*)

TABLE 10.—CHARACTERISTICS OF THE STEEL OF VARIOUS TUBE TYPES

Types of steel tubes	Tensile strength (min.)		Yield point (min.)		Elongation (min.)			Carbon	Manganese	Phosphorus (max.)	Sulphur (max.)	Nickel	Chromium	Vanadium (min.)
	Lb. per sq. in.	Kg. per sq. mm.	Lb. per sq. in.	Kg. per sq. mm.	In 2″ = 50.8mm. per cent.	In 4″ = 10.8mm. per cent.	In 8″ = 203.2mm. per cent.							
Soft carbon steel, seamless	55,000	38.67	36,000	25.31			..	0.10–0.20	0.30–0.60	0.045	0.050			
Medium carbon steel, seamless	80,000	56.24	60,000	42.18	..		..	0.30–0.40	0.50–0.80	0.045	0.050			
Alloy steel, seamless—temper No. 2	85,000	59.76	60,000	42.18	25		10	0.15–0.25	0.30–0.60	0.040	0.045	3.25–3.75		
Alloy steel, seamless—temper No. 1	110,000	77.33	90,000	63.27	15	..	10	0.15–0.25	0.30–0.60	0.040	0.045	....	0.60–0.90	0.15
Welded steel		(Used for unstressed parts only)						0.15–0.25	0.30–0.60	0.045	0.050			
Alloy steel, seamless	200,000	140.60	.....	.....	5	3		0.25–0.35	0.50–0.80	0.040	0.045	3.25–3.75		
Alloy steel, axle	200,000	140.60	.....	.....	5	3		0.25–0.35	0.50–0.80	0.040	0.045	1.00–1.50	0.45–0.75	
								0.25–0.35	0.50–0.80	0.040	0.045	....	0.80–1.10	0.15

TABLE 11.—WEIGHTS IN POUNDS PER FOOT OF STANDARD SIZES OF STEEL TUBING

Outside diam., in.	Gauge and thickness														Outside diam., mm.
	6.349 mm.	5.556 mm.	4.762 mm.	3.969 mm.	3.402 mm.	3.175 mm.	2.778 mm.	2.385 mm.	2.108 mm.	1.651 mm.	1.245 mm.	0.889 mm.	0.711 mm.	0.559 mm.	
					No. 10		No. 12		No. 14	No. 16	No. 18	No. 20	No. 22	No. 24	
	1/4"	7/32"	3/16"	5/32"	0.134"	0.125"	0.109"	0.094"	0.083"	0.065"	0.049"	0.035"	0.028"	0.022"	
1/4												0.084	0.066	0.054	6.349
5/16												0.104	0.086	0.069	7.937
3/8										0.215	0.170	0.126	0.102	0.083	9.525
1/2										0.299	0.236	0.174	0.141	0.112	12.700
5/8										0.390	0.301	0.221	0.179	0.148	15.875
3/4									0.602	0.476	0.367	0.267	0.216	0.171	19.050
7/8								0.782	0.700	0.561	0.432	0.314	0.253	0.201	22.225
1							1.04	0.907	0.812	0.648	0.498	0.361	0.291	0.228	25.400
1 1/8						1.34	1.18	1.030	0.923	0.735	0.563	0.407	0.328	0.262	28.574
1 1/4					1.60	1.50	1.01	1.160	1.030	0.823	0.629	0.454	0.365	0.289	31.749
1 3/8				2.03	1.78	1.67	1.47	1.280	1.140	0.915	0.694	0.501	0.416	0.334	34.924
1 1/2			2.63	2.24	1.95	1.84	1.62	1.410	1.250	1.000	0.759	0.548	0.439	0.347	38.099
1 5/8			2.88	2.46	2.14	2.00	1.76	1.540	1.360	1.080	0.825	0.597	0.480	0.374	41.274
1 3/4			3.13	2.66	2.31	2.17	1.91	1.660	1.480	1.170	0.890	0.641	0.514	0.405	44.449
1 7/8	4.34	3.87	3.38	2.87	2.49	2.31	2.06	1.790	1.590	1.260	0.956	0.698	0.562		47.624
2	4.67	4.16	3.63	3.08	2.67	2.50	2.20	1.910	1.700	1.350	1.020	0.735			50.800
2 1/4	5.34	4.75	4.13	3.49	3.02	2.84	2.50	2.100	1.910	1.540	1.160				57.149

TABLE 12.—POLAR MOMENTS OF INERTIA IN IN.[4] OF STANDARD SIZES OF STEEL TUBING

Outside diam., in.	Gauge and thickness														Outside diam., mm.
	6.349 mm.	5.556 mm.	4.762 mm.	3.969 mm.	3.402 mm.	3.175 mm.	2.778 mm.	2.385 mm.	2.108 mm.	1.051 mm.	1.245 mm.	0.889 mm.	0.711 mm.	0.559 mm.	
					No. 10		No. 12		No. 14	No. 16	No. 18	No. 20	No. 22	No. 24	
	¼″	7⁄32″	3⁄16″	5⁄32″	0.134″	0.125″	0.109″	0.094″	0.083″	0.065″	0.049″	0.035″	0.028″	0.022″	
¼												0.000281	0.000242	0.000207	6.349
5⁄16												0.000619	0.000513	0.000428	7.937
⅜										0.00160	0.00137	0.00110	0.000934	0.000788	9.525
½										0.00430	0.00357	0.00278	0.00232	0.00189	12.700
⅝										0.00909	0.00737	0.00567	0.00439	0.00377	15.875
¾									0.01964	0.01656	0.01326	0.01011	0.00829	0.00668	19.050
⅞								0.03579	0.03284	0.02740	0.02177	0.01629	0.01345	0.01134	22.225
1							0.06137	0.05558	0.05067	0.04193	0.03309	0.02475	0.02023	0.01620	25.400
1⅛						0.09869	0.09172	0.08062	0.07326	0.06050	0.04796	0.03467	0.02900	0.02259	28.574
1¼					0.14867	0.14190	0.12845	0.11459	0.10395	0.08573	0.06559	0.04982	0.03640	0.03234	31.749
1⅜				0.22500	0.20313	0.19332	0.17552	0.15731	0.14062	0.11528	0.09060	0.06596	0.05380	0.04360	34.924
1½			0.33878	0.30245	0.27005	0.25826	0.23273	0.20575	0.18559	0.15221	0.11749	0.08641	0.06972	0.05794	38.099
1⅝			0.44587	0.39476	0.35450	0.33388	0.30343	0.26809	0.23928	0.19380	0.15221	0.10998	0.09034	0.07106	41.274
1¾			0.56956	0.50125	0.44877	0.42301	0.37930	0.33790	0.30638	0.24648	0.18854	0.13847	0.11212	0.08901	44.449
1⅞	0.86318	0.79334	0.71686	0.62946	0.55974	0.52930	0.47627	0.42226	0.37709	0.30254	0.23404	0.16989	0.13748		47.624
2	1.07187	0.98482	0.88466	0.77236	0.68611	0.65205	0.58429	0.51457	0.46252	0.36962	0.28547	0.21211			50.800
2¼	1.59471	1.51390	1.30046	1.13062	1.00753	0.94364	0.84348	0.73902	0.66317	0.53214	0.40984				57.149

Perimeter $= 6.02d$
Area $\cong 6.37\,dt$
$D = 3d$
$I_x \cong td^3$
$I_y \cong 5I_x$

TABLE 13

	Thickness		Perim. (approx.)		Area		Weight		I_x		W_x	
	mm.	in.	mm.	in.	mm.²	in.²	Kg. per lin. m.	Lb. per lin. ft.	cm.⁴	in.⁴	cm.³	in.³
D = 30 mm.	0.560	0.0220	66.2	2.64	35.65	0.0554	0.282	0.190	0.0560	0.00134	0.112	0.0069
= 1.18 in.	0.710	0.0279	66.2	2.64	45.10	0.0701	0.357	0.240	0.0710	0.00170	0.142	0.0087
d = 10 mm.	0.890	0.0350	66.2	2.64	56.70	0.0880	0.448	0.301	0.0890	0.00213	0.178	0.0109
= 0.394 in.	1.245	0.0490	66.2	2.64	79.30	0.1234	0.627	0.422	0.1245	0.00239	0.249	0.0152
D = 45 mm.	0.710	0.0279	99.3	3.91	67.8	0.105	0.538	0.362	0.240	0.00573	0.343	0.0210
= 1.72 in.	0.890	0.0350	99.3	3.91	85.0	0.132	0.673	0.453	0.300	0.00718	0.428	0.0262
d = 15 mm.	1.245	0.0490	99.3	3.91	119.4	0.185	0.945	0.637	0.420	0.01010	0.600	0.0367
= 0.591 in.	1.650	0.0650	99.3	3.91	158.0	0.245	1.250	0.842	0.558	0.01340	0.798	0.0488
	2.110	0.0830	99.3	3.91	202.0	0.313	1.600	1.078	0.710	0.01700	1.005	0.0615
D = 60 mm.	0.890	0.035	132.4	5.21	106.5	0.165	0.845	0.568	0.712	0.0171	0.712	0.0437
= 2.36 in.	1.245	0.049	132.4	5.21	159.0	0.247	1.263	0.850	0.995	0.0238	0.995	0.0608
d = 20 mm.	1.650	0.065	132.4	5.21	210.0	0.326	1.670	1.123	1.320	0.0316	1.320	0.0808
= 0.788 in.	2.110	0.083	132.4	5.21	268.0	0.416	2.130	1.434	1.688	0.0403	1.688	0.1032
D = 75 mm.	1.245	0.049	163.5	6.52	198.0	0.307	1.570	1.056	1.945	0.0465	1.555	0.0952
= 2.95 in.	1.650	0.065	163.5	6.52	262.5	0.407	2.080	1.400	2.580	0.0617	2.063	0.1263
d = 25 mm.	2.110	0.083	163.5	6.52	335.0	0.520	2.660	1.790	3.300	0.0788	2.640	0.1615
= 0.983 in.	2.385	0.094	163.5	6.52	380.0	0.590	3.020	2.034	3.720	0.0890	2.975	0.1820
	2.770	0.109	163.5	6.52	440.0	0.683	3.490	2.350	4.320	0.1032	3.460	0.2120

Perimeter = $6.02d$ $\quad$ $\leftarrow D = 3d \rightarrow$ $\quad$ $I_x \cong td^3$
Area $\cong 0.37\,dt'$ $\quad$ $I_y \cong 5I_x$

TABLE 14

	Thickness		Perim. (approx.)		Area		Weight		I_x		W_x	
	mm.	in.	mm.	in.	mm.^{2}	in.^{2}	Kg. per lm.	Lb. per l. ft.	cm.^{4}	in.^{4}	cm.^{3}	in.^{3}
	1.245	0.049	198.6	7.82	238.0	0.368	1.89	1.272	3.36	0.0803	2.24	0.137
D = 90 mm.	1.650	0.065	198.6	7.82	315.0	0.488	2.49	1.675	4.45	0.1065	2.97	0.182
= 3.54 in.	2.110	0.083	198.6	7.82	403.0	0.627	3.19	2.150	5.70	0.1364	3.90	0.237
d = 30 mm.	2.385	0.094	198.6	7.82	455.0	0.707	3.60	2.425	6.45	0.1542	4.30	0.263
= 1.18 in.	2.770	0.109	198.6	7.82	515.0	0.798	4.08	2.750	7.50	0.1793	5.00	0.306
	3.170	0.125	198.6	7.82	605.0	0.938	4.78	3.222	8.55	0.2045	5.70	0.349
	1.650	0.065	231.7	9.10	368.0	0.572	2.92	1.965	7.05	0.1688	4.03	0.247
D = 105 mm.	2.110	0.083	231.7	9.10	470.0	0.728	3.73	2.510	9.00	0.2155	5.15	0.315
= 3.94 in.	2.385	0.094	231.7	9.10	532.0	0.825	4.22	2.840	10.15	0.2425	5.82	0.356
d = 35 mm.	2.77	0.109	231.7	9.10	620.0	0.962	4.92	3.310	11.80	0.2820	6.25	0.382
= 1.38 in.	3.17	0.125	231.7	9.10	710.0	1.100	5.63	3.790	13.50	0.3230	7.73	0.473
	3.40	0.134	231.7	9.10	760.0	1.178	6.03	4.070	14.50	0.3470	8.30	0.508
	2.110	0.083	264.9	10.42	536.0	0.832	4.25	2.86	13.50	0.3230	6.75	0.413
D = 120 mm.	2.385	0.094	264.9	10.42	608.0	0.945	4.82	3.24	15.25	0.3630	7.62	0.467
= 4.72 in.	2.77	0.109	264.9	10.42	705.0	1.092	5.58	3.76	17.70	0.4230	8.85	0.542
d = 40 mm.	3.17	0.125	264.9	10.42	805.0	1.247	6.38	4.30	20.30	0.4850	10.15	0.618
= 1.57 in.	3.40	0.134	264.9	10.42	865.0	1.340	6.85	4.62	21.75	0.5180	10.88	0.662
	3.96	0.156	264.9	10.42	1010.0	1.568	8.00	5.38	25.30	0.6050	12.65	0.772

For turnbuckles as well as for bolts, the reader may easily procure from the respective firms, tables of standard measurements with indications of breaking strength.

B. VARIOUS METALS

Table 15 gives the physical and chemical characteristics of various metals most commonly used; that is, copper, brass, bronze, aluminum, duraluminum, etc.

Copper and brass are generally used for tanks, radiators, and the relative piping systems.

Aluminum is used rather exclusively to make the cowling which serves to cover the motor. Aluminum can also be used for the tanks.

High resistance bronzes are used for the barrels of turnbuckles.

Tempered aluminum alloys, have not become of general use at all, because their tempering is very delicate and it is easily lost if for any reason the piece is heated above 400°F.

We call especial attention to the untempered aluminum alloy which, not requiring any treatment, has a resistance and an elongation comparable to those of homogeneous iron, although its specific weight is 3/8 that of iron.

C. WOODS

Wood is extensively used in the construction of the airplane; either in solid form or in the form of veneer.

Tables 16 and 17 give the characteristics of the principal species of woods used in aviation.[1]

Cherry, mahogany, and walnut are used especially for manufacturing propellers. For the wing structure, yellow poplar, douglas fir, and spruce are especially used.

Yellow birch, yellow poplar, red gum, red wood, mahogany (true), African mahogany, sugar maple, silver maple, spruce, etc., are especially used in manufacturing veneers.

Great attention must be exercised in the selection of the

[1] This table has been compiled by the Forest Products Laboratory. U. S. Forest Service. Madison, Wisconsin.

TABLE 15

Types of metallic tubes	Tensile strength (min.)		Yield point (min.)		Per cent. Elongation (min.) In 2″ = 50.8 mm.	Specific gravity (max.)	Copper	Lead (max.)	Iron (max.)	Zinc	Aluminum (min.)
	Lb. per sq. in.	Kg. per sq. mm.	Lb. per sq. in.	Kg. per sq. mm.							
Seamless brass	1,000	0.70	......	.....	..	(apx.) 8.52	79.0–82.0	0.20	0.10	Remainder	
Seamless copper	1,000	0.70	......	.....	..	8.90					
Aluminum No. 1	22,000	15.47	......	.....	3	2.60	....	....	....		99.0
Aluminum No. 2	22,000	15.47	......	.....	3	2.60	....	....	....		98.0
Aluminum alloy, temper No. 1	55,000	38.67	35,000	24.61	12	2.85					
Aluminum alloy, temper No. 2	50,000	35.15	25,000	17.58	20	2.85					
	49,600	35.0	......	.....	15						
Aluminum alloy, no temper	to	to			to	3.00					
	78,000	55.0	......	.....	5						

TABLE 16.—PROPERTIES O
Strength Values at 15 Per Cent. Mo

Common and botanical names	Specific gravity based on volume and weight when oven dry		Specific weight at 15 per cent. moisture	Static	
				Fiber stress at elastic limit	Modulus of rupture
	Average	Minimum permitted	Lb. per cu. ft.	Lb. per sq. in.	Lb. per sq. in. (f_b)
Ash (commercial white) (*Fraxinus Americana; Fraxinus Lanceolata; Fraxinus Quadrangulata*)	0.62	0.56	40	7700	12700
Ash (black) (*Fraxinus Nigra*)	0.53	0.48	35	5800	10500
Basswood (*Tillia Americana*)	0.40	0.36	25	4700	7200
Beech (*Fagus Atropunicea*)	0.66	0.60	41	7400	12600
Birch (*Betula Lutea, Lenta*)	0.67	0.61	43	8400	13500
Cherry (black) (*Prunus Serotina*)	0.53	0.48	35	7300	10600
Cottonwood (*Populus Deltoides*)	0.43	0.39	28	4500	7000
Elm (rock) (*Ulmus Racemosa*)	0.66	0.60	44	6700	12500
Gum (red) (*Liquidambar Styraciflua*)	0.53	0.48	34	6700	10400
Hickory (true hickories) (*Higoria Glabra, Laciniosa, Alba, Ovata*)	0.81	0.73	50	8900	16300
Mahogany (true) (*Swietenia Mahagoni*)	0.54	0.50	36	7000	10000
Mahogany (African) (*Khaya Senegalensis*)	0.50	0.46	34	7100	10400
Maple (hard) (*Acer Saccharum*)	0.66	0.60	42	8100	12900
Oak (commercial white) (*Quercus Alba Macrocarpa; Minor, Michauxii*)	0.72	0.65	46	6700	12000
Poplar (yellow) (*Liriodendrum Tulipifera*)	0.42	0.38	28	4800	7500
Walnut (black) (*Juglans Nigra*)	0.56	0.52	38	7900	11900

timbers for aviation uses; they must be free from disease, homogeneous, without knots and burly grain, and above all they must be thoroughly dry. Artificial seasoning does not decrease the physical qualities of wood, but, on the contrary, it improves them if such seasoning is conducted at a temperature not above 100°F. and is done with proper precautions.

It is very important, especially for the long pieces, as for instance the beams, that the fiber be parallel to the axis of the piece, otherwise the resistance is decreased.

Furthermore, it is important to select by numerous laboratory tests the quality of the wood to be used, because between one stock of wood and another, great differences may usually be found.

As an example of the importance which the value of the density of wood has upon the major or minor convenience of its use in the manufacture of a certain part, let us suppose that we design the section of a wing beam which has to

F VARIOUS HARD WOODS
isture, for Use in Airplane Design

bending		Compression parallel to grain max. crushing strength	Compression perpendicular to grain fiber stress at elastic limit	Shearing strength parallel to grain	Hardness side load required to imbed 0.444 in. ball to one-half its diameter	$\frac{f_s}{f_b}$	$\frac{f_t}{f_b}$	$A_1 = \frac{f_b}{S.W.}$	$A_2 = \frac{E}{S.W.}$
Modulus of elasticity	Work to maximum load								
1000 lb. per sq. in. (E)	In. lb. per cu. in.	Lb. per sq. in. (f_c)	Lb. per sq. in.	Lb. per sq. in. (f_s)	Lb.				
1500	14.2	6000	1300	1750	1150	0.472	0.138	317.5	37500
1400	14.1	4900	800	1350	740	0.467	0.128	300.0	40000
1300	6.4	3800	400	880	340	0.503	0.122	288.0	52000
1500	13.3	5900	1100	1700	1060	0.468	0.135	307.4	36585
1800	17.6	6600	1060	1620	1070	0.489	0.120	314.0	41860
1400	12.0	5800	700	1500	830	0.548	0.141	302.8	40000
1200	7.3	3800	400	800	380	0.527	0.114	250.0	42855
1400	19.3	5800	1200	1650	1200	0.464	0.132	284.1	31818
1400	11.0	4900	700	1500	650	0.471	0.144	305.9	41777
1900	28.0	7300	1800	1800	....	0.448	0.110	326.0	38000
1300	9.1	5500	1000	1420	860	0.550	0.142	277.8	36111
1400	10.3	5100	900	1270	730	0.489	0.102	305.9	41777
1600	12.9	6500	1200	1990	1200	0.504	0.155	307.1	38095
1400	12.7	5900	1300	1760	1270	0.490	0.147	260.9	30435
1300	6.2	4100	400	900	370	0.546	0.120	267.9	46430
1500	13.1	6100	1000	1300	950	0.513	0.110	313.2	39474

resist, for example, to a bending moment of 20,000 lb.-inch; and let us suppose that the maximum space which it is possible to occupy with this section is that of a rectangle having a base equal to 2.2″ and a height equal to 2.8″. We shall make a comparison between the use of spruce and the use of douglas fir, for which the value of coefficient A_1 is about the same. Table 17 gives a modulus of rupture of 7900 lb. per sq. in. for the spruce with a weight per cu. ft. of 27 lb.; that is, 0.0156 lb. per cu. inch. Since the maximum bending moment is equal to 20,000 lb.-inch, the section modulus of the section will equal

$$W_s = \frac{20,000}{7,900} = 2.53 \text{ inch}^3$$

For fir, instead, we shall have

$$W_F = \frac{20,000}{9,700} = 2.06 \text{ in.}^3$$

with a density of 0.0197 lb. cu. in.

TABLE 17.—PROPERTIES
Strength Values at 15 Per Cent. M

Common and botanical names	Specific gravity based on volume and weight when oven dry		Specific weight at 15 per cent. moisture	Static	
				Fiber stress at elastic limit	Modulus of rupture
	Average	Minimum permitted	Lb. per cu. ft.	Lb. per sq. in.	Lb. per sq. in. (f_b)
Cedar (incense) (*Libocedrus Decurrens*)	0.36	0.32	26	4900	7100
Cedar (Port Orford) (*Chamaecyparis Lawsoniana*)	0.47	0.42	31	6200	10300
Cedar (western red) (*Thuja Plicata*)	0.34	0.31	23	4200	6400
Cedar (white northern) (*Thuja Occidentalis*)	0.32	0.29	22	4200	5800
Fir (Douglas)	0.52	0.47	34	6800	9700
Pine (sugar) (*Pinus Lambertiana*)	0.39	0.36	27	5300	7400
Pine (western white) (*Pinus Monticola*)	0.45	0.40	29	5100	7800
Pine (white) (*Pinus Strobus*)	0.39	0.36	27	5100	7400
Spruce (red, white, Sitka) (*Picea Rubens; Canadensis Sitchensis*)	0.41	0.36	27	5100	7900
Cypress (bald) (*Taxodium Distichum*)	0.47	0.42	31	5100	8800

Let us call x the thickness of the flange (Fig. 150a). Making the thickness of the web equal to 0.8x, the section modulus and the area of the section will be respectively

$$W = \frac{1}{6}\,[2.2'' \times 2.8''^2 - (2.2 - 0.8x)(3 - 2x)^2] \text{ cu. in.}$$

$$A = 2.2'' \times 2.8 - (2.2 - 0.8x) \times (3 - 2x) \text{ sq. in.}$$

For spruce

$$W = W_s = 2.53 \text{ in.}^3$$

from which we have

$$x = 0.9''$$
$$A = 4.37 \text{ sq. in.}$$

For fir we shall have analogously

$$x = 0.65''$$
$$A = 3.29 \text{ sq. in.}$$

Consequently, the spruce beam will weigh 4.37 × 0.0156 = 0.069 lb. per in. of length, while the fir beam will weigh 3.29 × 0.0197 = 0.064 lb. Supposing then for instance, that the total length of the beams be 150 ft., *i.e.*, 1800 in., the weight of the spruce beams would be 1800 × 0.069 = 124 lb., while the weight of the fir beams would be 1800 ×

OF VARIOUS CONIFERS
oisture, for Use in Airplane Design

bending									
Modulus of elasticity	Work to maximum load	Compression parallel to grain max. crushing strength	Compression perpendicular to grain fiber stress at elastic limit	Shearing strength parallel to grain	Hardness side load required to imbed 0.444 in. ball to one-half its diameter	$\frac{f_c}{f_b}$	$\frac{f_s}{f_b}$	$A_1 = \frac{f_b}{S.W.}$	$A_2 = \frac{E}{S.W.}$
1000 lb. per sq. in. (E)	In. lb. per cu. in.	Lb. per sq. in. (f_c)	Lb. per sq. in.	Lb. per sq. in. (f_s)	Lb.				
1000	6.0	4300	600	850	430	0.606	0.120	273.1	38464
1700	9.7	5300	700	1160	580	0.513	0.112	332.3	54840
1000	5.5	4000	400	790	300	0.625	0.123	278.3	43478
750	5.1	3400	350	800	300	0.586	0.138	263.6	34091
1780	7.2	6000	750	1020	580	0.619	0.104	285.3	52353
1100	5.0	4300	540	950	410	0.581	0.128	274.1	40740
1400	6.9	4800	480	670	360	0.615	0.086	260.0	48276
1200	6.1	4500	530	850	380	0.608	0.115	274.1	44444
1300	7.4	4300	500	920	430	0.544	0.117	292.6	48148
1300	6.8	5400	670	940	400	0.612	0.107	284.0	41936

0.064 = 115 lb.; that is, a gain of 9 lb., more than 7 per cent., would be obtained.

If we use elm, which has the same coefficient A_1 as the preceding woods, but a resistance of 12,500 lb. per sq. in. and a weight per cu. in. of 0.0255 lb. we would have (Fig. 159*b*)

$$x = 0.48''$$
$$A = 2.44 \text{ sq. in.}$$

with a weight per inch of 2.44 × 0.0255 = 0.062 and for 1800 in., a weight of 112 lb.; that is, a gain of about 10 per cent. over the spruce.

Let us now examine an inverse case, a case in which the piece is loaded only to compression and no limit fixed upon the space allowed its section; this for instance is the case of fuselage longerons. Then the product $E \times I$ (elastic modulus × moment of inertia), is of interest for the resistance of the piece.

Let us suppose that the longeron has a square section of side x. We then have

$$I = \frac{1}{12} x^4$$

Supposing that we have two kinds of wood of modulus E_1 and E_2 and specific weight W_1 and W_2 respectively; and suppose that coefficient A_2 be the same for both kinds, that is

$$A_2 = \frac{E_1}{W_1} = \frac{E_2}{W_2}$$

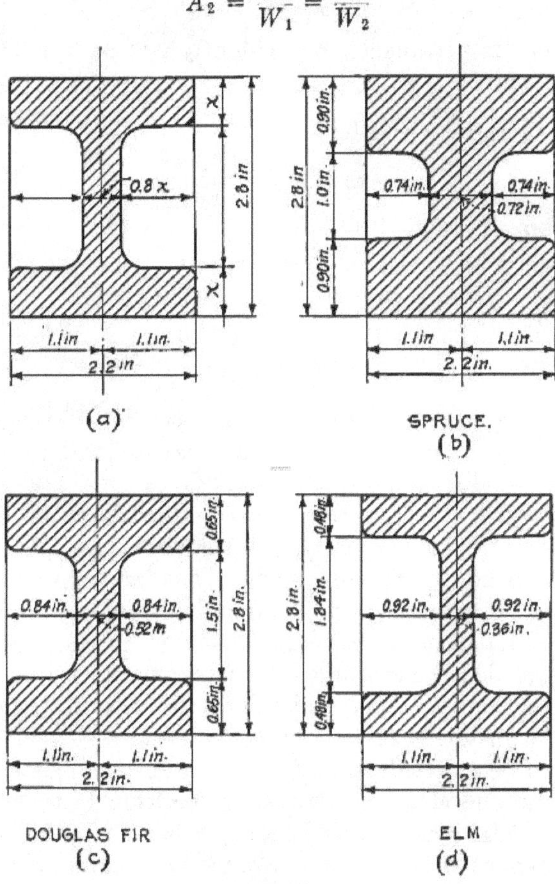

Fig. 150.

Let us call I_1 and I_2 the moments of inertia which the section must have respectively, according as to whether it is made of one or the other quality of wood. If we wish the piece to have the same resistance in both cases then

$$E_1I_1 = E_2I_2$$

that is

$$A_2 W_1 \frac{1}{12} x_1^4 = A_2 W_2 \frac{1}{12} x_2^4$$

from which

$$W_1 x_1^4 = W_2 x_2^4 \tag{1}$$

The weights per linear inch evidently will be in both cases

$$W_1 \times x_1^2 \text{ and } W_2 \times x_2^2$$

and their ratio w will be

$$w = \frac{W_1 \times x_1^2}{W_2 \times x_2^2}$$

But from (1)

$$\frac{W_1 \times x_1^2}{W_2 \times x_2^2} = \frac{x_2^2}{x_1^2}$$

consequently

$$w = \frac{x_2^2}{x_1^2}$$

that is, the piece having the greater section will weigh less, therefore it is convenient to use the material of smaller specific weight.

Let us now consider the veneers, which have become of very great importance in the construction of airplanes.

Wood is not, of course, homogeneous in all directions, as for instance, a metal from the foundry would be; its structure is of longitudinal fibers so that its mechanical qualities change radically according to whether the direction of the fiber or the direction perpendicular to the fiber is considered. Thus, for instance, the resistance to tension parallel to the fiber can be as much as 20 times that perpendicular to the fiber, and the elastic modulus can be from 15 to 20 times higher. *Vice versa*, for shear stresses we have the reverse phenomenon; that is, the resistance to shearing in a direction perpendicular to the fiber is much greater than in a parallel direction to the fiber. Now the aim in using veneer is exactly to obtain a material which is nearly homogeneous in two directions, parallel and perpendicular to the fiber.

Veneer is made by glueing together three or a greater odd number of thin plies of wood, disposed so that the fibers cross each other (Fig. 151). It is necessary that the number of plies be odd and that the external plies or faces have the same thickness and be of the same quality of wood, so that they may all be influenced in the same way by humidity, that is, giving perfect symmetrical deformations, thus avoiding the deformation of the veneer as a whole.

It is advisable to control the humidity of the plies during the manufacturing process, so that the finished panels may have from 10 to 15 per cent. of humidity. If we wish to

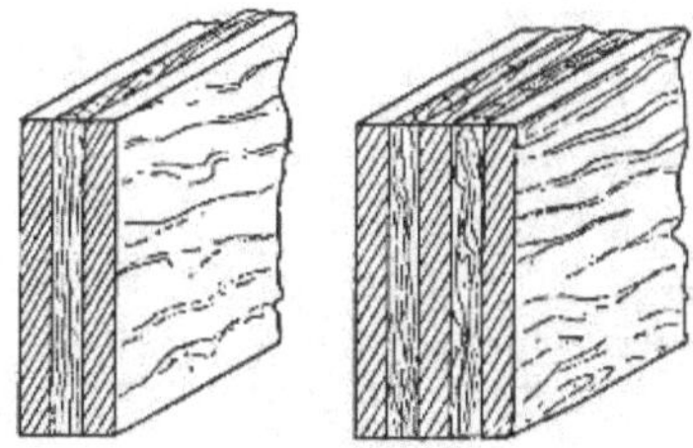

Fig. 151.

have the greatest possible homogeneity in both directions, it is advisable to increase the number of plies to the utmost, decreasing their thickness; this also makes the joining more easy by means of screws or nails, because the veneer offers a much better hold.

Considerations analogous to those given for the density of wood, lead to the conclusion that, wishing to attain a better resistance in bending, it is preferable to use plies of low density for the *core*. In fact, the weight being the same, the thickness of the panels will be inversely proportional to the density; but the moment of inertia, and consequently the resistance to column loads are proportional to the cube of the thickness; we see, therefore, the great advantage of having the *core* made of light thick material.

Light material would also be convenient for the faces, but they must also satisfy the condition of not being too soft, in order to withstand the wear due to external causes.

In Tables 18 and 19 we have gathered some of the tests

TABLE 18

Species	Average specific gravity of plywood tested	Per cent. moisture of plywood as tested	Column bending moment		Tensile strength		Split resistance per cent. of birch	Modulus of elasticity	
			Parallel to grain	Perpendicular to grain	Parallel to grain	Perpendicular to grain		Parallel to grain	Perpendicular to grain
			Lb. per sq. in.	Lb. per sq. in.	Lb. per sq. in.	Lb. per sq. in.		1000 lb. per sq. in.	1000 lb. per sq. in.
Ash (black)	0.49	9.1	7,760	1,779	6,180	3,940	73	1,073	96
Ash (commercial white)	0.60	10.2	9,930	2,620	6,510	4,350	71	1,420	143
Basswood	0.42	9.2	7,120	1,670	6,880	4,300	63	1,213	85
Beech	0.67	8.6	15,390	2,950	13,000	7,290	94	2,149	167
Birch (yellow)	0.67	8.5	16,000	3,200	13,200	7,700	100	2,259	197
Cedar (Spanish)	0.41	13.3	6,460	1,480	5,200	3,340	60	1,032	84
Cherry	0.56	9.1	12,260	2,620	8,460	5,920	80	1,627	152
Chestnut	0.43	11.7	5,160	1,110	4,430	2,600	74	744	75
Cottonwood	0.46	8.8	8,460	1,870	7,280	4,240	85	1,437	109
Cypress (bald)	0.47	10.3	7,830	1,820	6,560	4,390	69	1,144	91
Douglas fir	0.49	8.7	9,460	1,950	6,230	4,000	64	1,566	129
Elm (cork)	0.62	9.4	12,710	2,500	8,440	5,500	99	1,982	136
Elm (white)	0.52	8.9	8,630	1,970	5,860	3,990	75	1,224	109
Gum (black)	0.54	10.6	8,090	1,920	6,960	4,320	55	1,275	113
Gum (cotton)	0.50	10.3	7,760	1,580	6,260	3,760	60	1,300	111
Gum (red)	0.54	8.7	9,970	2,070	7,850	4,930	80	1,592	120

Forest Products Laboratory
Forest Service
United States Department of Agriculture

TABLE 19

Species	Average specific gravity of plywood tested	Per cent. moisture of plywood as tested	Column bending moment		Tensile strength		Split resistance per cent. of birch	Modulus of elasticity	
			Parallel to grain	Perpendicular to grain	Parallel to grain	Perpendicular to grain		Parallel to grain	Perpendicular to grain
			Lb. per sq. in.	Lb. per sq. in.	Lb. per sq. in.	Lb. per sq. in.		1000 lb. per sq. in.	1000 lb. per sq. in.
Hackberry	0.54	10.2	8,100	1,880	6,920	4,020	84	1,154	99
Hemlock (western)	0.47	9.7	9,250	1,960	6,800	4,580	63	1,581	112
Magnolia	0.59	9.9	9,830	2,340	10,000	5,740	98	1,704	135
Mahogany (African)	0.52	12.7	8,070	2,000	5,370	3,770	...	1,261	144
Mahogany (Philippine)	0.53	10.7	10,160	2,310	10,670	5,990	90	1,820	169
Mahogany (true)	0.48	11.4	8,500	1,940	6,390	3,780	...	1,252	117
Maple (soft)	0.57	8.9	11,540	2,420	8,180	5,380	106	1,752	145
Maple (sugar)	0.68	8.0	15,600	3,340	10,190	6,530	114	2,112	189
Oak (commercial red)	0.59	9.3	8,500	2,070	5,480	3,610	70	1,289	120
Oak (commercial white)	0.64	9.5	10,490	2,310	6,730	4,200	85	1,343	118
Pine (white)	0.43	10.2	7,920	1,770	5,640	3,870	52	1,274	99
Poplar (yellow)	0.50	9.4	8,860	1,920	7,390	4,720	51	1,544	115
Redwood	0.42	9.7	8,230	1,550	4,770	2,960	48	1,182	108
Spruce (Sitka)	0.43	8.4	7,640	1,680	5,600	3,250	82	1,395	105
Sycamore	0.56	9.2	11,040	2,340	8,030	5,220	77	1,628	130
Walnut (black)	0.59	9.1	12,660	2,770	8,250	5,260	77	1,736	141

FOREST PRODUCTS LABORATORY
FOREST SERVICE
United States Department of Agriculture

TABLE 20.—HASKELITE DESIGNING TABLE FOR THREE-PLY PANELS—NOT SANDED

Haskelite Research Laboratories—Report No. 109

Nominal thickness of panel	Faces	Core	Approximate weight	Approximate strength, lb. per in. of width	
	Thickness and kind of wood	Thickness and kind of wood	Lb. per 100 sq. ft.	Along face grain	Along core grain
	1/40 in.	1/40 in.			
	Spanish cedar	Spanish cedar	21	400	200
	Spanish cedar	Mex. mahogany	22	400	270
	Spanish cedar	Maple	24	400	300
	Spanish cedar	Birch	25	400	410
	Mex. mahogany	Spanish cedar	22	550	200
	Mex. mahogany	Mex. mahogany	23	550	270
	Mex. mahogany	Maple	26	550	300
0.075 in.	Mex. mahogany	Birch	26	550	410
	Maple	Spanish cedar	27	600	200
	Maple	Mex. mahogany	28	600	270
	Maple	Maple	31	600	300
	Maple	Birch	31	600	410
	Birch	Spanish cedar	28	820	200
	Birch	Mex. mahogany	28	820	270
	Birch	Maple	31	820	300
	Birch	Birch	31	820	410
	1/40 in.	1/20 in.			
	Spanish cedar	Basswood	26	400	500
	Spanish cedar	Spanish cedar	27	400	400
	Spanish cedar	Poplar	27	400	650
	Spanish cedar	Mex. mahogany	28	400	550
	Spanish cedar	Maple	33	400	600
	Spanish cedar	Birch	33	400	820
	Mex. mahogany	Basswood	27	550	500
	Mex. mahogany	Spanish cedar	28	550	400
	Mex. mahogany	Poplar	28	550	650
	Mex. mahogany	Mex. mahogany	29	550	550
	Mex. mahogany	Maple	34	550	600
0.100 in.	Mex. mahogany	Birch	35	550	820
	Maple	Basswood	32	600	500
	Maple	Spanish cedar	33	600	400
	Maple	Poplar	33	600	650
	Maple	Mex. mahogany	34	600	550
	Maple	Maple	39	600	600
	Maple	Birch	40	600	820
	Birch	Basswood	33	820	500
	Birch	Spanish cedar	33	820	400
	Birch	Poplar	34	820	650
	Birch	Mex. mahogany	35	820	550
	Birch	Maple	40	820	600
	Birch	Birch	40	820	820

TABLE 21.—HASKELITE DESIGNING TABLE FOR THREE-PLY PANELS—NOT SANDED

Haskelite Research Laboratories—Report No. 109

Nominal thickness of panel	Faces	Core	Approximate weight	Approximate strength, lb. per in. of width	
	Thickness and kind of wood	Thickness and kind of wood	Lb. per 100 sq. ft.	Along face grain	Along core grain
	1/28 in.	1/20 in.			
	Spanish cedar....	Basswood........	31	570	500
	Spanish cedar....	Spanish cedar....	32	570	400
	Spanish cedar....	Poplar...........	32	570	650
	Spanish cedar....	Mex. mahogany..	33	570	550
	Spanish cedar....	Maple...........	38	570	600
	Spanish cedar....	Birch............	38	570	820
	Mex. mahogany..	Basswood........	33	790	500
	Mex. mahogany..	Spanish cedar....	33	790	400
	Mex. mahogany..	Poplar...........	34	790	650
	Mex. mahogany..	Mex. mahogany..	35	790	550
	Mex. mahogany..	Maple...........	40	790	600
0.121 in.	Mex. mahogany..	Birch............	40	790	820
	Maple...........	Basswood........	40	860	500
	Maple...........	Spanish cedar....	41	860	400
	Maple...........	Poplar...........	41	860	650
	Maple...........	Mex. mahogany..	42	860	550
	Maple...........	Maple...........	47	860	600
	Maple...........	Birch............	48	860	820
	Birch............	Basswood........	40	1180	500
	Birch............	Spanish cedar....	41	1180	400
	Birch............	Poplar...........	42	1180	650
	Birch............	Mex. mahogany..	42	1180	550
	Birch............	Maple...........	47	1180	600
	Birch............	Birch............	48	1180	820

made at the "Forest Product Laboratory;" the veneers to which these tests refer were all three plies of the same thickness and the grain of successive plies was at right angles. All material was rotary cut. Perkins' glue was used throughout. Eight thicknesses of plies, from 3/30" to 3/6" were tested.

In Tables 20 to 29 are quoted the characteristics of three-ply panels of the Haskelite Mfg. Corp., Grand Rapids, Michigan.

TABLE 22.—HASKELITE DESIGNING TABLE FOR THREE-PLY PANELS—NOT SANDED

Haskelite Research Laboratories—Report No. 109

Nominal thickness of panel	Faces	Core	Approximate weight	Approximate strength, lb. per in. of width	
	Thickness and kind of wood	Thickness and kind of wood	Lb. per 100 sq. ft.	Along face grain	Along core grain
	1/28 in.	1/16 in.			
	Spanish cedar	Basswood	33	570	620
	Spanish cedar	Spanish cedar	34	570	500
	Spanish cedar	Poplar	35	570	810
	Spanish cedar	Mex. mahogany	36	570	690
	Spanish cedar	Maple	42	570	750
	Spanish cedar	Birch	43	570	1030
	Mex. mahogany	Basswood	35	790	620
	Mex. mahogany	Spanish cedar	36	790	500
	Mex. mahogany	Poplar	37	790	810
	Mex. mahogany	Mex. mahogany	38	790	690
	Mex. mahogany	Maple	44	790	750
0.133 in.	Mex. mahogany	Birch	45	790	1030
	Maple	Basswood	42	860	620
	Maple	Spanish cedar	43	860	500
	Maple	Poplar	44	860	810
	Maple	Mex. mahogany	45	860	690
	Maple	Maple	51	860	750
	Maple	Birch	52	860	1030
	Birch	Basswood	43	1180	620
	Birch	Spanish cedar	44	1180	500
	Birch	Poplar	44	1180	810
	Birch	Mex. mahogany	46	1180	690
	Birch	Maple	52	1180	750
	Birch	Birch	52	1180	1030

One of the best veneers for aviation is one obtained with spruce plies; this is easily understood if we consider the low density of spruce.

D. VARIOUS MATERIALS

(*a*) **Fabrics.**—Fabrics used for covering airplane wings are generally of linen or cotton, though sometimes of silk. The fabric is characterized by its resistance to tension and

TABLE 23.—HASKELITE DESIGNING TABLE FOR THREE-PLY PANELS—NOT SANDED

Haskelite Research Laboratories—Report No. 109

Nominal thickness of panel	Faces	Core	Approximate weight	Approximate strength, lb. per in. of width	
	Thickness and kind of wood	Thickness and kind of wood	Lb. per 100 sq. ft.	Along face grain	Along core grain
	1/28 in.	1/12 in.			
	Spanish cedar....	Basswood........	38	570	830
	Spanish cedar....	Redwood........	38	570	710
	Spanish cedar....	Spanish cedar....	39	570	670
	Spanish cedar....	Poplar...........	40	570	1080
	Spanish cedar....	Mex. mahogany..	41	570	920
	Spanish cedar....	Maple...........	50	570	1000
	Spanish cedar....	Birch............	50	570	1380
	Mex. mahogany..	Basswood........	40	790	830
	Mex. mahogany..	Redwood........	40	790	710
	Mex. mahogany..	Spanish cedar....	41	790	670
	Mex. mahogany..	Poplar...........	42	790	1080
	Mex. mahogany..	Mex. mahogany..	43	790	920
	Mex. mahogany..	Maple...........	51	790	1000
0.154 in.	Mex. mahogany..	Birch............	52	790	1380
	Maple...........	Basswood........	47	860	830
	Maple...........	Redwood........	47	860	710
	Maple...........	Spanish cedar....	48	860	670
	Maple...........	Poplar...........	49	860	1080
	Maple...........	Mex. mahogany..	50	860	920
	Maple...........	Maple...........	58	860	1000
	Maple...........	Birch............	59	860	1380
	Birch............	Basswood........	47	1180	830
	Birch............	Redwood........	47	1180	710
	Birch............	Spanish cedar....	49	1180	670
	Birch............	Poplar...........	49	1180	1080
	Birch............	Mex. mahogany..	51	1180	920
	Birch............	Maple...........	59	1180	1000
	Birch............	Birch............	60	1180	1380

TABLE 24.—HASKELITE DESIGNING TABLE FOR THREE-PLY PANELS—NOT SANDED

Haskelite Research Laboratories—Report No. 109

Nominal thickness of panel	Faces	Core	Approximate weight	Approximate strength, lb. per in. of width	
	Thickness and kind of wood	Thickness and kind of wood	Lb. per 100 sq. ft.	Along face grain	Along core grain
	1/20 in.	1/12 in.			
	Spanish cedar....	Basswood........	44	800	830
	Spanish cedar....	Redwood........	44	800	710
	Spanish cedar....	Spanish cedar....	46	800	670
	Spanish cedar....	Poplar...........	46	800	1080
	Spanish cedar....	Mex. mahogany..	48	800	920
	Spanish cedar....	Maple...........	56	800	1000
	Spanish cedar....	Birch............	57	800	1370
	Mex. mahogany..	Basswood........	47	1100	830
	Mex. mahogany..	Redwood........	47	1100	710
	Mex. mahogany..	Spanish cedar....	48	1100	670
	Mex. mahogany..	Poplar...........	49	1100	1080
	Mex. mahogany..	Mex. mahogany..	50	1100	920
	Mex. mahogany..	Maple...........	58	1100	1000
0.183 in.	Mex. mahogany..	Birch............	59	1100	1370
	Maple...........	Basswood........	57	1200	830
	Maple...........	Redwood........	57	1200	710
	Maple...........	Spanish cedar....	58	1200	670
	Maple...........	Poplar...........	59	1200	1080
	Maple...........	Mex. mahogany..	60	1200	920
	Maple...........	Maple...........	68	1200	1000
	Maple...........	Birch............	69	1200	1370
	Birch............	Basswood........	57	1650	830
	Birch............	Redwood........	57	1650	710
	Birch............	Spanish cedar....	59	1650	670
	Birch............	Poplar...........	60	1650	1080
	Birch............	Mex. mahogany..	61	1650	920
	Birch............	Maple...........	69	1650	1000
	Birch............	Birch............	70	1650	1370

TABLE 25.—HASKELITE DESIGNING TABLE FOR THREE-PLY PANELS—NOT SANDED

Haskelite Research Laboratories—Report No. 109

Nominal thickness of panel	Faces	Core	Approximate weight	Approximate strength, lb. per in. of width	
	Thickness and kind of wood	Thickness and kind of wood	Lb. per 100 sq. ft.	Along face grain	Along core grain
	1/16 in.	1/12 in.			
	Spanish cedar	Basswood	50	1000	830
	Spanish cedar	Redwood	50	1000	710
	Spanish cedar	Spanish cedar	51	1000	670
	Spanish cedar	Poplar	52	1000	1080
	Spanish cedar	Mex. mahogany	53	1000	920
	Spanish cedar	Maple	62	1000	1000
	Spanish cedar	Birch	62	1000	1370
	Mex. mahogany	Basswood	53	1370	830
	Mex. mahogany	Redwood	53	1370	710
	Mex. mahogany	Spanish cedar	54	1370	670
	Mex. mahogany	Poplar	55	1370	1080
	Mex. mahogany	Mex. mahogany	56	1370	920
	Mex. mahogany	Maple	65	1370	1000
0.208 in.	Mex. mahogany	Birch	65	1370	1370
	Maple	Basswood	65	1500	830
	Maple	Redwood	65	1500	710
	Maple	Spanish cedar	67	1500	670
	Maple	Poplar	67	1500	1080
	Maple	Mex. mahogany	69	1500	920
	Maple	Maple	77	1500	1000
	Maple	Birch	78	1500	1370
	Birch	Basswood	66	2060	830
	Birch	Redwood	66	2060	710
	Birch	Spanish cedar	68	2060	670
	Birch	Poplar	69	2060	1080
	Birch	Mex. mahogany	70	2060	920
	Birch	Maple	78	2060	1000
	Birch	Birch	79	2060	1370

TABLE 26.—HASKELITE DESIGNING TABLE FOR THREE-PLY PANELS—NOT SANDED

Haskelite Research Laboratories—Report No. 109

Nominal thickness of panel	Faces	Core	Approximate weight	Approximate strength, lb. per in. of width	
	Thickness and kind of wood	Thickness and kind of wood	Lb. per 100 sq. ft.	Along face grain	Along core grain
	1/16 in.	1/8 in.			
	Spanish cedar	Basswood	58	1000	1250
	Spanish cedar	Redwood	58	1000	1060
	Spanish cedar	Spanish cedar	61	1000	1000
	Spanish cedar	Poplar	62	1000	1620
	Spanish cedar	Mex. mahogany	64	1000	1370
	Spanish cedar	Maple	76	1000	1500
	Spanish cedar	Birch	77	1000	2060
	Mex. mahogany	Basswood	62	1370	1250
	Mex. mahogany	Redwood	62	1370	1060
	Mex. mahogany	Spanish cedar	64	1370	1000
	Mex. mahogany	Poplar	65	1370	1620
	Mex. mahogany	Mex. mahogany	67	1370	1370
	Mex. mahogany	Maple	79	1370	1500
0.250 in.	Mex. mahogany	Birch	80	1370	2060
	Maple	Basswood	74	1500	1250
	Maple	Redwood	74	1500	1060
	Maple	Spanish cedar	76	1500	1000
	Maple	Poplar	77	1500	1620
	Maple	Mex. mahogany	79	1500	1370
	Maple	Maple	92	1500	1500
	Maple	Birch	93	1500	2060
	Birch	Basswood	75	2060	1250
	Birch	Redwood	75	2060	1060
	Birch	Spanish cedar	77	2060	1000
	Birch	Poplar	78	2060	1620
	Birch	Mex. mahogany	80	2060	1370
	Birch	Maple	93	2060	1500
	Birch	Birch	94	2060	2060

TABLE 27.—HASKELITE DESIGNING TABLE FOR THREE-PLY PANELS—NOT SANDED

Haskelite Research Laboratories—Report No. 109

Nominal thickness of panel	Faces	Core	Approximate weight	Approximate strength, lb. per in. of width	
	Thickness and kind of wood	Thickness and kind of wood	Lb. per 100 sq. ft.	Along face grain	Along core grain
	1/12 in.	1/12 in.			
	Spanish cedar....	Basswood........	59	1330	830
	Spanish cedar....	Redwood........	59	1330	710
	Spanish cedar....	Spanish cedar....	61	1330	670
	Spanish cedar....	Poplar...........	61	1330	1080
	Spanish cedar....	Mex. mahogany..	63	1330	920
	Spanish cedar....	Maple...........	71	1330	1000
	Spanish cedar....	Birch............	72	1330	1370
	Mex. mahogany..	Basswood........	63	1830	830
	Mex. mahogany..	Redwood........	63	1830	710
	Mex. mahogany..	Spanish cedar....	65	1830	670
	Mex. mahogany..	Poplar...........	65	1830	1080
	Mex. mahogany..	Mex. mahogany..	67	1830	920
	Mex. mahogany..	Maple...........	75	1830	1000
0.250 in.	Mex. mahogany..	Birch............	76	1830	1370
	Maple...........	Basswood........	80	2000	830
	Maple...........	Redwood........	80	2000	710
	Maple...........	Spanish cedar....	81	2000	670
	Maple...........	Poplar...........	82	2000	1080
	Maple...........	Mex. mahogany..	83	2000	920
	Maple...........	Maple...........	92	2000	1000
	Maple...........	Birch............	92	2000	1370
	Birch............	Basswood........	81	2750	830
	Birch............	Redwood........	81	2750	710
	Birch............	Spanish cedar....	83	2750	670
	Birch............	Poplar...........	83	2750	1080
	Birch............	Mex. mahogany..	85	2750	920
	Birch............	Maple...........	93	2750	1000
	Birch............	Birch............	94	2750	1370

TABLE 28.—HASKELITE DESIGNING TABLE FOR THREE-PLY PANELS—NOT SANDED

Haskelite Research Laboratories—Report No. 109

Nominal thickness of panel	Faces	Core	Approximate weight	Approximate strength, lb. per in. of width	
	Thickness and kind of wood	Thickness and kind of wood	Lb. per 100 sq. ft.	Along face grain	Along core grain
	1/12 in.	1/8 in.			
	Spanish cedar....	Basswood........	68	1330	1250
	Spanish cedar....	Redwood........	68	1330	1060
	Spanish cedar....	Spanish cedar....	70	1330	1000
	Spanish cedar....	Poplar...........	71	1330	1620
	Spanish cedar....	Mex. mahogany..	73	1330	1370
	Spanish cedar....	Maple...........	86	1330	1500
	Spanish cedar....	Birch............	87	1330	2060
	Mex. mahogany..	Basswood........	72	1830	1250
	Mex. mahogany..	Redwood........	72	1830	1060
	Mex. mahogany..	Spanish cedar....	74	1830	1000
	Mex. mahogany..	Poplar...........	75	1830	1620
	Mex. mahogany..	Mex. mahogany..	77	1830	1370
	Mex. mahogany..	Maple...........	90	1830	1500
	Mex. mahogany..	Birch............	91	1830	2060
0.291 in.	Maple...........	Basswood........	89	2000	1250
	Maple...........	Redwood........	89	2000	1060
	Maple...........	Spanish cedar....	91	2000	1000
	Maple...........	Poplar...........	92	2000	1620
	Maple...........	Mex. mahogany..	94	2000	1370
	Maple...........	Maple...........	106	2000	1500
	Maple...........	Birch............	107	2000	2060
	Birch............	Basswood........	90	2750	1250
	Birch............	Redwood........	90	2750	1060
	Birch............	Spanish cedar....	92	2750	1000
	Birch............	Poplar...........	93	2750	1620
	Birch............	Mex. mahogany..	95	2750	1370
	Birch............	Maple...........	108	2750	1500
	Birch............	Birch............	109	2750	2060

TABLE 29.—HASKELITE DESIGNING TABLE FOR THREE-PLY PANELS—NOT SANDED

Haskelite Research Laboratories—Report No. 109

Nominal thickness of panel	Faces	Core	Approximate weight	Approximate strength, lb. per in. of width	
	Thickness and kind of wood	Thickness and kind of wood	Lb. per 100 sq. ft.	Along face grain	Along core grain
	1/8 in.	1/8 in.			
	Spanish cedar....	Basswood........	87	2000	1250
	Spanish cedar....	Redwood........	87	2000	1060
	Spanish cedar....	Spanish cedar....	89	2000	1000
	Spanish cedar....	Poplar...........	90	2000	1620
	Spanish cedar....	Mex. mahogany..	92	2000	1370
	Spanish cedar....	Maple...........	104	2000	1500
	Spanish cedar....	Birch............	105	2000	2060
	Mex. mahogany..	Basswood........	93	2750	1250
	Mex. mahogany..	Redwood........	93	2750	1060
	Mex. mahogany..	Spanish cedar....	95	2750	1000
	Mex. mahogany..	Poplar...........	96	2750	1620
	Mex. mahogany..	Mex. mahogany..	98	2750	1370
	Mex. mahogany..	Maple...........	111	2750	1500
0.375 in.	Mex. mahogany..	Birch............	112	2750	2060
	Maple...........	Basswood........	118	3000	1250
	Maple...........	Redwood........	118	3000	1060
	Maple...........	Spanish cedar....	120	3000	1000
	Maple...........	Poplar...........	121	3000	1620
	Maple...........	Mex. mahogany..	123	3000	1370
	Maple...........	Maple...........	136	3000	1500
	Maple...........	Birch............	137	3000	2060
	Birch............	Basswood........	120	4120	1250
	Birch............	Redwood........	120	4120	1060
	Birch............	Spanish cedar....	122	4120	1000
	Birch............	Poplar...........	123	4120	1620
	Birch............	Mex. mahogany..	125	4120	1370
	Birch............	Maple...........	138	4120	1500
	Birch............	Birch............	139	4120	2060

TABLE 30

	Material	Weight, oz. per sq. yd. max. "W"	Av. breaking load, lb. per yd. of width (min.)		Elongation				$\frac{S'}{W}$	$\frac{S''}{W}$
			Warp—S'	Filling—S"	Tension, lb. per yd. of width	In warp, per cent.	Tension, lb. per yd. of width	In filling, per cent.		
1	Linen—A	4.50	2160	2160	....	....	....	....	480	480
2	Linen—B	4.50	1870	1870	....	....	....	....	416	416
					288	12"— 5.4	288	12"— 2.7		
3	Mercerized cotton—A	4.50	2300	2300	576	12"— 6.7	576	12"— 3.3	512	512
					2020	12"—10.0	2020	12"— 5.3		
					288	12"— 5.8	288	12"— 3.7		
4	Mercerized cotton—B	4.00	2100	2100	576	12"— 7.2	576	12"— 4.3	525	525
					1870	12"— 9.7	1870	12"— 6.5		
5	Balloon cloth—AA	2.10	1080	1080	....	....	....	....	515	515
6	Balloon cloth—BB	2.65	1620	1620	....	....	....	....	612	612
7	Balloon cloth—DD	4.60	2340	2340	....	....	....	....	508	508
8	Balloon cloth—EE	5.00	3240	3240	....	....	....	....	648	648
9	Silk—A	2.36	2020	1825	2020	8"—26.0	1825	8"—16.5	856	775
10	Silk—B	2.12	2000	1735	2000	8"—28.0	1735	8"—22.0	944	820

to tearing, both in the direction of the woof and the warp, and by its weight per square foot.

Table 30 gives the characteristics of several types of fabric. In this table we find for various types the weight per square yard, the resistance in pounds per square yard (referring to both woof and warp) and the ratio between the resistance and weight. We see that silk is the most convenient material for lightness; the cost of this material with respect to the gain in weight is so high as to render its use impractical.

Fabric must be homogeneous and the difference between the resistance in warp and woof should not exceed 10 per

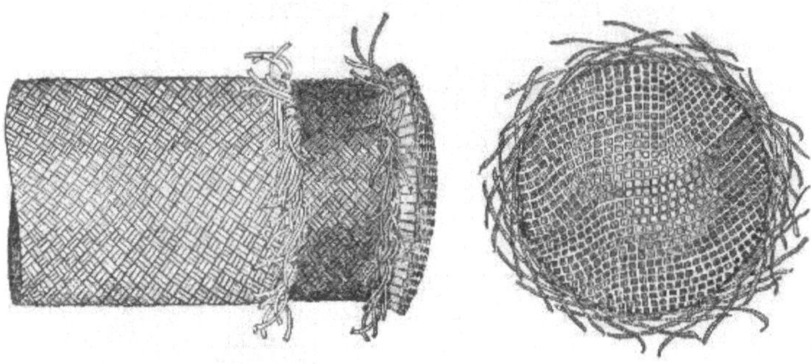

Fig. 152.

cent. of the total resistance; in fact the fabric on the wings is so disposed that the threads are at 45° to the ribs, thus working equally in both directions and having consequently the same resistance: in the calculations, therefore, the minor resistance should be taken as a basis; the excess of resistance in the other direction resulting only in a useless weight.

(*b*) **Elastic Cords.**—For landing gears the so-called elastic cord is universally adopted as a shock absorber.

It is made of multiple strands of rubber tightly incased within two layers of cotton braid (Fig. 152). Both the inner and outer braids are wrapped over and under with three or four threads. The rubber strands are square and are made of a compound containing not less than 90 per cent.

of the best Para rubber. The size of a single strand is between 0.05 and 0.035 inch.

The rubber strands are covered with cotton while they are subjected to an initial tension, in order to increase the

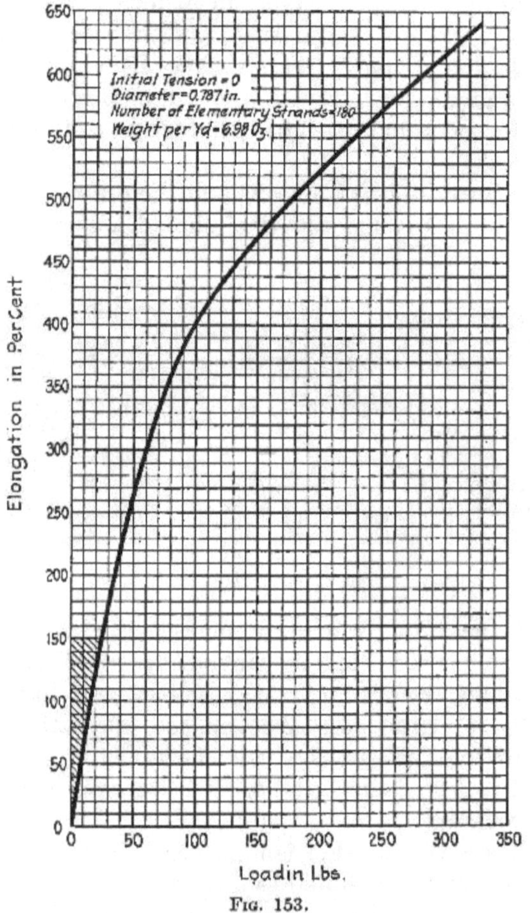

FIG. 153.

work that the elastic can absorb. The diagrams of Figs. 153 and 154 show this clearly.

Fig. 153 gives the diagram of work of a mass of rubber strands without cotton wrapping and without initial tension.

Fig. 154 gives the diagram of the same mass of rubber strands with an initial tension of 127 per cent., and with the cotton wrapping.

In general, the elongation is limited for structural reasons; let us suppose for instance, that an elongation of 150 per cent. be the maximum possible. It is then interesting to calculate the work which can be absorbed by 1 lb. of elastic cord having initial tension and cotton wrapping and to compare it to that which can be absorbed by 1 lb. of elastic cord without initial tension and without cotton

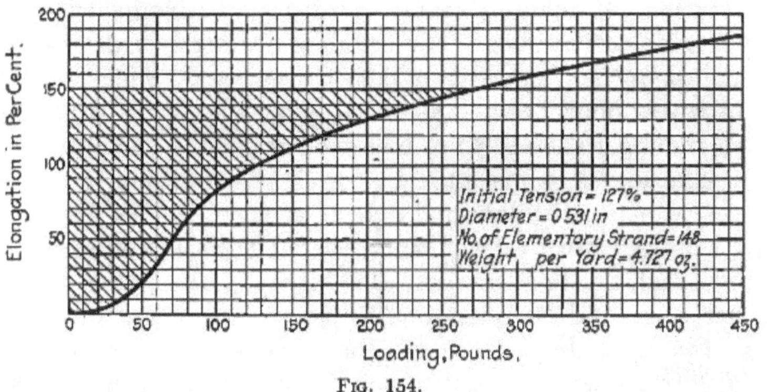

Fig. 154.

wrapping. The work can be easily calculated by measuring the shaded areas in Figs. 153 and 154. Naturally to do this it is necessary to translate the per cent. scale of elongation into inches, which is easy when the weight per yard is known.

For 150 per cent. of elongation the work absorbed by 1 lb. of elastic cord without initial tension and without cotton wrapping is 1280 lb.-in.; while that absorbed by elastic cord with 127 per cent. of initial elongation is equal to 20,200 lb.-in.; that is, in the second case a work about 16 times greater can be absorbed with the same weight. This shows the great convenience in using elastic cords with a high initial tension.

(*c*) **Varnishes.**—Varnishes used for airplane fabrics are divided into two classes: stretching varnishes (called "dope"), and finishing varnishes.

The former are intended to give the necessary tension to the cloth and to make it waterproof, increasing at the same time its resistance. The finishing varnishes which are applied over the stretching varnishes have the scope of protecting these latter from atmospheric disturbances, and of smoothing the wing surfaces so as to diminish the resistance due to friction in the air.

The stretching varnishes are generally constituted of a solution of cellulose acetate in volatile solvents without chlorine compounds. The cellulose acetate is usually contained in the proportion of 6 to 10 per cent. The solvents mixtures must be such as not to alter the fabrics and not to endanger the health of men who apply the varnish.

The use of gums must be absolutely excluded because they conceal the eventual defects of the cellulose film. A good stretching varnish must render the cloth absolutely oil proof, and will increase the weight of the fabric by 30 per cent. and its resistance by 20 to 30 per cent.

Finally it should be noted that it is essential for the varnish to increase the inflammability of the fabric as little as possible; precisely for this reason the cellulose nitrate varnish is used very seldom, notwithstanding its much lower cost when compared with cellulose acetate.

In general for linen and cotton fabrics three to four coats of stretching varnish are sufficient; for silk instead, it is preferable to give a greater number of coats, starting with a solution of 2 to 3 per cent. of acetate and using more concentrated solutions afterward.

The finishing varnishes are used on fabric which have already been coated with the stretching varnishes. These have as base linseed oil with an addition of gum, the whole being dissolved in turpentine.

A good finishing varnish must be completely dry in less than 24 hours, presenting a brilliant surface after the drying, resistant to crumpling, and able to withstand a wash with a solution of laundry soap.

(*d*) **Glues.**—Glues are greatly used both in manufacturing propellers and veneers.

Beside having a resistance to shearing superior to that of wood, a good glue must also resist humidity and heat. There are glues which are applied hot (140°F.), and those which are applied cold.

A good glue should have an average resistance to shearing of 2400 lb. per sq. in.

CHAPTER XVII

PLANNING THE PROJECT

When an airplane is to be designed, there are certain imposed elements on the basis of which it is necessary to conduct the study of the other various elements of the design in order to obtain the best possible characteristics.

Airplanes can be divided into two main classes: war airplanes and mercantile airplanes.

In the former, those qualities are essentially desired which increase their war efficiency, as for instance: high speed, great climbing power, more or less great cruising radius, possibility of carrying given military loads (arms, munitions, bombs, etc.), good visibility, facility in installing armament, etc.

For mercantile airplanes, on the contrary, while the speed has the same great importance a high climbing power is not an essential condition; but the possibility of transporting heavy useful loads and great quantities of gasoline and oil, in order to effectuate long journeys without stops, assumes a capital importance.

Whatever type is to be designed, the general criterions do not vary. Usually the designer can select the type of engine from a more or less vast series; often though, the type of motor is imposed and that naturally limits the fields of possibility.

Rather than exposing the abstract criterions, it is more interesting to develop summarily in this and the following chapters, the general outline of a project of a given type of airplane, making general remarks which are applicable to each design as it appears. In order to fix this idea, let us suppose that we wish to study a fast airplane to be used for sport races.

The future aviation races will certainly be marked by imposed limits, which may serve to stimulate the designers of airplanes as well as of engines towards the increase of efficiency and the research of all those factors which make flight safer.

For instance, for machines intended for races the ultimate factor of safety, the minimum speed, the maximum hourly consumption of the engine, etc., can be imposed.

The problem which presents itself to the designer may be the following: to construct an airplane having the maximum possible speed and also embodying the following qualities:

1. A coefficient of ultimate resistance equal to 9.
2. Capable of sustentation at the minimum speed of 75 m.p.h.
3. Capable of carrying a total useful load of 180 lb. (pilot and accessories), beside the gasoline and oil necessary for three hours flight.
4. An engine of which the total consumption in oil and gasoline does not surpass 180 lb. per hour when running at full power.

Let us call W the total weight in pounds of the airplane at full load, A its sustaining surface in sq. ft., W_u the useful load in pounds, P the power of the motor in horsepower, and C the total specific consumption of the engine in oil and gasoline.

Remembering that in normal flight

$$W = 10^{-4} \lambda A V^2$$

since the condition is imposed that the airplane sustain itself for $V = 75$ m.p.h., we must have

$$\frac{W}{A} < 0.56 \lambda_{max.}$$

that is, the load per square foot of wing surface will have to equal $\frac{56}{100}$ of the maximum value $\lambda_{max.}$ which it is possible to obtain with the aerofoil under consideration.

The total useful load will equal

$$W_u = 180 + 3cP$$

Let us call W_p the weight of the motor including the propeller, W_R the weight of the radiator and water, W_A the weight of the airplane.
Then

$$W = W_u + W_p + W_R + W_A \qquad (1)$$

Calling p the weight of the engine propeller group per horsepower we will have

$$W_p = pP$$

The weight of the radiator and water, by what we have said in Chapter V, can be assumed proportional to the power of the engine and inversely proportional to the speed.

$$W_R = b\,\frac{P}{V}$$

As to the weight of the airplane, for airplanes of a certain well-studied type and having a given ultimate factor of safety, it can be considered proportional to the total weight; we can therefore write

$$W_A = aW$$

Then (1) can be written

$$W = 180 + 3cP + pP + b\,\frac{P}{V} + aW$$

that is

$$W = \frac{180}{1-a} + \frac{P}{1-a}\left(3c + p + \frac{b}{V}\right) \qquad (2)$$

The machine we must design is of a type analogous to the single-seater fighter. Consequently in the outline of the project we can use the coefficients corresponding to that type.

For these, the value of a is about 0.34; also, expressing V in m.p.h. we can take $b = 45$.

Remembering the imposed condition that cP must not exceed 180 lb., we will have to select an engine having the minimum specific consumption c, in order to have the maximum value of P; at the same time the weight p per horsepower must be as small as possible.

Let us suppose that four types of engines of the following characteristics are at our disposal:

TABLE 31

Type	P H.P.	p lb. per H.P.	c lbs. per H.P.	pP lbs.	cP lbs.
I	250	2.3	0.54	575	135
II	300	2.2	0.53	660	159
III	350	2.1	0.56	735	196
IV	400	2.0	0.59	800	236

It is clearly visible that engines No. III and No. IV should without doubt be discarded since their hourly consumption is greater than the already imposed, 180 lbs. Of the other engines the more convenient is undoubtedly type II for which the value of p is lower.

Then formula (2), making $a = 0.34$, $p = 300$, $c = 0.53$, $p = 2.2$, $b = 45$, becomes

$$W = 1992 + \frac{20{,}400}{V} \qquad (3)$$

To determine W as a first approximation, let us remember that the formula of total efficiency gives

$$r = 0.00248 \frac{WV}{P} \qquad (4)$$

and that for a machine of great speed we can take $r = 2.8$; then making $P = 300$ we have

$$\frac{1}{V} = \frac{0.00248}{840} W$$

and substituting in (3)

$$W\,(1 - 0.06) = 1992$$

that is,

$$W = 2130$$

Then $V = 159$ m.p.h.

Consequently we can claim; in the first approximation, that the principal characteristics of our airplane will be

$W = 2130$ lbs.
$V_{max.} = 159$ m.p.h.
$V_{min.} = 75$ m.p.h.
$P = 300$ H.P.

Let us now determine the sustaining surface.

We have seen that we must have

$$\frac{W}{A} \lesseqgtr 0.56\ \lambda_{max.}$$

where $\lambda_{max.}$ is the maximum value it is practical to obtain.

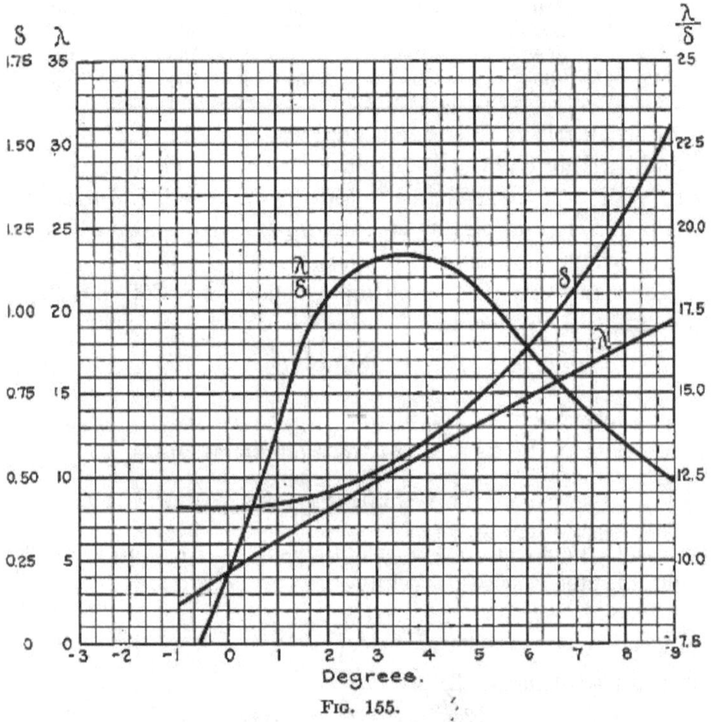

FIG. 155.

From the aerofoils at our disposition, let us select one which, while permitting the realization of the above condition, at the same time gives a good efficiency at maximum speed.

Let us suppose that we choose the aerofoil having the characteristics given in the diagram of Fig. 155.

Then as $\lambda_{max.} = 14.4$, we must have

$$\frac{W}{A} \lesseqgtr 8$$

4.17 In.
65.67 In.
61.50 In.
320.47 In.
160.235 In.
45.67 In.
89.76 In.
44.88 In.
44.88 In.
49.61 In.
24.805 In.
24.805 In.
44.88 In.
89.76 In.
44.88 In.
160.235 In.
45.67 In.
37.40 In.
89.96 In.
144.29 In.
16.93 In.
16.93 In.
288.58
89.96 In.
144.29 In.
37.40 In.

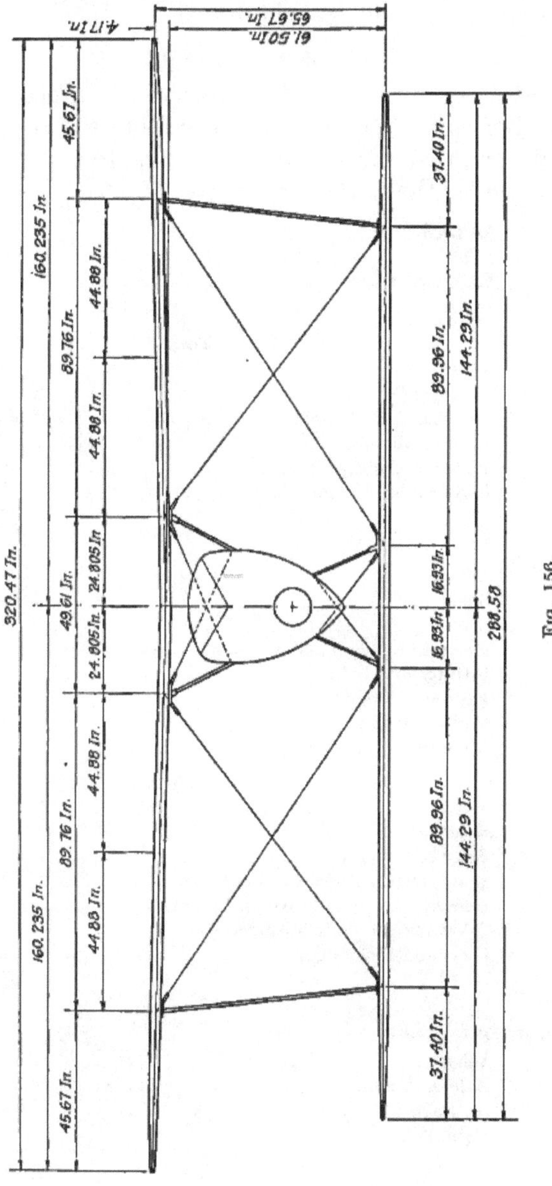

Fig. 156.

For $\frac{W}{A} = 8$ and $W = 2130$ lb.

$$A \doteq 265 \text{ sq. ft.}$$

Let us select a type of biplane wing surface adopting a chord of 65″. The scheme will be that shown in Fig. 156.

We can then compile the approximate table of weights, considering the following groups:

1. *Useful Load*	
Pilot	180 lb.
Gasoline and oil	477 lb.
Instruments	11 lb.
Total	668 lb.
2. *Engine Propeller Group*	
Dry engine and propeller	660 lb.
Exhaust pipes	6 lb.
Water in the engine	30 lb.
Radiator and water	125 lb.
Total	821 lb.
3. *Wing Truss*	
Spars	100 lb.
Ribs	26 lb.
Horizontal struts and diagonal bracings	20 lb.
Fittings and bolts	30 lb.
Fabric and varnish	25 lb.
Vertical struts	40 lb.
Main diagonal bracing	35 lb.
Total	276 lb.
4. *Fuselage*	
Body of fuselage	155 lb.
Seat, control stick, and foot bar	25 lb.
Gasoline tanks and distributing system	40 lb.
Oil tanks and distributing system	6 lb.
Cowl and finishing	25 lb.
Total	251 lb.
5. *Landing Gear*	
Wheels	32 lb.
Axle and spindle	25 lb.
Struts	15 lb.
Cables	4 lb.
Total	76 lb.

6. *Controls and Tail Group*

Ailerons	12 lb.
Fin	2 lb.
Rudder	6 lb.
Stabilizer	8 lb.
Elevator	10 lb.
Total	38 lb.

We can then compile the following approximate table:

TABLE 32

Denomination	Weight in lb.	Per cent. of total weight
1. Useful load	668	31.0
2. Engine propeller group	821	38.5
3. Wing truss	276	13.0
4. Fuselage	251	12.0
5. Landing gear	76	3.5
6. Controls	38	2.0
Total	2130	100.0

A schematic side view of the machine is then drawn in order to find the center of gravity as a first approximation.

In determining the length of the airplane, or better, the distance of tail system from the center of gravity, we have a certain margin, since it is possible to easily increase or decrease the areas of the stabilizing and control surfaces. For machines of types analogous to those which we are studying, the ratio between the wing span and length usually varies from 0.60 to 0.70. Since we have assumed the wing span equal to 26.6 ft., we shall make the length equal to 18 ft.; that is, we shall adopt the ratio 0.678. The side view (Fig. 157) shows the various masses, with the exception of the wings and landing gear; these are separately drawn in Figs. 158 and 159. Then with the usual methods of graphic statics we determine separately the centers of gravity of the fuselage (with all the loads), of the wing truss, and of the landing gear.

It is then easy to combine the three drawings so that the following conditions be satisfied:

1. That the center of gravity of the whole machine be on

Rudder 6 lbs.
Elevator 10 lbs.
C.G. of Tail Weights
Fin 2 lbs.
Stabilizer 8 lbs.
Fuselage and Controls 180 lbs.
Pilot 180 lbs.
Instrument Boards 11 lbs.
Gasoline and Gas Tank 477 lbs.
Cowl 25 lbs.
Motor, Exhaust-Tubes and Water in Jackets 661 lbs.
Auxiliary Radiator 20 lbs.
Propeller 35 lbs.
Total Fuselage Weights 1615 lbs.
178.75 in.
19.5 in.
8.5 in.
Scale of Lengths
0 20 40 Inches
Scale of Weights
0 200 400 LBS.
26 LBS. 538 LBS. 661 LBS. 180 LBS. 180 LBS. 180 LBS. 35 LBS.

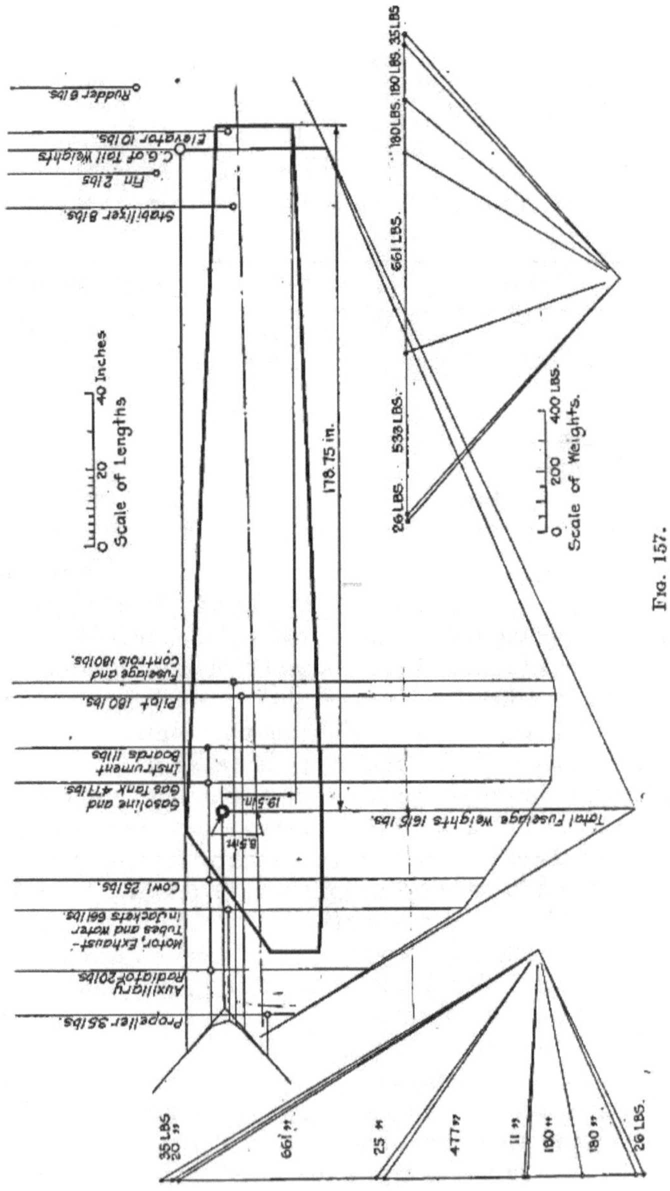

FIG. 157.

Ribs 13 lbs.
Fabric and Varnish 12.5 lbs
Ailerons, 6 lbs
63.58 in.
Total Wing Weights 439 lbs.
19.5 in.
15.6 in.
Weight of Struts, Wires, and Fittings 113 lbs.
Radiator 171 lbs.
Ribs 13 lbs
Fabric and Varnish 12.5 lb.
Weight of Struts, Wires and Fittings 92 lb.
Ailerons 6 lbs.
0 10 20 Inches.
Scale of Lengths
0 100 200 Pounds.
Scale of Weights.
Pounds.
113.0
171.0
26.0
25.0
92.0
12.0
25.5
6.0
113.0
92.0
171.0
25.5
6.0
Pounds.

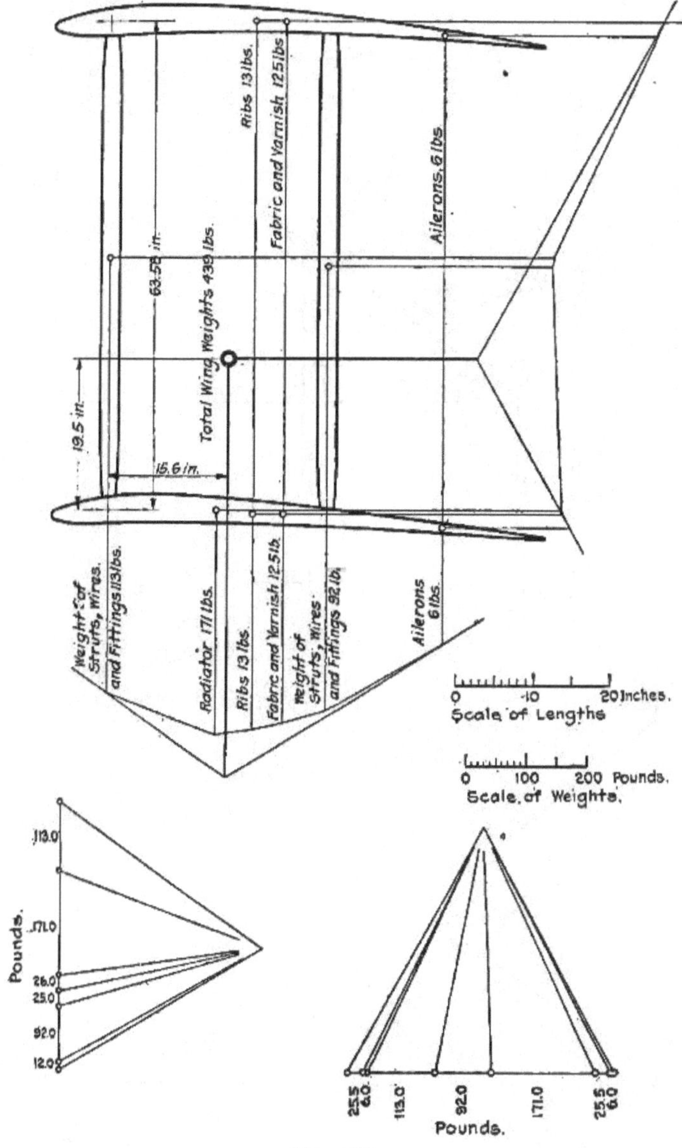

Fig. 158.

63.23 in
6.61 in
38.30 in
38.35 in
2.50
4.70 in
57 lbs
8 lbs
11 lbs
57 lbs
Totals Landing Gear weights 76 lbs
0 10 20 Inches.
Scale of Lengths.
0 10 20 Pounds.
Scale of Weights
11
57
8
Pounds
19
57
Pounds

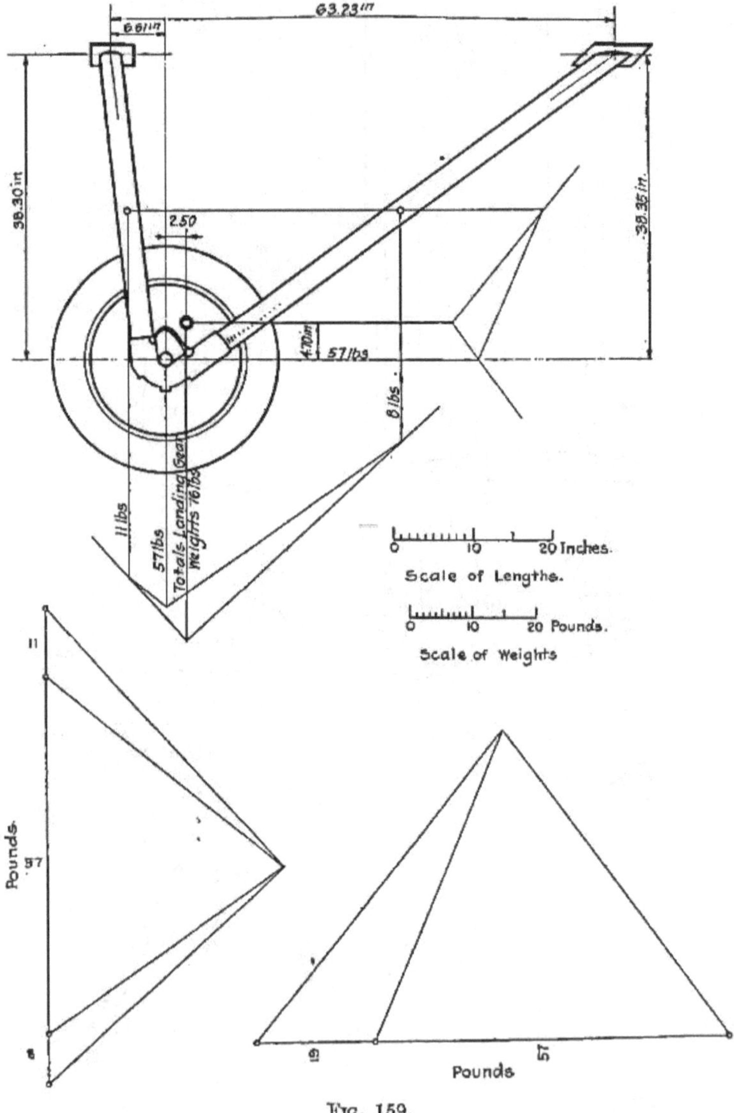

FIG. 159.

0
20
40 Inches
Scale of Lengths
0
500
1000 Pounds
Scale of Weights.
76
439
1615
Pounds
13.4 in.
178.0 in
Propeller Axis
2.5 in.
Wing 439 lbs
Fuselage 1615 lbs.
Total Weights 2130 lbs
Landing Gear 76 lbs.
76
1615
439
Pounds

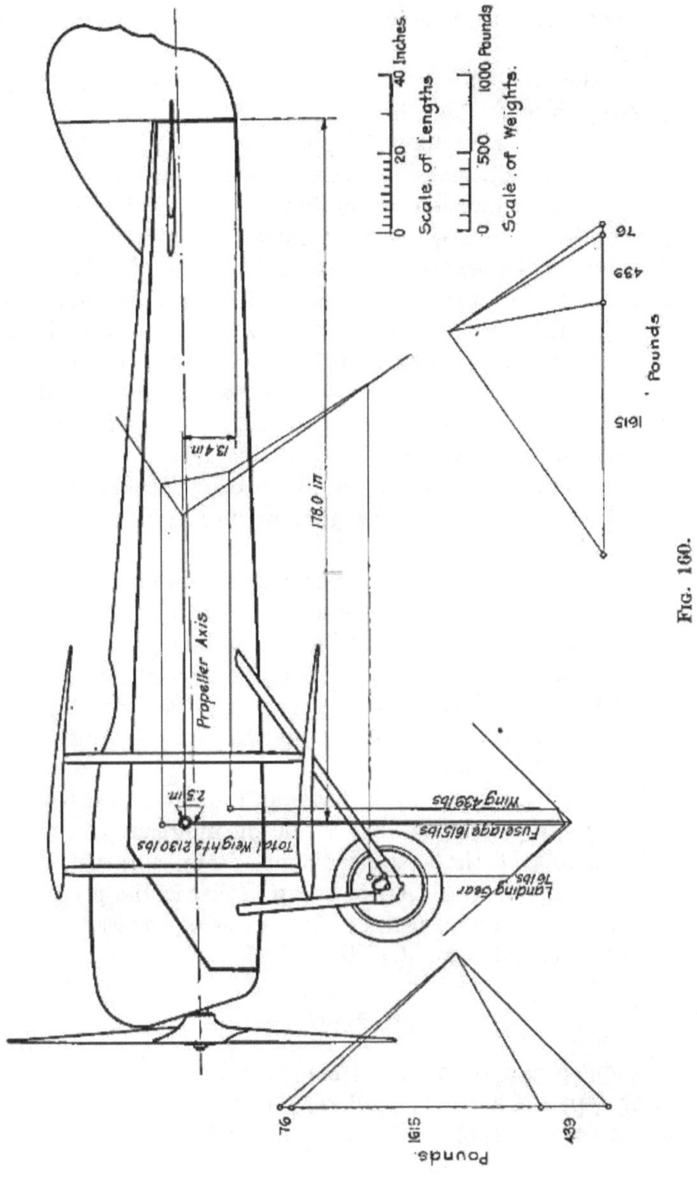

Fig. 160.

the vertical line passing by the center of pressure of the wings.

2. That the axis of the landing gear be on a straight line passing through the center of gravity and inclined forward by 14°; that is, by about 25 per cent.

The superimposing has been made in Fig. 160.

The ideal condition of equilibrium is that the center of gravity, thus found, not only must be on the vertical line passing by the center of pressure, but must also be on the axis of thrust; if it falls above the axis of thrust it is advisable that its distance from it be not greater than 4 or 5 inches at the maximum; if instead it falls below the axis of thrust, we have a greater margin as the conditions of stability improve. This shall be seen in Chapter XXI. In our case, it falls 2.5 in. above the propeller axis.

The center of gravity having been approximately determined we can draw the general outline (Figs. 161, 162 and 163).

It is then necessary to calculate the dimensions of the stabilizer, fin, rudder, and elevator. To do this, it would be essential to know the principal moments of inertia of the airplane. The graphic determination of these moments is certainly possible but it is a long and laborious task because of the great quantity and shape of masses which compose the airplane.

Practically a sufficient approximation is reached by considering the weight W instead of the moment of inertia. Then calling M the static moment of any control surface whatever about the center of gravity (that is, the product of its surface by the distance of its center of thrust from the center of gravity) we shall have

$$M = a \times \frac{W}{V^2}$$

Value a can be assumed constant for machines of the same type. Then, having determined a based on machines which have notably well chosen control surfaces, it is easy to determine M. Value a in our case can be taken equal

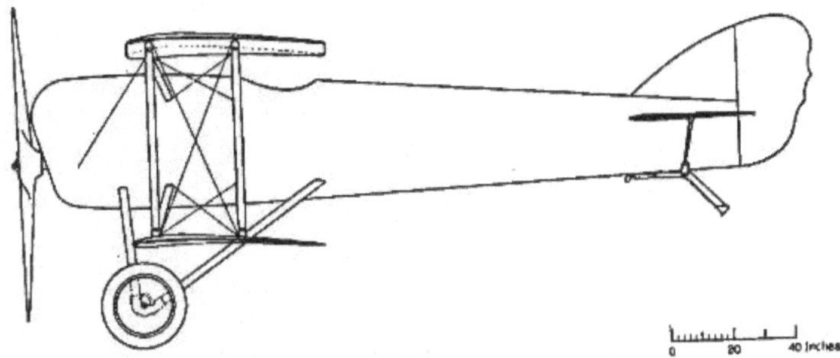

FIG. 161.

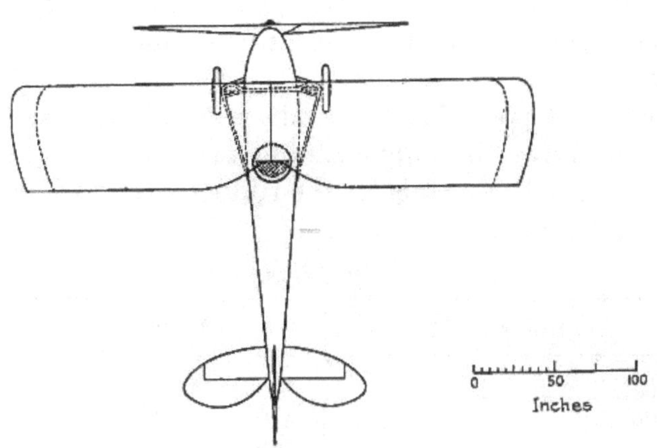

FIG. 162.

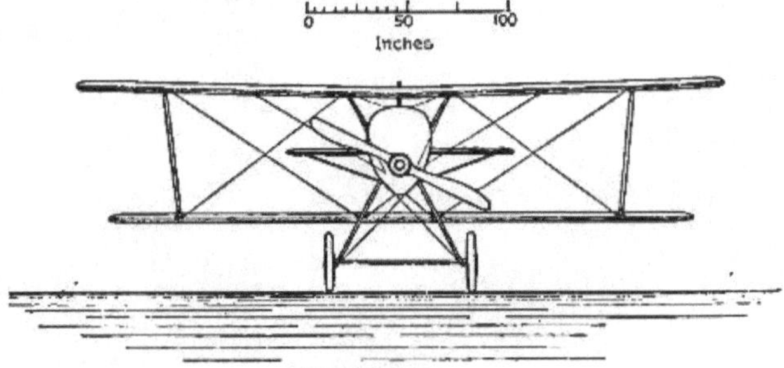

FIG. 163.

to 3900 for the ailerons, 2100 for the elevator, and 2500 for the rudder, taking as the units of measure pounds for W and feet per second for V.

Then it is possible to compile the following table where a and M have the above significance, l is the lever arm in feet and S is the surface of the rudder elevator and ailerons in square feet.

TABLE 33

Controls	a	M (cu. ft.)	l (feet)	S (sq. ft.)
Ailerons	3900	172	8.3	21.0
Elevator	4100	178	14.8	12.0
Rudder	2400	105	15.6	6.7

CHAPTER XVIII

STATIC ANALYSIS OF MAIN PLANES AND CONTROL SURFACES

Owing to the broadness of the discussion we shall limit ourselves to summarily resume the principal methods used in analyzing the various parts, referring to the ordinary treaties on mechanics and resistance of materials for a more thorough discussion.

In this chapter the static analysis of the wing truss and of the control surfaces is given.

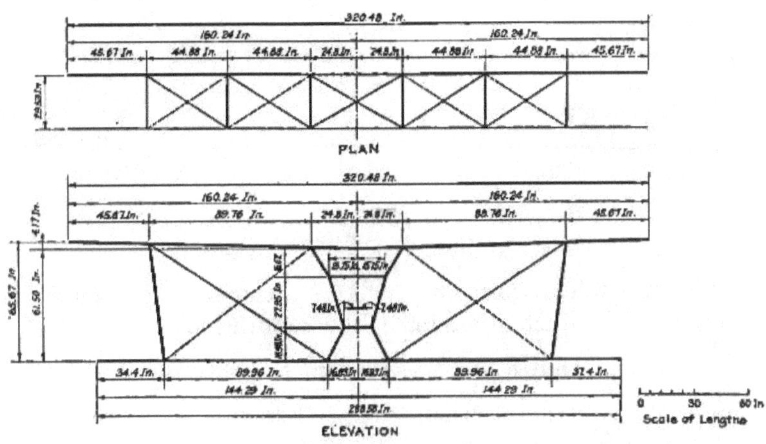

Fig. 164.

Fig. 164 shows that the structure to be calculated is composed of four spars, two top and two bottom ones, connected to one another by means of vertical and horizontal trussings.

For convenience the analysis of the vertical trussings is usually made separately from the analysis of the horizontal ones, and upon these calculations the analysis of the main beams can be made.

First of all it is necessary to determine the system of the acting forces. An airplane in flight is subjected to three kinds of forces: the weight, the air reaction and the propeller thrust.

The weight is balanced by the sustaining component L, of the air reaction; the propeller thrust is balanced by the drag-component D. The weight and the propeller thrust are forces which for analytical purposes can be considered as applied to the center of gravity of the airplane. The components L and D instead, are uniformly distributed on the wing surface. Practically, the ratio $\frac{D}{L}$ assumes as many

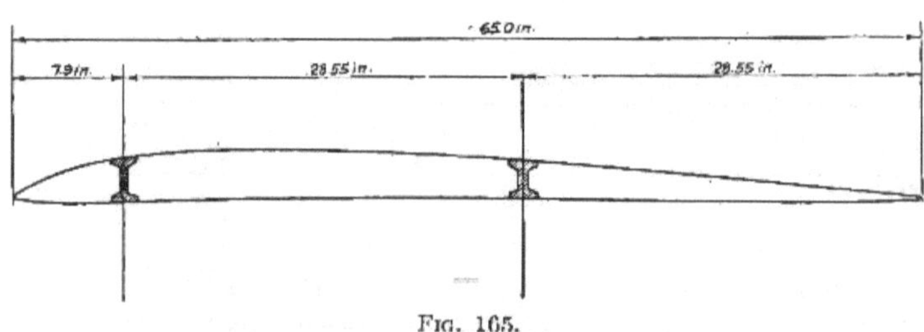

FIG. 165.

different values as there are angles of incidence. The maximum value, which is assumed in computations, is, usually, $\frac{D}{L} = 0.25$. Thus it will be sufficient to study the distribution of L, because, when this is known the horizontal stresses can immediately be calculated.

Let us suppose that the aerofoil be that of Fig. 165 and that the relative position of the spars be that indicated in this figure. The first step is to determine the load per linear inch of the wing. Fig. 164 shows that the linear wing development of the upper wing is 320.48 inches while that of the lower wing is 288.58 inches.

We know that the two wings of a biplane do not carry equally because of the fact that they exert a disturbing influence on each other; in general the lower wing carries less than the upper one; usually in practice the load per unit length of lower wing is assumed equal to 0.9 of that of the upper wing. Then evidently the load per linear inch of the upper wing is given by

$$\frac{2130}{320.48 + 0.9 \times 288.58} = 3.66 \text{ lb. per inch}$$

and for the lower wing it is given by

$$0.9 \times 3.66 = 3.29 \text{ lb. per inch}$$

From these linear loads we must deduct the weight per linear inch of the wing truss, because this weight, being

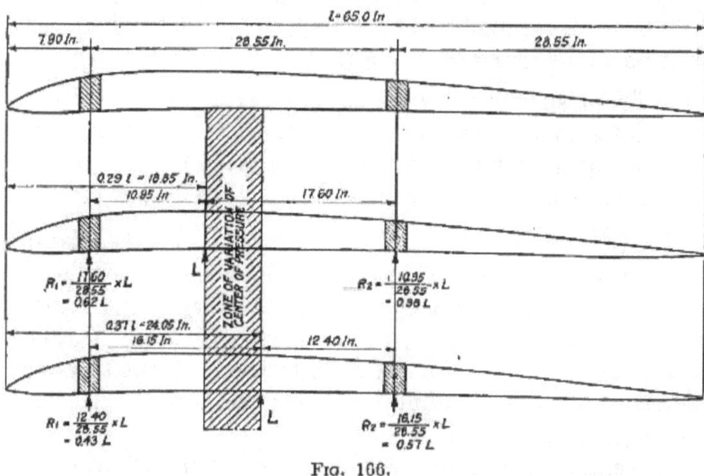

FIG. 166.

applied in a directly opposite direction to the air reaction, decreases the value of the reaction. In our case the figured weight of the wing truss is 276 lb.; thus the weight per linear inch to be subtracted from the preceding values will be 0.45 lb. per linear inch.

We shall then have ultimately:

Upper wing loading.......3.21 lb. per linear inch
Lower wing loading.......2.86 lb. per linear inch

Knowing these loads, it is possible to calculate the distribution of loading upon the front spars and upon the rear spars. For this it is necessary to know the law of variation of the center of thrust.

It is easily understood that when the center of thrust is displaced forward, the load of the front spar increases, and that of the rear spar decreases; and that the contrary happens when the center of thrust is displaced backward. We shall suppose that in our case the center of thrust has a displacement varying from 29 per cent. to 37 per cent. of the wing cord (Fig. 166). In the first case the front spar will support 0.62 of the total load and the rear spar will support 0.38; in the second case these loads will be respectively 0.43 and 0.57 of the total load.

Thus the normal loads per linear inch of the four spars can be summarized as follows:

Front spar upper wing..........1.98 lb. per inch
Rear spar upper wing...........1.82 lb. per inch
Front spar lower wing..........1.75 lb. per inch
Rear spar lower wing...........1.62 lb. per inch

Practically it is convenient to make the calculations using the breaking load instead of the normal load; in fact there are certain stresses which do not vary proportionally to the load but follow a power greater than unity, as we shall see presently. In our case, as the coefficient must be equal to 10, the breaking load must be equal to 10 times the preceding values.

We can then initiate the calculation of the various trusses which make up the structure of the wings. We shall proceed in the following order, computing:

(*a*) bending moments, shear stresses and spar reactions at the supports. Determination of the neutral curve of the spars

(*b*) front and rear vertical trusses

(*c*) upper and lower horizontal trusses

(*d*) unit stresses in the spars.

(*a*) The spars can be considered as uniformly loaded continuous beams over several supports. In our case there are four supports for the upper spars as well as for the lower ones; the uniformly distributed loadings are the preceding.

Let us note first, that in our case as in others, the distribution of the spans of the rear spars is equal to that of the spans of the front spars; thus the only difference between the front and rear spars is in the load per unit of length. It suffices then to calculate the bending moments, the shear stresses and the reactions at the supports for the front spars; the same diagrams, by a proper change of scales, can be used for the rear spars. In our case, the unit loading for the rear spars is equal to 0.92 of that for the front spars.

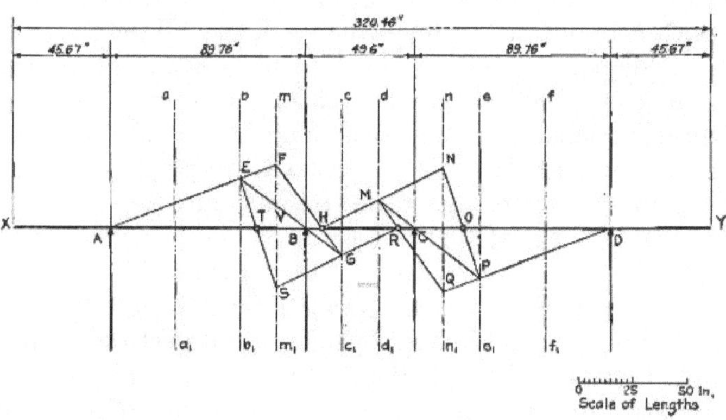

FIG. 167.

With this premise we shall give the graphic analysis based upon the theorem of the three moments, but we shall not explain the reason of the successive operations, referring the reader to treaties on the resistance of materials. First consider the upper front spar (Fig. 167); let XY be its length and A, B, C, D, its supports, made by the struts. Let each span be divided into three equal parts by means of trisecting lines aa_1, bb_1, cc_1, etc. For each support with the exception of the first and last ones, the difference between the third parts of its adjacent spans shall be determined; and that difference is layed off starting from the support, toward the bigger span. In our case we subtract the third part of span BC from the third part of span AB, and the difference is layed off starting from B toward A. Thus V is obtained. The line mm_1 drawn through V perpendicular to XY is called counter vertical of support. Analogously one-third of BC is subtracted from one-third of CD, and its difference is laid off from C toward D, fixing a second counter vertical of support nn_1.

Starting from A (Fig. 167) let us draw any straight line that will cut the trisecant bb_1, and the first counter vertical of support mm_1 in the points E and F respectively.

Draw the straight line EB which prolonged will cut the first triscecant of the second span cc_1 in the point G. Join G with F by a straight line which will cut XY at the point H. This point is called the *right-hand point* of support B. Starting from H we draw any straight line that will meet the second triscecant of the second span dd_1 and the second diagonal nn_1 at the points M and N respectively. Find the point P by prolonging the straight line between M and C. Point O, the right-hand point of the second support, is given by the intersection of line NP and line XY. In order to find the left-hand points for the supports C and B, draw the straight line PD which will interest the counter vertical nn_1 at point Q. Point R where the lines MQ and XY intersect each other will be the left-hand point of support C. Starting from R draw the line RG which will cut the first counter diagonal at point S. Point T, the point of intersection of lines SE and XY will be the left-hand point of support B.

The right-hand and left-hand points being known, we shall suppose that we load one span at a time, determining the bending moments which this load produces on all the supports. Summing up at every support the moments due to the separate loads, we shall obtain the moments originated by the whole load.

The moment on the external supports is equal to that given by the load on the cantilever ends, as it cannot be influenced by the loads on the other spans, owing to the fact that the cantilever beam can rotate around its support. The load on the cantilever spans however affects the other spans. To determine this effect we proceed in the following manner: Consider support A (Fig. 168); the moment at this support is equal to $\frac{wl^2}{2}$, calling w the load in lb. per linear inch and l the length of the span in inches. Lay off, to any scale, the segment $AA' = \frac{wl^2}{2}$.

Let us then draw the straight line $A'T$; it will intersect the vertical line through support B at point 1; the segment $1B$ measures, to the scale of moments, the moment that the load on the cantilevered span produces on support B.

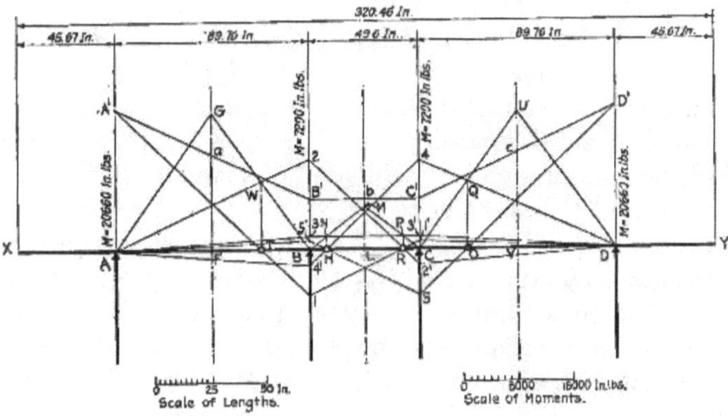

FIG. 168.

Then draw the straight line $1R$; it will meet the vertical line through support C at $1'$; the segment $1'C$ measures, always to the scale of moments, the moment originated on support C by the load of the cantilevered span. The moment in D cannot be influenced by the cantilever load on $\times A$.

Let us now determine the effect of the load on span AB, on the moment of the various supports. Draw FG perpendicular bisectrix of AB and lay off, to the scale of moments, a segment FG equal to $\frac{w \times l^2}{8}$; that is, equal to the moment which would be obtained at the center point of AB, by a unit load w, if AB were a free-end span supported at the extremities. From T, the left-hand point of support B, raise a perpendicular which cuts line GB at W. Draw line AW to meet the perpendicular through support T at point 2. The segment $B2$ read to scale, will give the moment on support B due to the load on AB.

Point 2′ is obtained by prolonging line $2R$ until it meets the perpendicular through C at 2′. Segment C-2′ represents to the scale of moments, the moment on support C due to the load on AB.

In order to find the effect of the load of span BC on the other spans, proceed analogously; that is lay off ML on the bisectrix of BC, equal to scale, to the moment $ML = \frac{wl^2}{8}$.

Let us find points N and P as indicated in the figure and let us draw the line NP which prolonged will meet the perpendiculars on supports B and C at points 3 and 3′. Segments B-3 and C-3′ read to the scale of moments, will give the moments produced by the load of span BC on the supports B and C respectively.

Proceeding as for spans XA and AB we obtain the moments originated on BC by the loads on spans CB and BY. The construction is clearly indicated in Fig. 168.

Resuming, we shall have the moment originated by cantilever loads on the supports A and D, and the moment originated by the loads on all the different spans, on the supports B and C.

For the point of support B the moment due to the cantilever load is equal, read to the scale of moments, to distance B-1, the moment due to the load on AB is equal to B-2, the moment due to the load on BC is equal to B-3, the moment due to the load on CD is equal to B-4′ and that due to the cantilever load on DY is equal to B-5′. If we assume that the distances above the axis XY are positive and those below are negative, the total moment BB' on support B will be equal to the algebraic sum of the moments B-1, B-2, B-3, B-4′, and B-5′.

Analogously the algebraic sum CC' will represent the total moment on C. The total moment on the external supports will naturally remain the one due to cantilevers, and consequently equal to AA' and DD'. In order to find the variations of the bending moment on all the spans, the load being uniformly distributed, we must draw the parabolæ of the bending moments as though the spans were simply supported (Fig. 169).

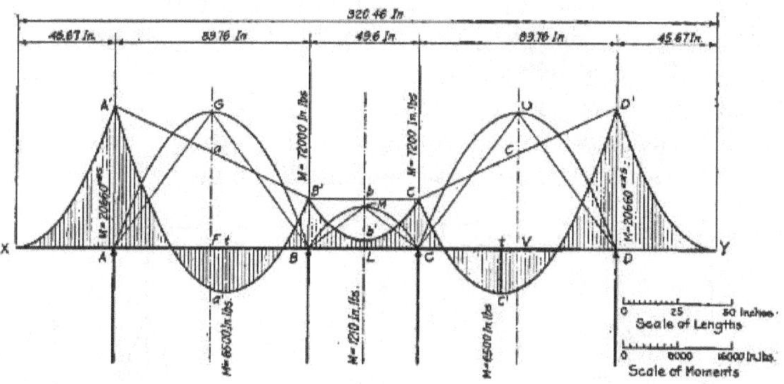

Fig. 169.

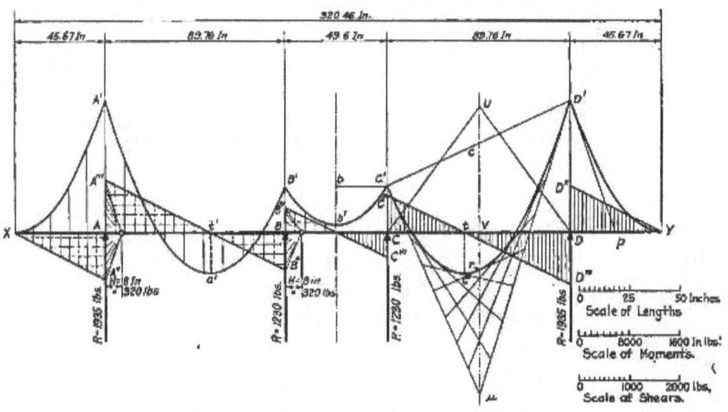

Fig. 170.

Then the difference between the ordinates of the parabolæ and those of the diagram $AA'\ B'\ C'\ D'\ D$ give us the diagram $XA'\ a'\ B'\ b'\ C'\ c'\ D'\ YX$ which represents the diagram of the bending moment (Fig. 169).

Knowing the diagram of the bending moments, it is easy

through a process of derivation applying the common methods of graphic statics, to find the diagram of the shearing stresses, and consequently the reactions on the supports (Fig. 170). The scale of forces is obtained by multiplying the basis H of the derivation, by the ratio between the scale of moments and that of the lengths. In Fig. 170 the scale of forces has been drawn, and on the supports the corresponding numerical values of the reactions have been marked.

Furthermore, from the diagram of bending moments we can obtain the *elastic curve*, which will be needed later.

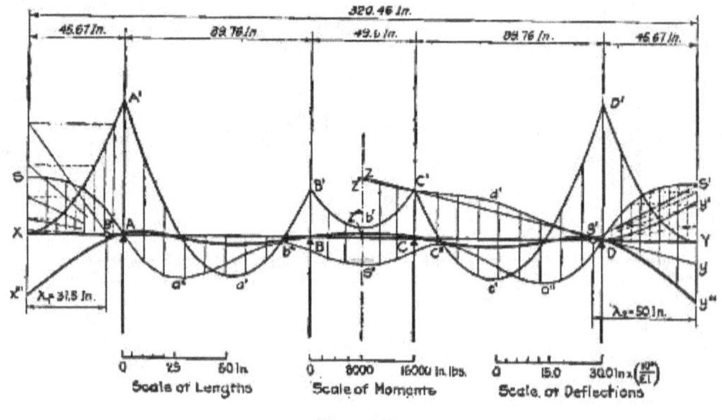

FIG. 171.

In fact let us remember that the analytic expression of the bending moment is given by

$$M_B = E \times I \times \frac{d^2y}{dx^2}$$

and consequently

$$y = \frac{1}{E\,I} \int dx \int M_B \; dx^2$$

that is, by double integration of the diagram of M_B we obtain the deflections y, that is, we obtain the form which the neutral axis of the spar assumes, and which is called *elastic curve* (Fig. 171).

We shall not pause in the process of graphic integration, as it can be found in treaties on graphic statics.

We shall make use of the elastic curve for the determination of the supplementary moments produced on the spars by the compression component of the vertical and horizontal trussings.

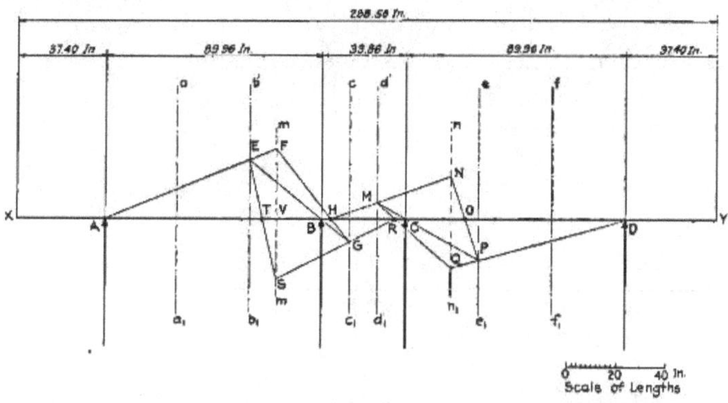

Fig. 172.

Figs. 167, 168, 169, 170 and 171 refer to the calculation of the upper front spar. In Figs. 172, 173, 174, 175 and 176 instead, the graphic analysis of the lower front spar is developed.

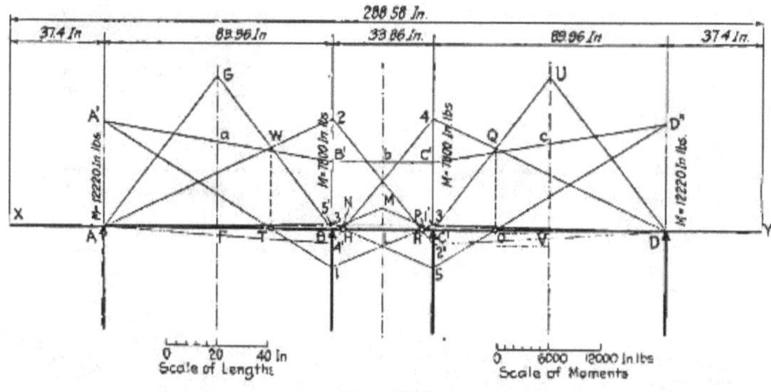

Fig. 173.

On these figures, beside the unit loads which are already known, the scale of the moments, of the lengths and of the forces are also indicated.

The preceding diagrams also give the bending moments,

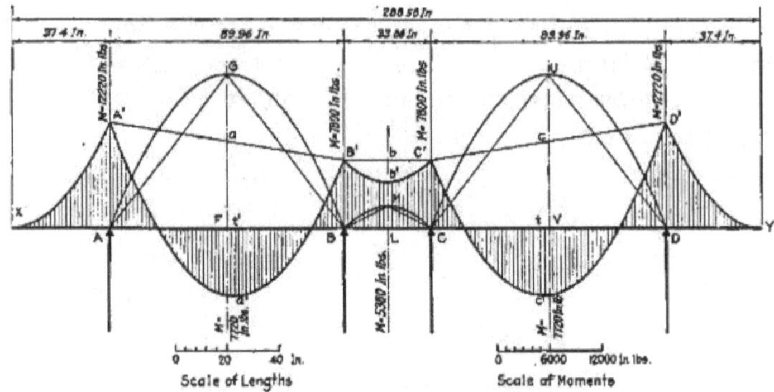

FIG. 741.

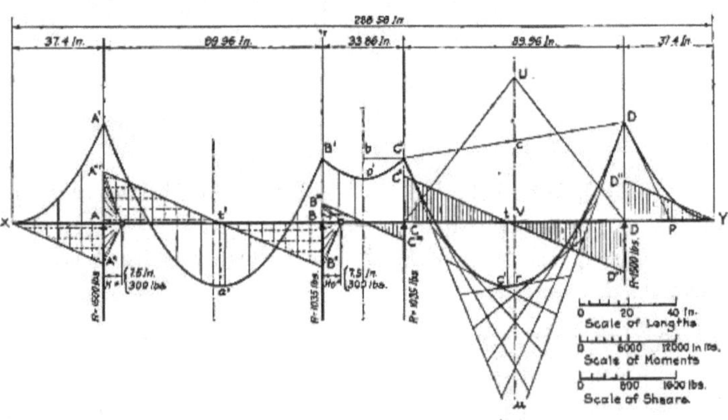

FIG. 175.

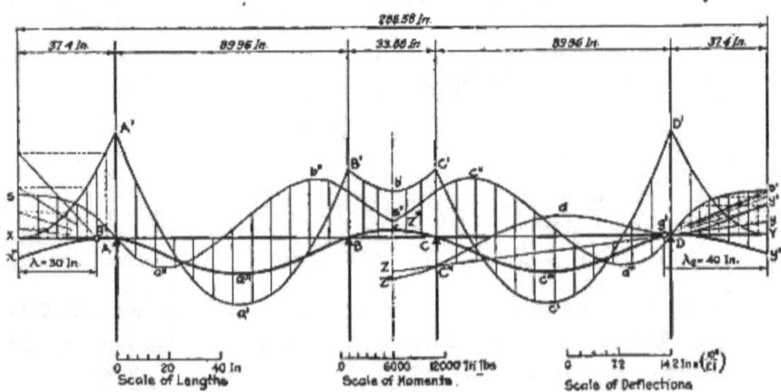

FIG. 176.

the shearing stresses and the reactions on the supports for the rear spars; in fact it suffices to multiply both the values of the forces and those of the moments by 0.92, as the spans are the same, and the loads per linear inch of the rear spars are equal to 0.92 of the loads of the front spars.

A special note should be made of the scales of ordinates for the elastic curve; these are inversely proportional to the product $E \times I$, the elastic modulus by the moment of inertia, and consequently they vary from spar to spar. But we shall return to this in speaking of the unit stresses in spars.

(*b*) Knowing the reactions upon the supports, it is possible to calculate the vertical trussings. Since the front trussing has the same dimensions as the rear one, and since the reactions on the supports are in the ratio 0.92, it suffices to calculate only the first.

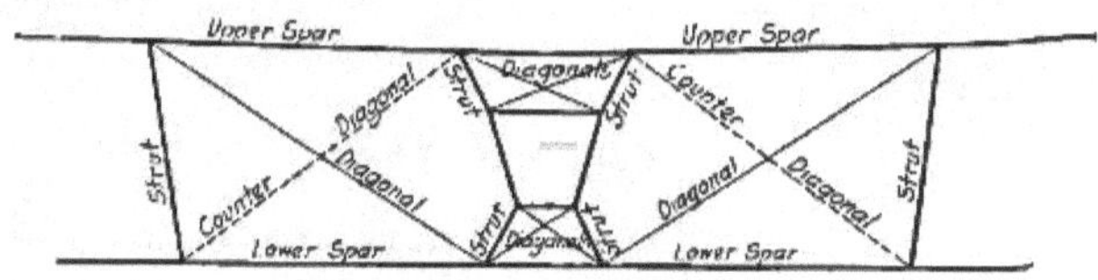

Fig. 177.

The vertical trussing is composed of two spars, one above, and the other below, connected by struts capable of resisting compression, by bracings called diagonals, which must resist tension, and by bracings called counter diagonals which serve to stiffen the structure (Fig. 177). In flight, the counter diagonals relax and consequently do not work; for the purpose of calculation we can consequently consider the vertical trussing as though it were made only of spars, struts, and diagonals; furthermore, because of the symmetry of the machine, for simplicity we shall consider only one-half of it, as evidently the stresses are also symmetrical (Fig. 178); the plane of symmetry will naturally have to be considered as a plane of perfect fixedness.

With that premise let us remember that for equilibrium it is first of all necessary that the resultant of the external forces be equal to zero. The reactions upon the supports are all vertical and directed from bottom to top; their sum is equal to 5695 lb.; now, this force is balanced by that part of the weight of the machine which is supported at point *A* and which is exactly equal to 5695 lb. Moreover it is necessary that in any case the applied external force (reaction at support), be in equilibrium with the internal reaction; that is, as it is usually expressed in graphic statics, it is essential that the polygon of the external forces and of

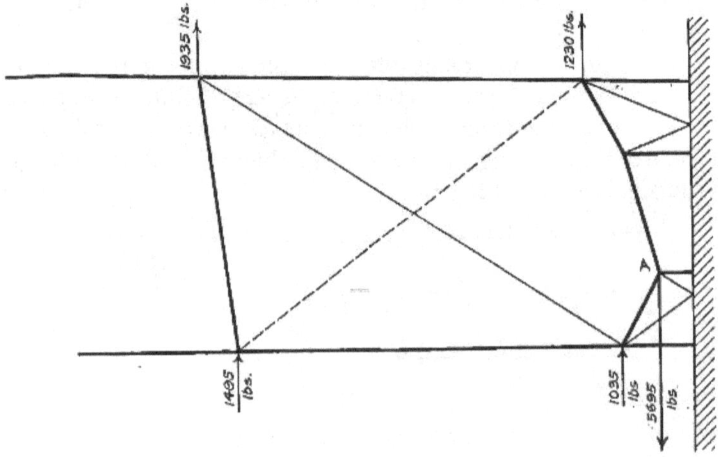

Fig. 178.

the internal reactions close on itself. This consideration enables the determination of the various internal reactions through the construction of the stress diagram, illustrated, for our example, in Fig. 179.

Referring to treaties on graphic statics for the demonstration of the method, we shall here illustrate, for convenience, the various graphic operations.

The values of the reactions on the supports individuated by zones *ab, bc, cd,* and *de* are laid off to a given scale on *AB*, *BC*, *CD*, and *DE* (Fig. 179); from *B* and *C* we draw two parallels to the truss members determined by the zones *bh* and *ch* respectively; in *BH* we shall have the

160.24 In
45.67 In.
89.76 In.
24.30 In
4.17 In.
Center Line
1935 lbs.
d
-5120 lbs
1230 lbs.
e
-5780 lbs.
f
+1400 lbs.
16.62 In.
g
+6380 lbs.
65.67 In
61.50 In.
c
-1500 lbs.
h
27.95 In
16.93 In.
-200 lbs.
-5200 lbs
i
+7600 lbs
b
a
1495 lbs.
1035 lbs.
37.40 In.
89.96 In.
16.93 In
144.29 In.
TRUSS DIAGRAM
0
22
44 In
Scale of Lengths.
F
E
G
D
W
C
STRESS DIAGRAM
B
H
I
A
0
1000
2000 lbs.
Scale of Forces

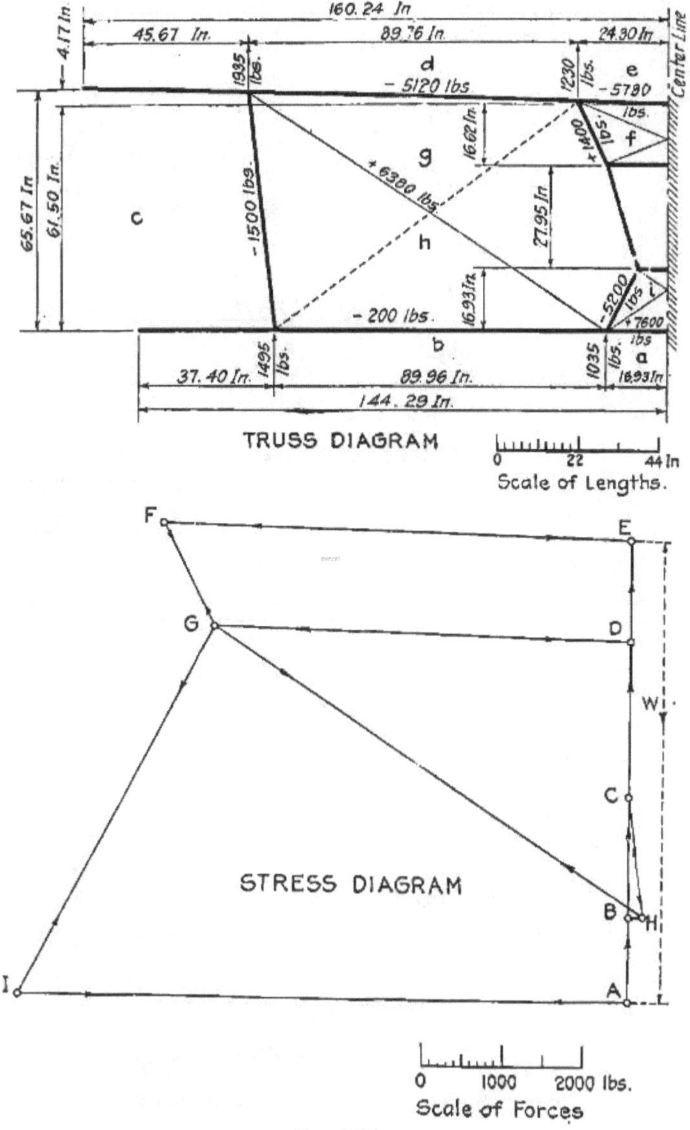

Fig. 179.

144.29 In.
37.40 In.
89.96 In.
16.93 In.
Center Line
748 lbs.
-2330 lbs.
518 lbs.
-2650 lbs
16.93 In.
+620 lbs
+2800 lbs.
27.95 In.
65.67 In
61.50 In.
-990 lbs.
16.62 In.
-2800 lbs
+3750 lbs
-130 lbs
4.17 In
-45.67 In.
968 lbs.
89.76 In.
615 lbs.
24.80 In
160.24 In.
TRUSS DIAGRAM
0 22 44 In
Scale of Lengths
STRESS DIAGRAM
0 500 100 lbs.
Scale of Forces

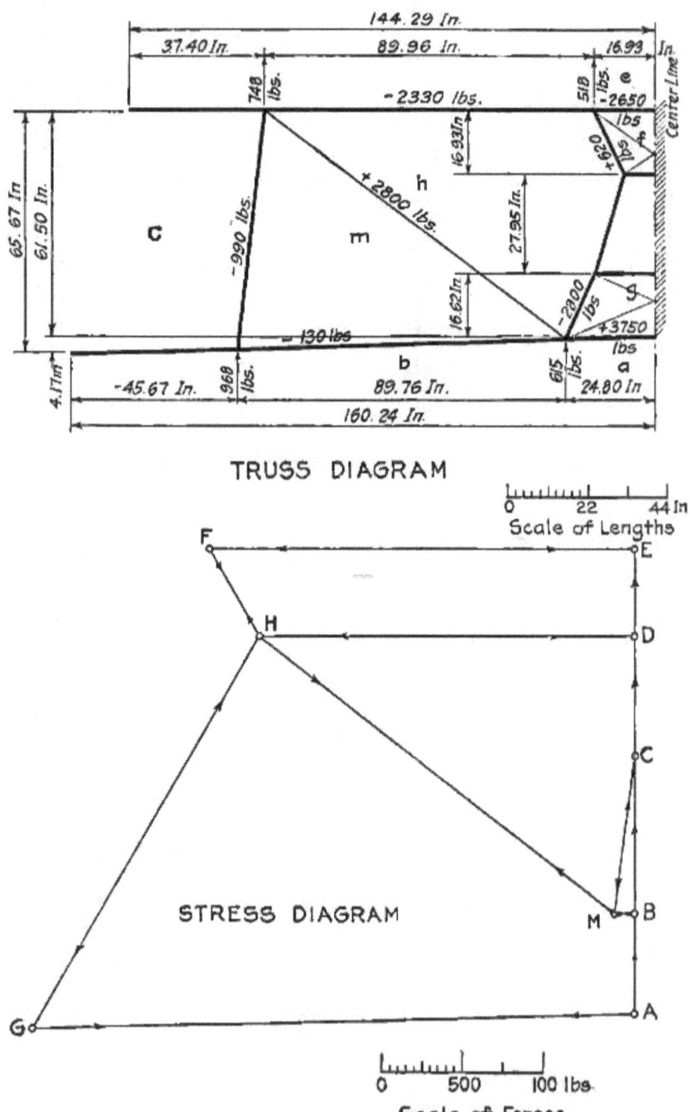

Fig. 180.

stress corresponding to member *bh*, and in *CH* that corresponding to the member *ch*. From points *H* and *D* we draw the parallels to the members *gh* and *gd*; in *HG* and *DG* we shall have the stresses in *hg* and *dg*; from points *E* and *G* we draw the parallels to the members determined by zones *ef* and *gf*; in *EF* and *GF* we shall obtain the stresses in these members; finally from points *G* and *A* we draw the parallels to the members individuated by zones *gi* and *ai*, obtaining the corresponding stresses in *GI* and *AI*. The arrows of the stress diagram enable the easy determination of which parts of the truss are subjected to tension and which to compression.

In Fig. 179, beside marking the scales of lengths and of forces, we have marked the lengths and the stresses corresponding to the various parts, adopting + signs for tension stresses, and − signs for compression stresses. By multiplying these stresses by 0.92 we shall obtain the values of the stresses of the rear trussing.

The counter diagonals which do not work in normal flight, function only in case of flying with the airplane upside down. For this case, which is absolutely exceptional, a resistance equal to half of that which is had in normal flight is generally admitted. The determination of stresses is analogous to that made for normal flight and is shown in Fig. 180.

Based upon the values found in the preceding construction, Table 34 can be compiled. That table permits the calculation of the bracings and struts.

The calculation of the bracings presents no difficulties; it is sufficient to choose cables or wires having a breaking strength equal to or greater than that indicated in the table; naturally the turnbuckles and attachments must have a corresponding resistance. Table 35 gives the dimensions of the cables selected for our example. For the principal bracings we have adopted double cables, as is generally done in order to obtain a better penetration; in fact not only does the diameter of the cable exposed to the wind

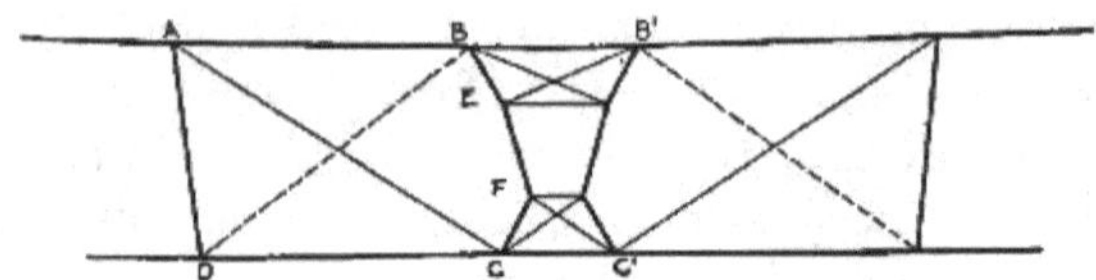

TABLE 34

Member	Front vertical truss				Rear vertical truss			
	Normal position		Inverted position		Normal position		Inverted position	
	Tension	Compression	Tension	Compression	Tension	Compression	Tension	Compression
A–B	....	**5120**	130	....	....	**4710**	120	
B–E	1400	....	....	**2800**	1290	....	....	**2580**
F–C	....	**5200**	620	....	....	**4780**	570	
A–C	**6380**	....	....	....	**5870**			
D–C	....	200	....	**2330**	....	180	....	**2140**
A–D	....	**1500**	....	990	....	**1380**	....	910
B–B′	....	**5780**	3750	....	....	**5320**	3450	
C–C′	**7600**	....	....	2650	**6990**	....	....	2440
B–D	....	....	**2800**	....	....	....	**2580**	

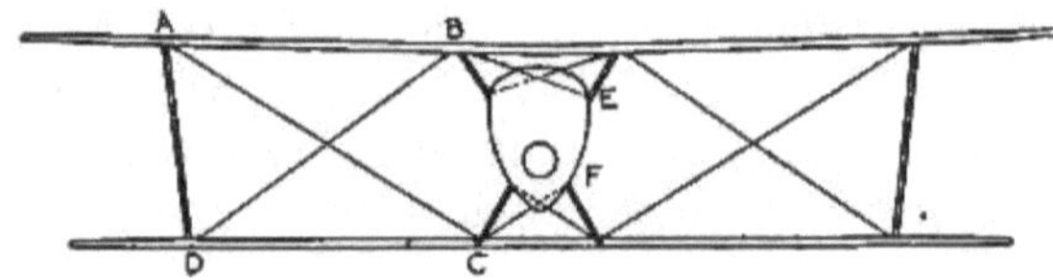

TABLE 35

	Member	Stress coef. 10 lb.	Number of cables used	Diameter of cable, in.	Ultimate strength of cable, lb.	Total ultimate strength, lb.	Coef. of safety
Front truss	A–C	+6380	2	3/16	4200	8400	13.1
	B–D	+2800	1	1/8	2000	2000	7.1
	B–E	+4700	1	5/32	2800	2800	5.9
	C–F	+9200	1	7/32	5600	5600	6.1
Rear truss	A–C	+5870	2	3/16	4200	8400	14.3
	B–D	+2580	1	1/8	2000	2000	7.7
	B–E	+4320	1	5/32	2800	2800	6.5
	C–F	+8500	1	7/32	5600	5600	6.6

result smaller, but it becomes possible to streamline the two cables by means of wooden faring.

For the struts, which can be considered as solids under compression, it is necessary to apply Euler's formula which gives the maximum load W that a solid of length l with a section having a moment of inertia I can support

$$W = \alpha \frac{EI}{l^2}$$

In that formula α is a numerical coefficient and E is the elastic modulus of the material of which the solid is made.

The theory gives the value 10 for coefficient α. We shall quickly see that practically it will be convenient to adopt a smaller coefficient in consideration of practical unforeseen factors.

Let us remember that the struts, being exposed to the wind, present a head resistance which must be reduced to a minimum by giving them a shape of good penetration as well as by reducing their dimensions to the minimum. This last consideration shows, by what has been said in Chapter XVI, that for struts it is convenient to use materials which even having high coefficients A_1 and A_2 have a high specific weight.

Then the best material for struts is steel. In Chapter XVI a table has been given of oval tubes normally used for struts, with the most important characteristics, such as weight per unit of length, area of section, relative moment of inertia, etc.

Let us apply Euler's formula to these tubes, remembering that for them $I = td^3$, where t is the thickness and d is the smaller axis. We shall have

$$W = \alpha \frac{Etd^3}{l^2}$$

Remembering then that the area of these struts is given with sufficient approximation by the expression $A = 6.37td$ the preceding formula can be written as follows

$$\frac{W}{A} = \frac{\alpha \times E}{6.37} \times \frac{1}{\left(\frac{l}{d}\right)^2}$$

where

$\frac{W}{A}$ = unit stress of the material

$\frac{l}{d}$ = ratio between that portion of the length which can be considered as free ended, and the minimum dimension of the strut.

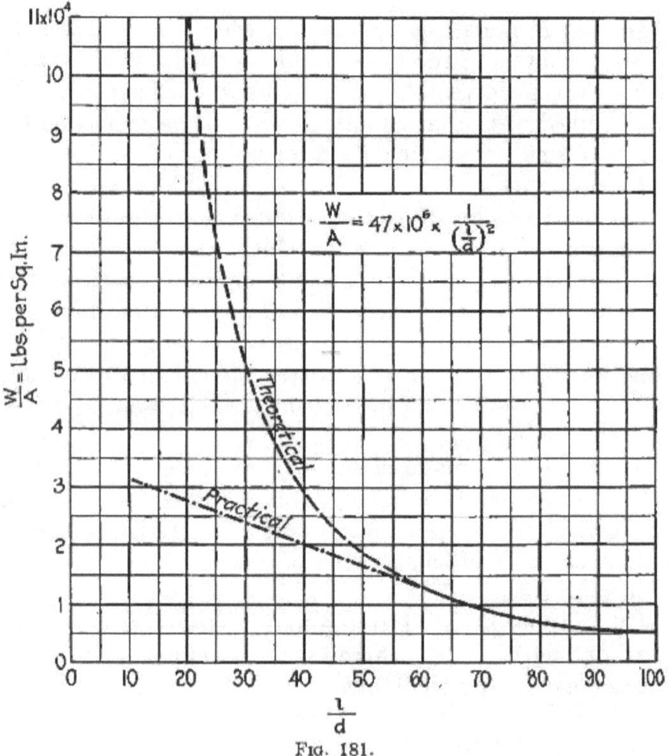

Fig. 181.

Adopting pounds and inches as the unit, we have $E = 3 \times 10^7$ and consequently

$$\frac{W}{A} = 47 \times 10^5 \times \alpha \times \frac{1}{\left(\frac{l}{d}\right)^2} \qquad (1)$$

Naturally this formula can be applied only for high values of the ratio $\frac{l}{d}$; practically below the value $\frac{l}{d} = 60$ this formula can no longer be relied upon. In Fig. 181 the diagram corresponding to the preceding formula is given, drawing the diagram with a dotted instead of a full line for the values of $\frac{l}{d} < 60$. For these values the practical diagram is shown by a dot and dash line.

In Tables 36 to 39 we have tabulated the results of some of the many tests on metal struts which have been made at our works. In these tables the practical value of coefficient α of Euler's formula has been calculated; it is seen that while in some tests α has a value higher than 10, in general it gives lower values. That depends upon the struts being partly manufactured by hand and partly rolled, and also upon the thickness of the sheet and the dimensions of the sections being not always uniform. Based on average values we can therefore assume that for properly manufactured struts a coefficient $\alpha = 8$ can be adopted for computation purposes.

With this premise it is simple, when the ultimate stress which a strut must withstand, and its length, are known, to determine its dimensions.

Moreover infinite solutions exist, since formula (1) when W and l are given, can be satisfied by infinite couples of values A and d.

Evidently by increasing d, the value of A becomes smaller and consequently the weight of the strut diminishes; from that point of view it would be convenient to use struts having large dimensions and small thicknesses. However, the increase of d increases the head resistance of the airplane, and increases the power necessary to fly.

Therefore it becomes necessary to adopt that solution which requires the minimum power expension.

If β is the weight per horsepower lifted by the airplane, γ is the weight of one foot of strut of width d, k its coefficient of head resistance as was definitely stated in Chapter

TABLE 36

D in.	d in.	t in.	l in.	$\frac{l}{d}$	$\left(\frac{l}{d}\right)^2$	A in.²	W lbs.	Δ in.	$\frac{W}{A}$ lbs./in.²	a	
2.40	0.83	0.110	61.50	74.2	5,505	0.58	6,500	0.39	11,200	13.0	Welded.
2.36	0.83	0.079	61.50	74.2	5,505	0.41	2,360	0.59	5,720	6.8	Welded.
2.38	0.77	0.071	57.50	75.0	5,625	0.38	3,020	0.90	8,770	10.7	Seamless drawn—poplar filling.
2.36	0.79	0.062	62.60	79.5	6,320	0.31	1,870	0.78	6,040	8.3	Seamless drawn.
2.36	0.79	0.059	65.00	82.7	6,839	0.30	1,510	0.86	5,100	7.6	Welded.
2.36	0.79	0.059	62.60	79.5	6,320	0.30	2,040	0.98	6,920	9.5	Seamless drawn.
2.39	0.79	0.059	63.00	79.4	6,300	0.30	1,595	1.25	5,350	7.3	Welded.
2.38	0.80	0.059	64.20	80.7	6,512	0.30	1,585	1.18	5,250	7.4	Seamless drawn.
2.36	0.81	0.059	61.00	75.5	5,700	0.30	2,690	0.63	8,840	10.8	Welded.
2.36	0.79	0.059	61.00	77.4	5,990	0.30	2,670	0.65	9,030	11.2	Welded.
2.36	0.79	0.039	62.60	79.5	6,320	0.20	990	0.78	5,080	6.9	Seamless drawn.
2.36	0.79	0.039	59.20	75.4	5,685	0.20	1,360	0.78	6,940	8.6	Welded.
2.36	0.79	0.039	59.20	75.4	5,685	0.20	1,150	0.86	5,600	6.9	Seamless drawn.
2.36	0.79	0.039	61.00	77.3	5,975	0.20	1,360	0.59	6,940	9.0	Welded.
2.36	0.79	0.031	62.60	79.5	6,320	0.16	880	0.78	5,570	7.7	Seamless drawn.

TABLE 37

D in.	d in.	t in.	l in.	$\frac{l}{d}$	$\left(\frac{l}{d}\right)^2$	A in.²	W lbs.	Δ in.	$\frac{W}{A}$ lbs./in.²	α	A=6.37 dt
2.92	0.97	0.083	67.80	70.0	4,900	0.51	3,370	0.985	6,680	7.20	Seamless drawn.
2.96	1.01	0.083	68.00	67.8	4,596	0.53	3,890	1.300	7,170	7.20	Welded.
2.96	0.95	0.081	68.00	72.0	5,184	0.49	3,190	0.235	6,540	7.40	Seamless drawn.
2.88	1.01	0.079	68.00	67.8	4,596	0.51	3,610	0.985	7,170	7.20	Welded.
2.94	1.01	0.079	67.60	67.3	4,529	0.51	3,360	1.100	6,680	6.60	Welded.
2.94	0.97	0.079	68.00	70.5	4,950	0.48	3,150	0.590	6,540	7.10	Welded.
2.96	1.01	0.079	68.00	67.7	4,583	0.51	3,450	1.420	6,900	6.90	Welded.
2.96	1.01	0.079	67.50	68.8	4,733	0.48	3,410	0.865	6,920	7.10	Welded.
2.93	1.01	0.079	69.80	69.5	4,830	0.51	3,520	1.180	7,000	7.40	Welded.
2.92	1.01	0.079	68.00	67.7	4,583	0.51	3,190	1.460	6,340	6.30	Welded.
2.96	0.99	0.071	67.50	68.8	4,733	0.44	3,290	1.180	7,400	7.60	Welded.
2.96	0.99	0.071	68.00	69.2	4,788	0.44	2,850	.....	6,400	6.70	Seamless drawn.
2.96	0.99	0.071	67.00	68.0	4,624	0.44	3,080	.....	6,940	6.90	Seamless drawn.
2.96	1.02	0.071	69.00	67.4	4,542	0.46	3,740	0.590	8,080	7.90	Welded.
2.96	0.95	0.059	74.80	79.2	6,272	0.36	3,120	.....	15,080	20.50	Welded.

TABLE 38

D	d	t	l	$\frac{l}{d}$	$\left(\frac{l}{d}\right)^2$	A	W	Δ	$\frac{W}{A}$	α	$A = 6.37\ dt$
2.96	0.99	0.059	69.80	75.0	5625.00	0.37	2030	.....	5560	6.80	Welded.
2.92	0.99	0.059	74.80	76.0	5776.00	0.37	2625	.....	7110	9.10	Welded.
2.96	0.99	0.059	71.00	75.2	5655.04	0.37	2170	0.590	5870	7.20	Welded.
2.96	0.99	0.059	71.00	75.2	5655.04	0.37	2620	.....	7150	8.80	Seamless drawn.
2.92	0.99	0.059	68.00	69.3	4802.49	0.37	2860	0.590	7750	8.10	
2.96	0.97	0.059	67.50	70.2	4928.02	0.37	3300	0.196	9120	3.90	Seamless drawn—walnut filling.
2.96	0.99	0.059	67.00	68.3	4664.89	0.37	4180	0.310	.....	11.50	Seamless drawn—poplar filling.
2.96	0.99	0.039	72.00	73.2	5358.24	0.25	1880	0.985	7470	8.75	Seamless drawn—poplar filling.
2.96	0.82	0.039	72.00	83.2	6922.24	0.22	880	0.865	4060	6.10	Welded.
2.96	0.99	0.039	71.70	75.4	5685.16	0.25	1360	0.305	5500	6.80	Welded.
2.96	0.99	0.039	71.70	75.4	5685.16	0.25	1680	.....	6820	8.40	Seamless drawn.
2.96	0.99	0.039	67.00	68.3	4664.89	0.25	2090	.....	8480	8.60	Seamless drawn.
2.88	0.93	0.039	74.80	80.8	6528.64	0.23	1360	0.925	5900	8.40	Welded.
2.92	0.95	0.039	76.30	80.8	6528.24	0.24	1360	0.790	5810	8.25	Welded.
2.92	0.95	0.039	74.30	78.6	6177.96	0.24	1760	0.790	7420	10.00	Welded.

TABLE 39

D	d	t	l	$\frac{l}{d}$	$\left(\frac{l}{d}\right)^2$	A	W	Δ	$\frac{W}{A}$	α	$A = 6.37\ dt$
3.58	1.22	0.059	45.20	37.0	1,369.00	0.44	11,000	0.060	24,000	7.10	Welded.
3.55	1.18	0.039	39.30	33.4	1,115.56	0.23	15,000	0.310	67,200	12.00	Seamless drawn.
3.55	1.12	0.039	39.30	35.1	1,232.01	0.28	6,600	0.118	23,800	6.40	Seamless drawn.
3.55	1.18	0.039	39.30	33.4	1,115.56	0.30	14,400	0.158	48,400	11.70	Seamless drawn—poplar filling.
3.55	1.18	0.039	47.20	40.0	1,600.00	0.30	6,600	0.785	22,200	7.75	Seamless drawn.
3.55	1.18	0.039	55.60	47.0	2,209.00	0.30	8,580	0.590	28,800	13.80	Seamless drawn.
3.55	1.18	0.039	55.60	47.0	2,209.00	0.30	9,030	.....	34,000	14.60	Seamless drawn—fir filling.
3.55	1.19	0.039	39.30	33.1	1,095.61	0.30	9,600	0.118	32,100	7.65	Welded.
3.55	1.18	0.039	47.20	37.3	1,391.29	0.30	7,920	0.310	26,600	8.10	Seamless drawn.

VII, V the speed of the airplane in m.p.h., and ρ the propeller efficiency, the total power p absorbed by a foot of strut will be equal to

$$p = \frac{\gamma}{\beta} + \frac{1}{\rho} \times 267 \times 10^{-9} k \frac{d}{12} V^3$$

Now the weight γ is equal to

$$\gamma = 12 \times A \times 0.280 \text{ lb.} = 3.36A \text{ lb.}$$

where A is expressed in square inches.

In Chapter III we have seen that $k = 3.5$ for struts of the type which we are studying. Then, taking an average value $\rho = 0.75$ we shall have

$$p = \frac{3.36A}{\beta} + 103.6 \times 10^{-9} dV^3$$

Formula (1) permits expressing A as function of d

$$A = \frac{W}{47 \times 10^5 \times \alpha} \times \left(\frac{l}{d}\right)^2$$

consequently we shall have

$$p = \frac{715 \times 10^{-9} Wl^2}{\alpha\beta d^2} + 103.6 \times 10^{-9} dV^3$$

Supposing W, l, α, β and V to be known, the preceding equation gives the expression of total power (that is, the resultant of the weight and head resistance), absorbed by one foot of strut as function of the minor axis d of its section.

Evidently the designer's interest is to find the value of d that makes p minimum; but that value is the one which makes the derivative of the second term of the preceding equation equal to zero, that is, the one which satisfies the equation

$$-2 \times \frac{715 \times 10^{-9} Wl^2}{\alpha\beta d^3} + 103.6 \times 10^{-9} V^3 = 0$$

from which

$$d^3 = 13.8 \times \frac{W \times l^2}{\alpha \times \beta \times V^3}$$

Let us remember that the symbols have the following significance:

W = maximum braking load which a strut must support,
l = length of strut,

α = coefficient of Euler's formula,

β = ratio between the total weight and power of the airplane,

V = speed of the airplane,

For our example the weight of the airplane is 2130 lb. and its power is 300 H.P.; then $\beta = 7.1$; the foreseen speed is about 158 m.p.h. Furthermore for α we can adopt the value 8.

Then the preceding formula becomes:

$$d^3 = 61.5 \times 10^{-9} Wl^2 \qquad (2)$$

Euler's formula, for $\alpha = 8$, gives

$$\frac{W}{A} = 3.76 \times 10^{-7} \times \frac{1}{\left(\frac{l}{d}\right)^2} \qquad (3)$$

Equations (2) and (3) enable obtaining d and A, when W and l are known; then since

$$A = 6.37td$$

the thickness t of the tube is easily obtained.

The computations of the struts for the airplane in our example, Table 40, have been made with these criterions.

Before passing to the calculation of the horizontal trussings it is necessary to mention the vertical transversal trussings which serve to unite the front and rear struts (Fig. 182). The scope of these bracings is that of stiffening the wing truss and at the same time of establishing a connection between the diagonals of the principal vertical trussings. Their calculation is usually made by admitting that they can absorb from ½ to ⅔ of the load on the struts.

(c) The horizontal trussings have the scope of balancing the horizontal components of the air reaction. As we have seen, it is sufficient for the calculation, to assume for these horizontal components 25 per cent. of the value of the vertical reactions.

As an effect of the stresses in the vertical trussings, a certain compression in the spars of the upper wings and a certain tension in the spars of the lower wings are developed.

As an effect of the stresses in the horizontal trussings we have a certain tension in the front spars and a certain compression in the rear spars.

TABLE 40

Member		Stress coef. 10 W, lb.	Length l, in.	Diameter d (theoretical), in.	Thickness t (theoretical), in.	Diameter d (actual), in.	Thickness t (actual), in.
Front truss	I	−1500	65	0.731	0.068	0.788	0.065
	II	−2800	20	0.883	0.068	0.983	0.065
	III	−5200	20	1.085	0.068	1.180	0.065
Rear truss	I	−1380	65	0.710	0.068	0.788	0.065
	II	−2580	20	0.860	0.068	0.983	0.065
	III	−4780	20	1.057	0.066	1.180	0.065

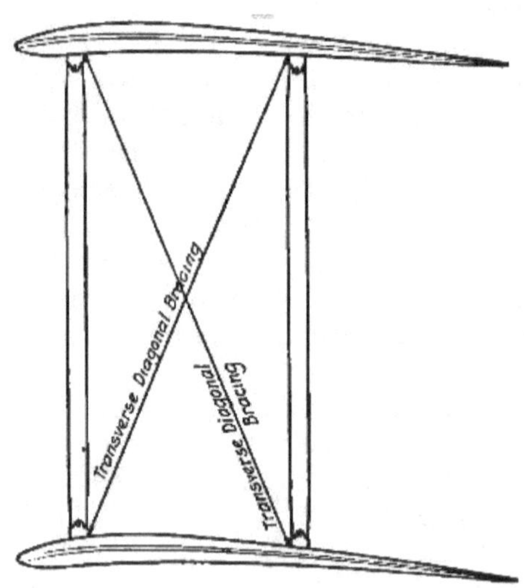

FIG. 182.

Consequently in the various spars there is a distribution of stresses as shown in Table 41.

TABLE 41

Spar	Effect of vertical trussing	Effect of horizontal trussing
Upper front	Compression	Tension
Upper rear	Compression	Compression
Lower front	Tension	Tension
Lower rear	Tension	Compression

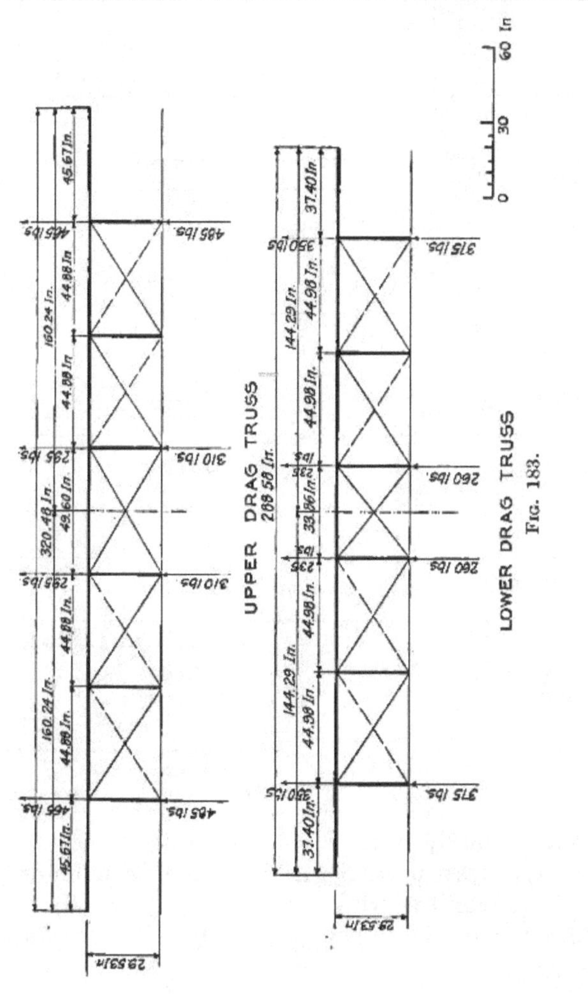

FIG. 183.

We see then that while there is partial compensation of stresses in the upper-front and lower-rear spars, in the other two spars instead the stresses add to each other. The spar which is in the worst condition is the upper-rear one,

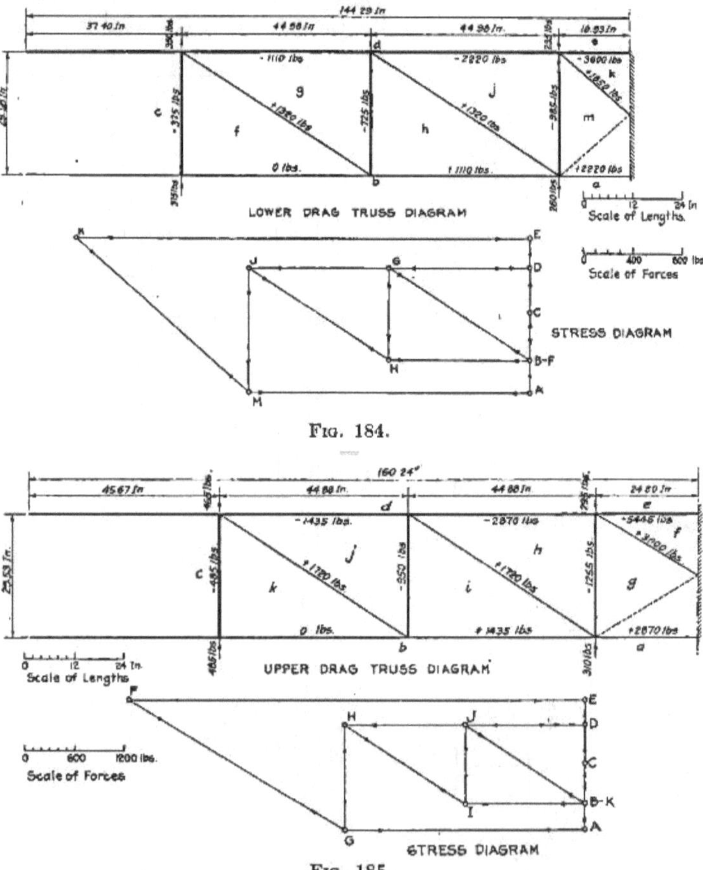

Fig. 184.

Fig. 185.

which is doubly compressed. In order to take the stress from it, at least partially, it is practical to adopt drag cables which anchor the wings horizontally. Usually these drag cables anchor the upper wings only. Sometimes also the lower ones.

In Fig. 183 the schemes of the horizontal trussings for the lower and upper wing are given. They are made of spars, a certain number of horizontal transversal struts, and of steel wire cross bracing. As we have already seen, in Fig. 183 the acting forces have been indicated equal to 25 per cent. of the vertical components. In Figs. 184 and 185 the graphic analysis of the horizontal trussings of the lower and upper wings have been given; as they are entirely analogous to those described for the vertical trussing, we need not discuss them.

(*d*) **Analysis of the Unit Stresses in the Spars.**—This analysis is usually made following an indirect method, that is, under form of verification. We fix certain sections for the spars and determine the unit load corresponding to the ultimate load of the airplane.

After various attempts, the most convenient section is determined.

Let us suppose that in our case the sections be those indicated in Fig. 186.

The areas and the moment of inertia are determined first. The areas are determined either by the planimeter or by drawing the section on cross-section paper. The moment of inertia is determined either by mathematical calculation or graphically by the methods illustrated in graphic statics. Fig. 187 gives this graphic construction for the upper rear spar.

Practically two principal methods of verification are used:

A. The elastic curve method.

B. The Johnson's formula method.

A. This method consists of determining the total unit stress f_T by adding the three following stresses:

1. Stresses of tension or of pure compression $f_c = \frac{P_T}{A}$ where P_T is the sum of the stresses P_L and P_D originated in the considered part of the spar by vertical and horizontal load, and A is the area of the section.

2. Stress due to bending moments $f_M = \frac{M}{Z}$ where M is the

UPPER FRONT SPAR
DOUGLAS FIR.
UPPER REAR SPAR
DOUGLAS FIR.
LOWER FRONT SPAR
SPRUCE.
LOWER REAR SPAR
SPRUCE.
320.48"
288.58
160.24"
89.76"
45.67"
144.29"
89.96"
37.40"
24.80"
65.67"
61.5"
4.17"
1.97"

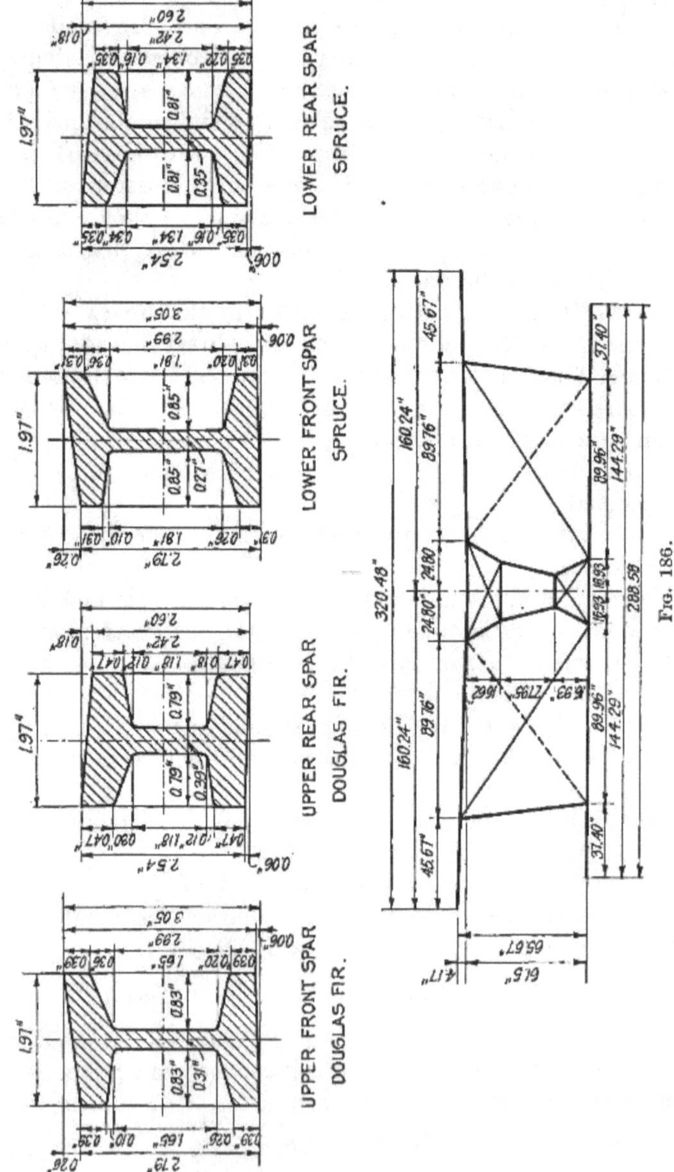

Fig. 186.

2.745 Sq. In.
H = 0.686 Sq. In.
0 0.4 0.8 Sq. In.
Scale of Area Forces.
H' = 1.5 In.
0 0.5 1.0 in.
Scale of Moment Forces.
Center of Gravity
y = 2.13 In.
1.32 In.
1.28 In.
2.60 In.
0.18 In.
0.47 In.
0.12 In.
1.18 In.
0.18 In.
0.47 In.
1.97 In.
0.985 In.
0.985 In.
0.39
0 0.5 1.0 In.
Scale of Spar-Section
0.47 In.
0.30 In.
0.30 In.
1.18 In.
0.12 In.
0.47 In.
0.06 In.
2.60 In.
I = H x H' x y
= .686 x 1.5 x 2.13 = 2.19 In.4

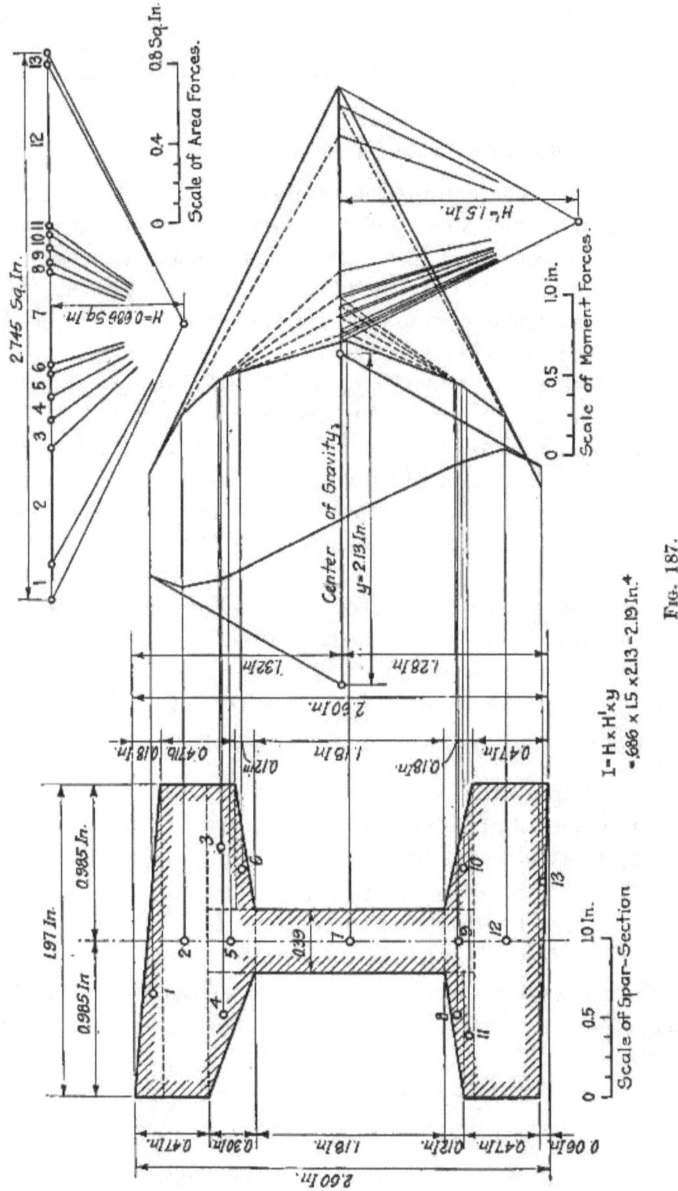

FIG. 187.

bending moment and Z is the section modulus. We shall remember that this modulus is obtained by dividing the moment of inertia I by the distance of the farthest fiber from the neutral axis.

3. Bending stress due to the compression stress $f_{\Delta} = \frac{P_T \times \Delta}{Z}$ where P_T is the compression stress and Δ is the maximum deflexion of the span which is obtained from the elastic curve. In order to know Δ it is necessary to know the elastic modulus E of the material because this modulus enters into the equation which gives the scale of the elastic curve (see Figs. 171 and 176).

By adding the values f_c, f_M and f_{Δ} we obtain f_T, which is the total unit stress, in our case corresponding to a load equal to ten times the normal flying load. If we wish to determine the factor of safety of the section it is necessary to know the modulus of rupture of the material; this modulus of rupture divided by $\frac{1}{10} f_T$ gives the factor of safety.

We have given in Chapter XVI the moduli of rupture to bending for various kinds of wood. For combined stresses of bending and compression stresses, it is necessary to adopt an intermediate modulus of rupture. Fig. 188 shows diagrams giving the modulus of rupture as function of ratio $\frac{f_c}{f_T}$ for the four following kinds of wood; Douglas fir, port-orford, spruce and poplar.

In Table 42 all the preceding data for the sections of the spars most stressed has been collected. In this table

P_L = stress due to vertical trussings.
P_D = stress due to horizontal trussings.
$P_T = P_L + P_D$ = total stress due to both trussings.

For these stresses the $-$ sign has been adopted when they are compression stresses and the $+$ sign when they are tension stresses.

A = area of the section.
$f_c = \frac{P_T}{A}$.
E = elastic modulus of the material.

Modulus of Rupture - Lbs. per Sq. In.
10500
10000
9500
9000
8500
8000
7500
7000
6500
6000
5500
5000
4500
4000
Ratio of $\frac{Fc}{Ft}$
0.0
0.1
0.2
0.3
0.4
0.5
0.6
0.7
0.8
0.9
1.0
Port Or Ford
Douglas Fir
Spruce
Poplar
Douglas Fir
Port or Ford

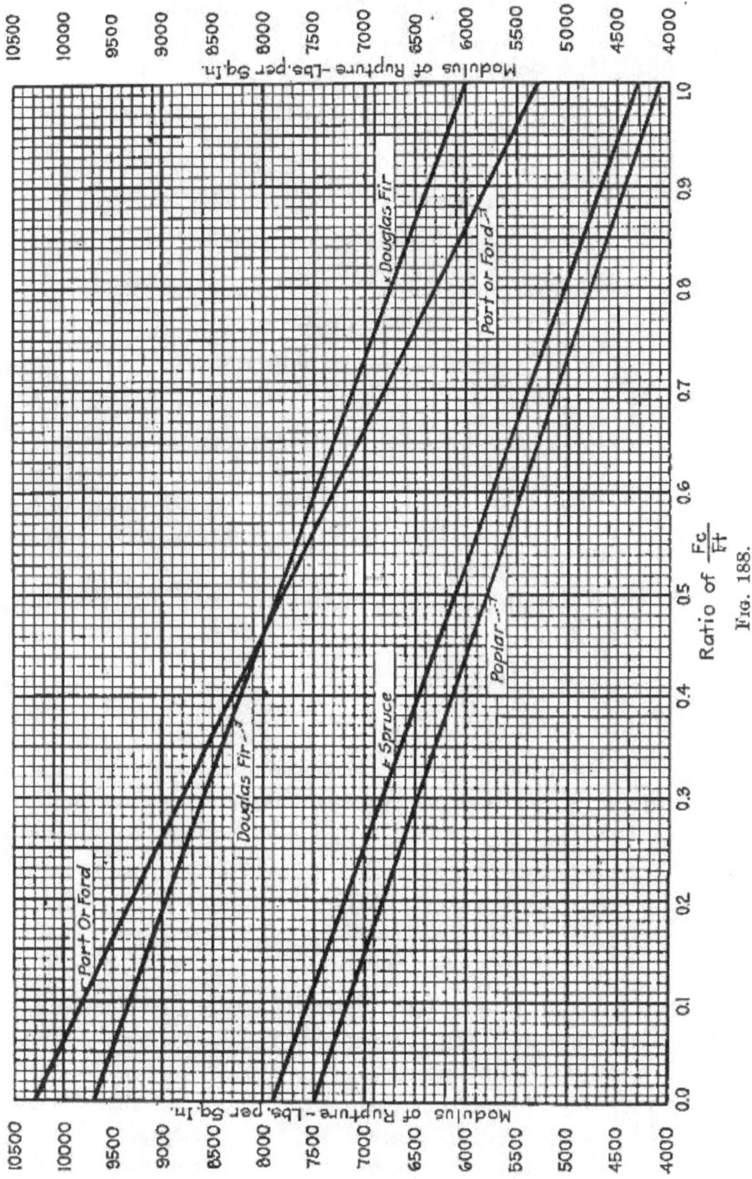

Fig. 188.

I = moment of inertia of the section.
Z = section modulus.
M = bending moment due to air pressure.
$f_M = \frac{M}{Z}$ unit stress due to this bending moment.
Δ = maximum deflexion of the span.
P_T = moment due to compression stress P_T.
$f_\Delta = \frac{P_T \times \Delta}{A}$ = unit stress due to the moment originated by the compression stress.
S = total shearing stress.
$s = \frac{S}{A}$ = unit stress to shearing.
$f_T = f_c + f_M + f_\Delta$.
f_c/f_T = ratio between the compression stress and total stress. By using the diagrams of Fig. 188, this ratio enables us to determine the modulus of rupture, thence the factor of safety.

B. The Johnson's formula method is based upon Johnson's formula:

$$f_T = \frac{P_T}{A} + \frac{M}{Z \times \left(1 - \frac{P_T l^2}{KEI}\right)}$$

where l is the length of the span, K is a numerical coefficient and the other symbols are those of the preceding method.

The value of coefficient K is dependent on end conditions and is

= 10 for hinged ends
= 24 for one hinged, one fixed
= 32 for both ends fixed

In Table 43 all the values of the quantities necessary for calculating the factor of safety by the Johnson's formula method have been collected.

We see that the factors of safety are about equal to those found by the preceding method, with the exception of that corresponding to point B of the upper-rear-spar. This

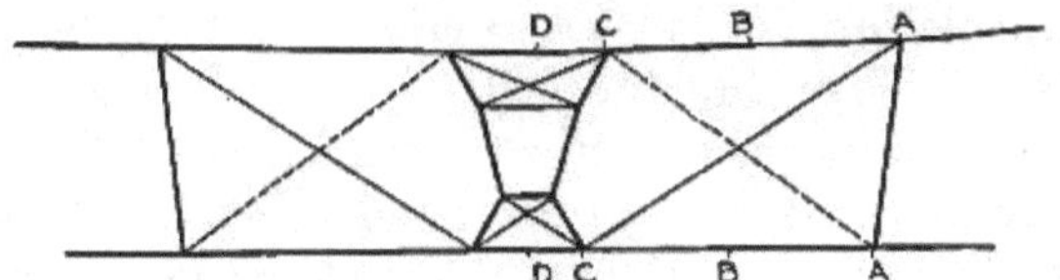

TABLE 42

Member	Sec.	P_L, lb.	P_D, lb.	P_T, lb.	A, sq. in.	$f_c = P_T/A$, lb. per sq. in.	E, lb. per sq. in.	I, in.^{4}	Z, in.^{3}	M, in. lb.
Upper front spar.....	A	0	0	0	5.70	0	1.78×10^6	4.02	2.62	20660
	B	−5120	0	− 5120	2.34	2190	1.78×10^6	2.83	1.84	6500
	C	−5120	+1435	− 3685	5.56	645	1.78×10^6	3.06	2.00	7200
	D	−5780	+2850	− 3020	2.34	1290	1.78×10^6	2.83	1.84	1210
Upper rear spar......	A	0	0	0	4.88	0	1.78×10^6	2.51	1.96	19000
	B	−4700	−1435	− 6135	2.58	2380	1.78×10^6	2.10	1.64	5980
	C	−4700	−2865	− 7565	4.88	1550	1.78×10^6	2.51	1.96	6630
	D	−5320	−5445	−11765	2.58	4370	1.78×10^6	2.10	1.64	1110
Lower front spar......	A	0	0	0	5.70	0	1.30×10^6	3.06	2.00	12220
	B	− 200	0	− 200	1.98	100	1.30×10^6	2.52	1.84	7720
	C	− 200	+1110	+ 910	5.56	165	1.30×10^6	3.06	2.00	7800
	D	+7600	+2220	+ 9820	1.98	4960	1.30×10^6	2.52	1.84	5380
Lower rear spar......	A	0	0	0	4.88	0	1.30×10^6	1.92	1.49	11300
	B	− 185	−1110	− 1295	2.11	615	1.30×10^6	1.84	1.40	7140
	C	− 185	−2220	− 2405	4.88	495	1.30×10^6	1.92	1.49	7270
	D	+7000	−3600	+ 3400	2.11	1610	1.30×10^6	1.84	1.40	4970

* No bolt holes.

discrepancy occurs because the coefficient K for this point should have been 32 instead of 24, as was assumed. In fact, from an examination of the elastic curve of the upper spars (Fig. 171), it is seen that point A is to be considered as an actual fixed point, and consequently for this point the coefficient 32 should have been taken.

With this single exception, the two methods are practically equivalent.

Before leaving the calculation of the wing truss, the calculation of the shearing stresses and of the bending moments which are developed in the ribs should be mentioned.

This calculation, which is usually made graphically is illustrated in Figs. 189 and 190.

The rib can be considered as a small beam with two supports and 3 spans; the supports being made by the spars.

Diagram (a) of Fig. 189 gives the values of the pressures

TABLE 42—(*Continued*)

$f_M = M/Z$, lb. per sq. in.	Δ, in.	$P_T.\Delta$, in. lb.	$f_\Delta = P_T.\Delta/Z$ lb. per sq. in.	S, lb.	$s = \frac{S}{A}$, lb. per sq. in.	$f_t = f_c + f_M f_\Delta$ lbs. per sq. in.[3]	f_c/f_T	Modulus, lb. per sq. in.	Factor safety	Sec.
7880	.....	....	....	1040	185	7880	0.000	9700	12.3	■•
3530	0.578	2960	1600	900	385	7320	0.299	8550	11.6	I
3600	0.578	2130	1065	720	125	5310	0.121	9250	17.4	■
660	0.106	320	175	350	150	2125	0.606	7450	35.0	I
9700	.....	....	....	950	195	9700	0.000	9700	10.0	■•
3640	0.716	4390	2670	825	320	8600	0.270	8680	10.0	I
3380	0.716	5420	2840	660	133	7770	0.197	8900	11.5	■•
680	0.131	1540	940	320	125	5990	0.723	7000	11.7	I
6110	.....	....	....	1030	180	6110	0.000	7900	12.9	■
4200	1.43	290	160	900	455	4460	0.025	7850	17.5	I
3900	.....	....	....	900	160	4065	0.000	7000	19.4	■
2930	.....	....	....	300	150	7890	0.000	7900	10.0	I
7600	.....	....	....	955	195	7600	0.000	7900	10.4	■
5100	1.55	2010	1435	830	395	7150	0.086	7600	10.6	I
4840	1.55	3730	2500	830	170	7835	0.062	7800	10.0	■
4350	.....	....	....	280	130	5960	0.000	7900	13.2	I

along the entire rib; the integration of this diagram gives diagram (b) of Fig. 189 whose ordinates correspond to the shearing stresses.

In Fig. 190, diagram (a) represents diagram (b) of Fig. 189. In order to render this diagram more clearly it has been redrawn in Fig. 190 (b) referring it to a rectilinear axis and adopting a doubled scale for the shearing stresses.

The integration of this diagram gives the diagram of the bending moments, Fig. 190 (c).

The distributions of the shearing stresses and bending

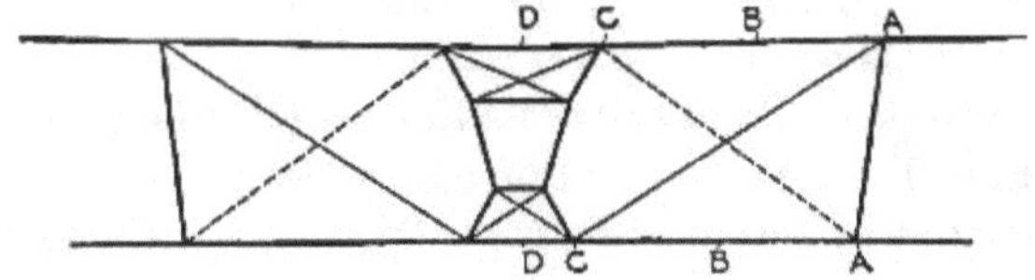

TABLE 43

Member	Sec.	P_L, lb.	P_D, lb.	$P = P_L + P_D$ lb.	A, sq. in.	$f_c = P/A$, lb. per sq. in.	l, in.	Pl^2	Z, in.³	K	E, lb. per sq. in.	I, in.⁴
Upper front spar	A	0	0	0	5.70	0	.....	....	2.62	..	1.78×10^6	4.02
	B	−5120	0	− 5120	2.34	2190	89.76	41.2×10^6	1.84	24	1.78×10^6	2.83
	C	−5120	+1435	− 3685	5.56	645	89.76	29.7×10^6	2.00	32	1.78×10^6	3.06
	D	−5780	+2850	− 3020	2.34	1290	48.18	7.0×10^6	1.84	32	1.78×10^6	2.83
Upper rear spar	A	0	0	0	4.88	0	.....	....	1.96	..	1.78×10^6	2.51
	B	−4700	−1435	− 6135	2.58	2380	89.76	49.4×10^6	1.64	24	1.78×10^6	2.10
	C	−4700	−2865	− 7565	4.88	1550	89.76	61.0×10^6	1.96	32	1.78×10^6	2.51
	D	−5320	−6445	−11765	2.58	4370	48.18	27.3×10^6	1.64	32	1.78×10^6	2.10
Lower front spar	A	0	0	0	5.70	0	.....	....	2.00	..	1.30×10^6	3.06
	B	− 200	0	− 200	1.98	100	89.96	1.62×10^6	1.84	24	1.30×10^6	2.52
	C	− 200	+1110	+ 910	5.56	165	89.96	....	2.00	32	1.30×10^6	3.06
	D	+7600	+2220	+ 9820	1.98	4960	33.86	....	1.84	32	1.30×10^6	2.52
Lower rear spar	A	0	0	0	4.88	0	.....	....	1.49	..	1.30×10^6	1.92
	B	− 185	−1110	− 1295	2.11	615	89.96	10.5×10^6	1.40	24	1.30×10^6	1.84
	C	− 185	−2220	− 2405	1.88	495	89.96	19.5×10^6	1.49	32	1.30×10^6	1.92
	D	+7000	−3600	+ 3400	2.11	1610	33.86	....	1.40	32	1.30×10^6	1.84

* No bolt holes.

moments being known the dimensions of the web and of the rib flanges can easily be determined.

In Fig. 191 a general view of a very light type of rib is given.

We shall now pass to the calculation of the tail system and the control surfaces. Figs. 192 and 193 give respectively the assembly of the fin-rudder group and the stabilizer-elevator group. The calculation of their frame is very easy when the distribution of the loads on the surface is known. Consequently only the procedure for the calculation of these loads will be indicated.

Let us first of all consider the fin-rudder group (Fig. 194). In normal flight as well as during any maneuver whatever, the distribution of the pressures on these surfaces is very

TABLE 43—(*Continued*)

KEI	M, in. lb.	$\frac{Pl^2}{KEI}$	$1-\frac{Pl^2}{KEI}$	$\frac{M}{z\left(1-\frac{Pl^2}{KEI}\right)}$, lb.	S, lb.	$S=\frac{S}{A}$ lb. per sq. in.	f_t, lb. per sq. in.	f_c/f_t	Mod. rupt., lb. per sq. in.	Factor safety	Sec.
....	20600	.....	.....	7890	1040	180	7890	.....	9700	12.3	■*
121.0×10^6	6500	0.340	0.060	5350	900	385	7540	0.291	8600	11.4	I
174.2×10^6	7200	0.170	0.830	4330	720	125	4975	0.130	9200	18.5	■
161.0×10^6	1210	0.044	0.956	690	350	150	1950	0.651	7300	37.0	I
....	19000	.....	.....	9700	950	195	9700	.....	9700	10.5	■*
189.5×10^6	5980	0.552	0.448	8130	825	320	10510	0.226	8850	8.5	I
143.0×10^6	6630	0.427	0.573	5900	660	135	7450	0.203	8900	11.9	■*
119.3×10^6	1110	0.228	0.772	880	320	125	5250	0.883	6600	12.6	I
....	12220	.....	.....	6110	1030	180	6110	.....	7900	12.9	■
78.6×10^6	7720	0.021	0.979	4300	900	455	4400	0.021	7850	17.8	I
....	7800	.....	.....	3900	900	160	4065	.....	7900	19.4	■
....	5380	.....	.....	2930	300	150	7890	.....	7900	10.0	I
....	11300	.....	.....	7600	955	195	7600	.....	7900	10.4	■
57.4×10^6	7140	0.183	0.817	6240	830	395	6865	0.090	7600	11.0	I
80.0×10^6	.7210	0.244	0.756	6410	830	170	6905	0.070	7650	11.1	■
....	4970	.....	.....	3550	280	130	5160	.....	7900	15.3	I

complex and varies according to their profile and their form.

Practically, though, such high factors of safety are assumed for them, that it suffices to follow any loading hypothesis even if only approximate.

For instance, as it is usually done in practice, the hypothesis illustrated in the diagram of Fig. 194 (*c*) can be adopted. We suppose that the unit load decreases linearly on the fin as well as on the rudder; in the fin it decreases from a maximum value u in the front part to a minimum

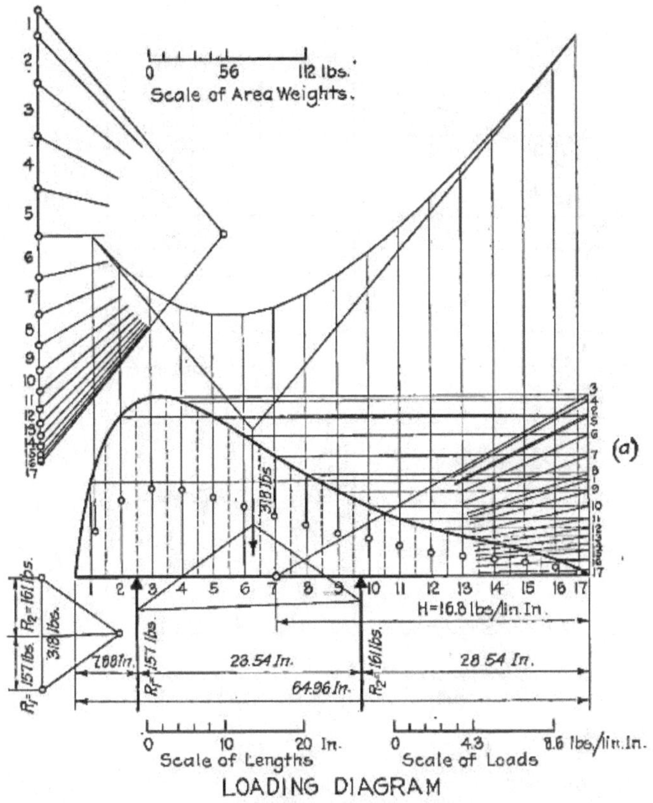

TABLE OF AREA WEIGHTS IN POUNDS

Area	1	8.2	3	4	5	6	7	8	9	10	11	12	13	14	15	16	17	Σ
Load	18.0	33.5	38.3	37.3	34.0	30.0	25.8	21.6	18.0	14.9	12.3	10.1	8.3	6.9	5.3	3.2	0.5	318

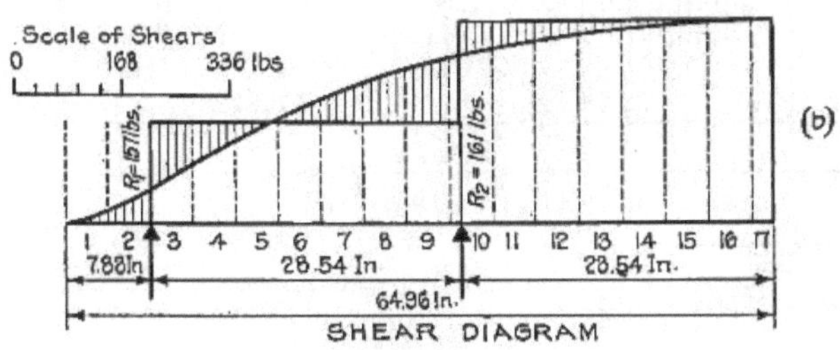

Fig. 189.

(a)
7.88 In.
28.54 In.
28.54 In.
64.96 In.
SHEAR DIAGRAM (1)
0
10
20 in
Scale of Lengths
0
168
336 lbs.
Scale of Shears
(b)
H=55 lbs.
H=55 lbs
7.88 In
28.54 In.
28.54 In
64.96 In.
SHEAR DIAGRAM (2)
0
84
168 lbs.
Scale of Shears.
(c)
7.88 In
28.54 In.
28.54 In.
64.96 In.
MOMENT DIAGRAM
0
550
1100 In lbs
Scale of Moments.

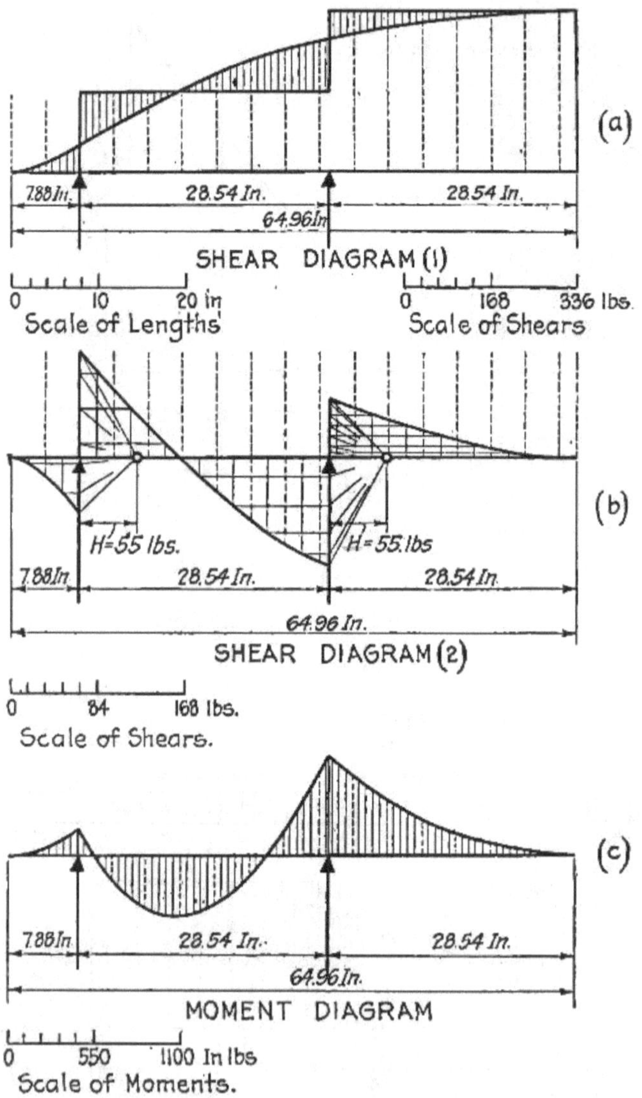

Fig. 190.

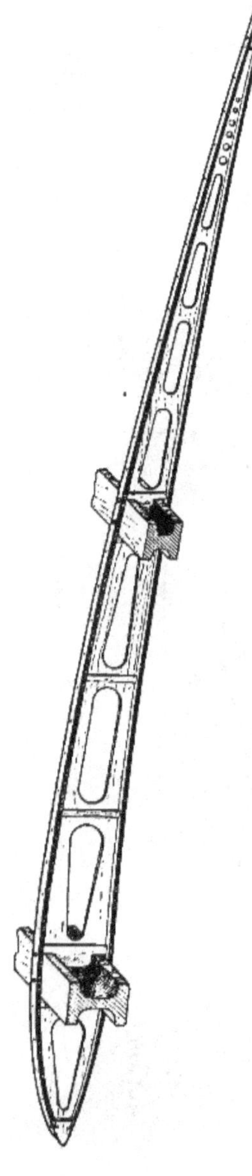

Fig. 191.

value equal to 0.5 u in the rear part. In the rudder instead, the unit load decreases from u to zero.

In order to determine the numerical value of u the average value u_m of the unit load of the surfaces is usually given. This average value is assumed so much greater, as the airplane is faster; practically for speeds between 100 and 200 m.p.h. we can assume

$$u_m = 0.16V$$

expressing u_m in pounds per square foot.

In our case we shall have about $u_m = 25$ lbs. per sq. ft. Then the surfaces of the fin and rudder are divided into sections (Fig. 194 (*a*)), and their areas are determined. In our case they are as given in the table of Fig. 194 (*b*); let us call a one of these areas and ku the corresponding unit load; the load upon it will be evidently aku.

If A is the total area, we have

$$\Sigma a \times k \times u = Au_m$$

that is

$$u = \frac{A \times u_m}{\Sigma a \times k}$$

The value u having been determined, we have all the

Section E-E
Section A-A
Section B-B
Section C-C
Section D-D
0
8
16 inches
Scale of Inches

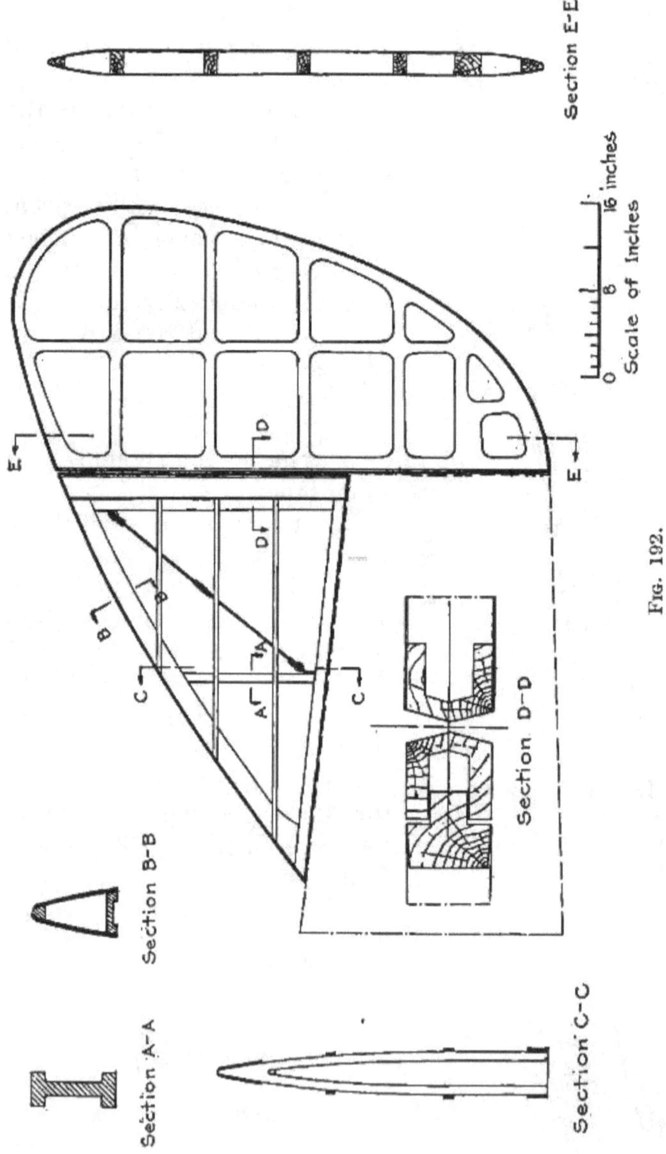

FIG. 192.

Section B-B
Section A-A
Section C-C
0
8
16 inches.
Scale of Lengths.

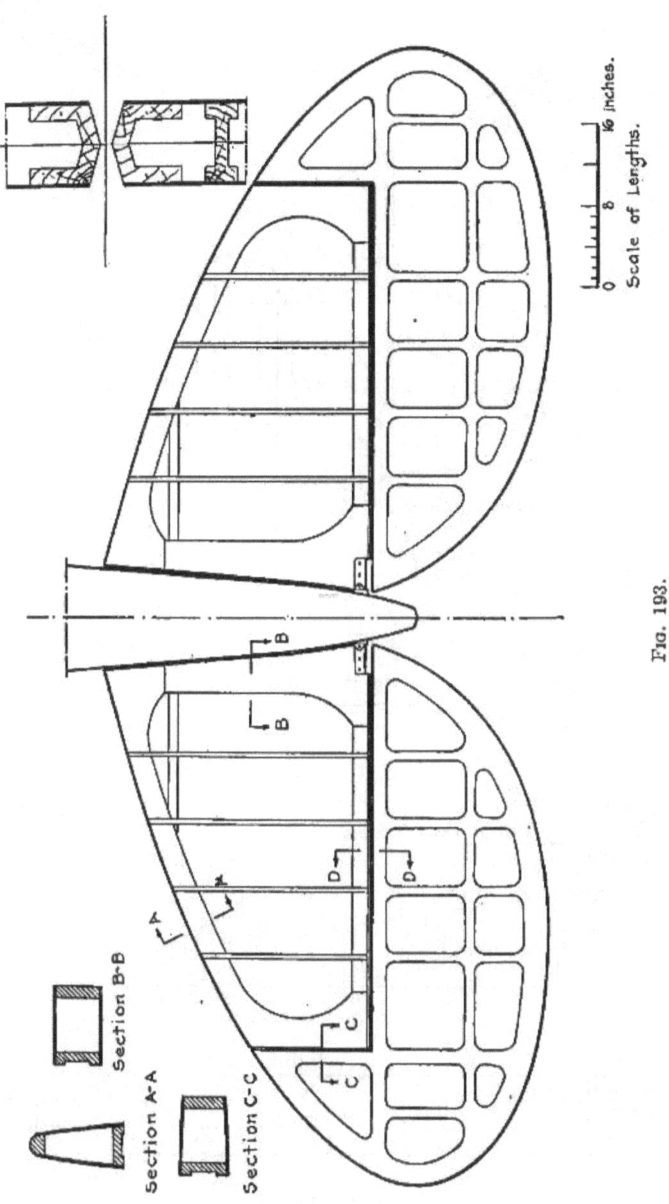

FIG. 193.

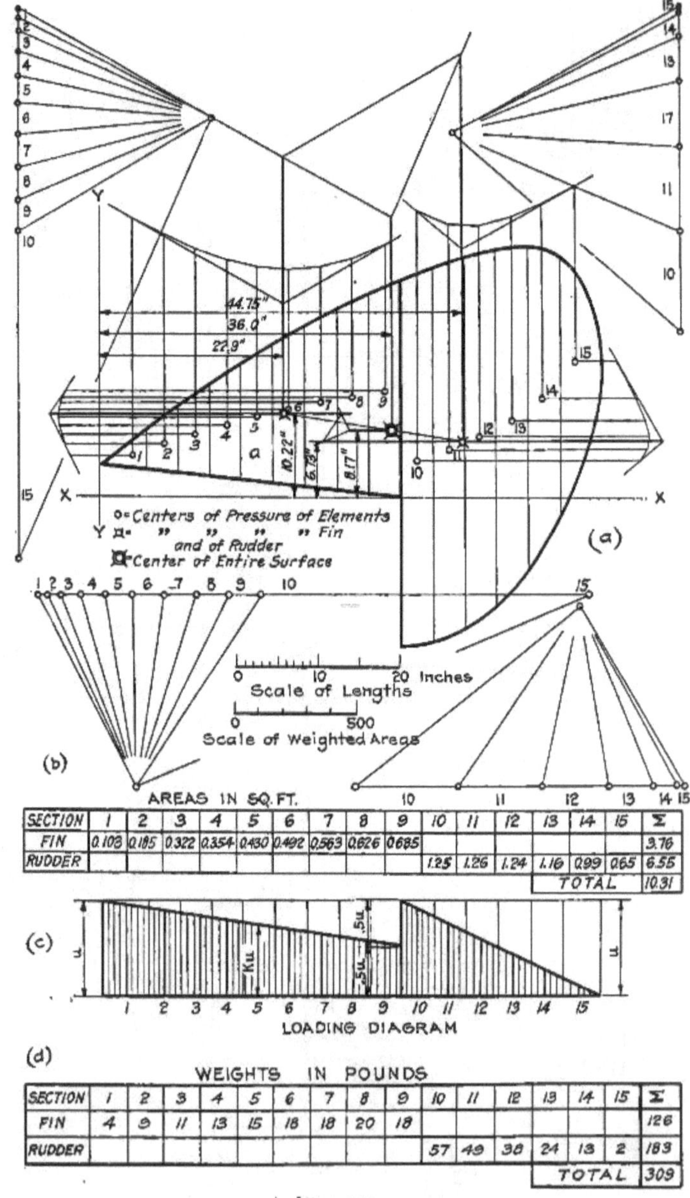

AREAS IN SQ. FT.

SECTION	1	2	3	4	5	6	7	8	9	10	11	12	13	14	15	Σ
FIN	0.103	0.185	0.322	0.354	0.430	0.492	0.563	0.626	0.685							3.76
RUDDER										1.25	1.26	1.24	1.16	0.99	0.65	6.55
													TOTAL			10.31

WEIGHTS IN POUNDS

SECTION	1	2	3	4	5	6	7	8	9	10	11	12	13	14	15	Σ
FIN	4	9	11	13	15	18	18	20	18							126
RUDDER										57	49	38	24	13	2	183
													TOTAL			309

FIG. 194.

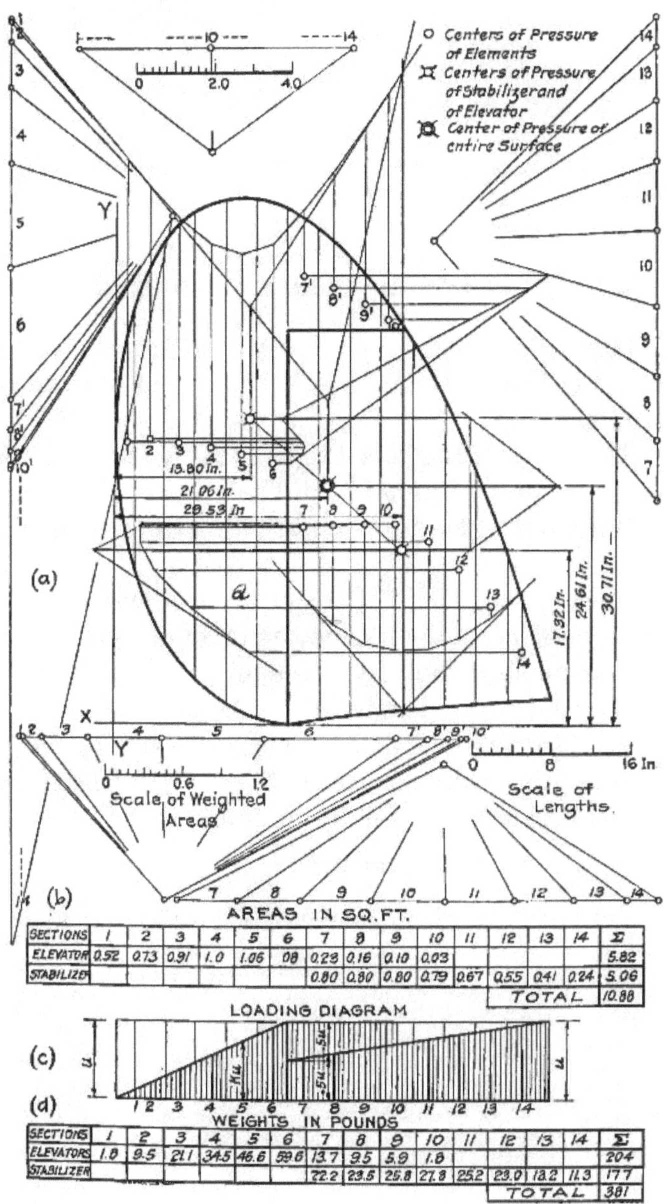

AREAS IN SQ. FT.

SECTIONS	1	2	3	4	5	6	7	8	9	10	11	12	13	14	Σ
ELEVATOR	0.52	0.73	0.91	1.0	1.06	08	0.23	0.16	0.10	0.03					5.82
STABILIZER							0.80	0.80	0.80	0.79	0.67	0.55	0.41	0.24	5.06
												TOTAL			10.88

WEIGHTS IN POUNDS

SECTIONS	1	2	3	4	5	6	7	8	9	10	11	12	13	14	Σ
ELEVATORS	1.8	9.5	21.1	34.5	46.6	59.6	13.7	9.5	5.9	1.8					204
STABILIZER							22.2	23.5	25.8	27.8	25.2	23.0	18.2	11.3	177
												TOTAL			381

Fig. 195.

elementary values aku, which in our case are as given in the table of Fig. 194 (d). These loads being obtained, we easily determine:

(a) the center of loads of the fin, that is, what is usually termed the center of pressure of the fin,

(b) the center of loads or center of pressure of the rudder, and

(c) the center of loads of the entire system.

It is then possible to determine the reactions on the various structures and consequently to make the calculation of their dimensions, following the usual methods.

In Fig. 195 all the operations previously described are repeated for the stabilizer-elevator group, noting, however, that for this group we usually assume

$$u_m = 0.22 \times V$$

that is, in our case u_m = 35 lbs. per sq. ft.

CHAPTER XIX

STATIC ANALYSIS OF FUSELAGE, LANDING GEAR AND PROPELLER

A. Analysis of Fuselage.—Let us consider the following particular cases:

(*a*) Stresses in normal flight.
(*b*) Stresses while maneuvering the elevator.
(*c*) Stresses while maneuvering the rudder.
(*d*) Maximum stresses in flight.
(*e*) Stresses while landing.

(*a*) In normal flight the fuselage should be considered as a beam supported at the points where the wings are attached to it and loaded at the various joints of the trussing which make the frame of the fuselage. In these conditions it is easy to determine the shearing stresses and the bending moments when the weight of the various parts composing the fuselage or contained in it are known.

Let us consider the case of a fuselage made of veneer. As we have seen in the first part of this book, such a fuselage has a frame of horizontal longerons connected by wooden bracings; this frame is covered with veneer, glued and nailed to the longerons and bracings. Let us suppose the frame to be the one shown in Fig. 196*a*.

First the reactions of the various weights on the joints of the structure, and the reactions on the supports are calculated (Fig. 196*b*). It is then easy to draw the diagram of the shearing stresses (Fig. 196*c*), and of the bending moments (Fig. 196*d*), corresponding to the case of normal flight.

(*b*) When the pilot maneuvers the elevator, the fuselage is subjected to an angular acceleration, which is easily calculated if the moment of inertia of the fuselage is known.

(a)
215.7 In.
15.75" 15.75" 23.10" 24.50" 23.20" 21.80" 20.30" 18.90" 17.50" 16.30" 18.60"
265 lbs.
284 lbs.
324 lbs.
451 lbs.
162 lbs.
20 lbs.
18 lbs.
17 lbs.
16 lbs.
14 lbs.
23 lbs.
21 lbs.
2 3 4 5 6 7 8 9 10 11 12 13
M
N
31.5 In.
23.10 In.
161.10 In.
(b)
654 lbs. Wing Reaction
961 lbs. Wing Reaction
SPACE DIAGRAM.
0 30 60 In.
Scale of Lengths
(c)
H=167 lbs
5 6 7 8 9 10 11 12
2 3 4
0 250 500 lbs.
SHEAR DIAGRAM
Scale of Shears
(d)
0 5000 10000 lbs.In.
Scale of Moments.
2 3 4 5 6 7 8 9 10 11 12
MOMENT DIAGRAM

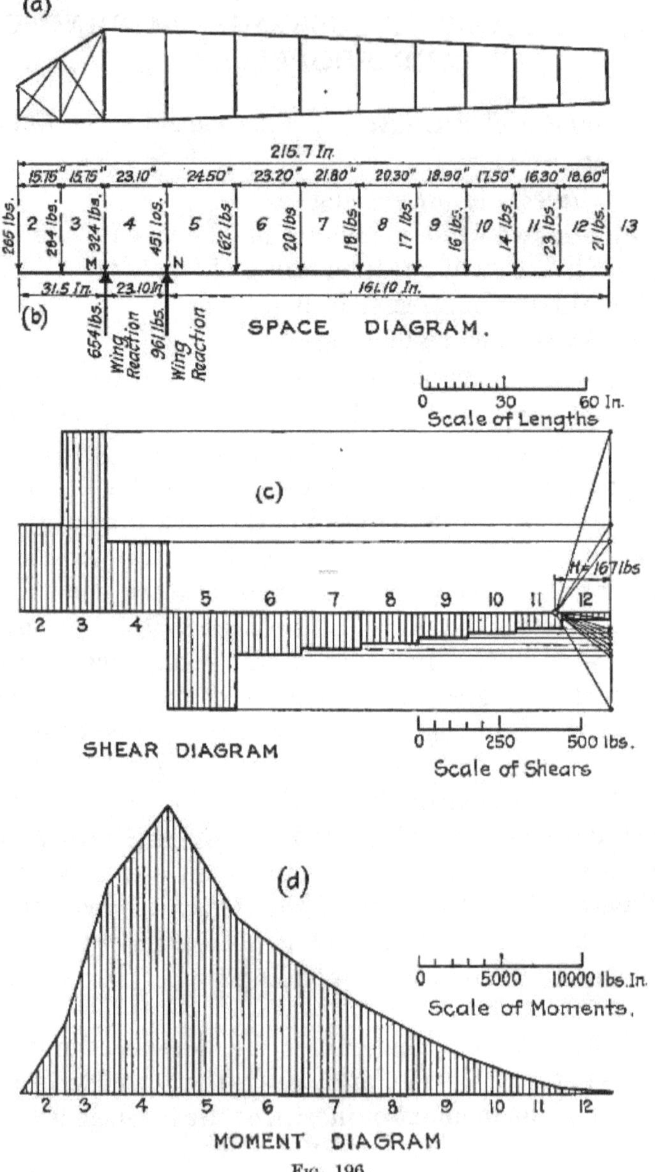

Fig. 196.

In Fig. 197 the graphic determination of this moment of inertia has been made; its result is $I = 97{,}000$ lb. $\times$ inch^2. We shall suppose that a force equal to 1000 lb. acts suddenly upon the elevator. Then remembering the equation of mechanics

$$C = I_p \times \frac{d\omega}{dt}$$

where

C = acting couple
I_p = polar moment of inertia
$\frac{d\omega}{dt}$ = angular acceleration

and as in our case

$C = 1000 \times 177 = 177{,}000$ lb. $\times$ inch
$I = 97{,}000$ lb. mass $\times$ inch^2

we shall have

$$\frac{d\omega}{dt} = \frac{177{,}000}{97{,}000} = 1.82 \ 1/\text{sec.}^2$$

This angular acceleration originates a linear acceleration in each mass proportional to its distance from the center of gravity and in a direction tending to oppose the rotation originated by the couple C. Thus, each mass will be subjected to a force, as illustrated for our example, in Fig. 198*a*. It is then easy to obtain the diagrams of the shearing stresses (Fig. 198*b*), and of the bending moments (Fig. 198*c*), originated by the forces of inertia which appear in the various masses of the fuselage, when a force of 1000 lb. is suddenly applied upon the elevator.

Let us note that the stresses thus calculated are greater than those had in practice; in fact for the calculation of the angular acceleration, the total moment of inertia of the airplane and not only that of the fuselage should have been introduced: therefore the angular acceleration found is greater than the effective one. However this approximation is admissible, since its results give a greater degree of safety.

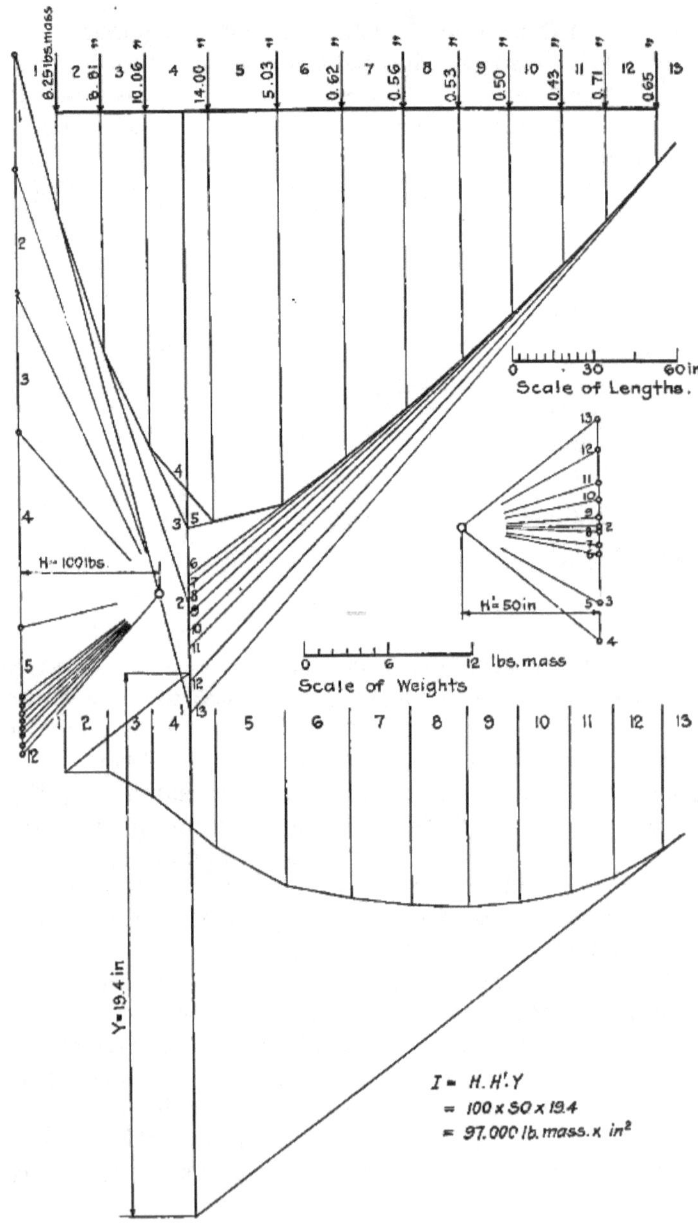

8.25 lbs. mass
8.81
10.06
14.00
5.03
0.62
0.56
0.53
0.50
0.43
0.71
0.65
H = 100 lbs.
Scale of Lengths.
0 30 60 in
H' = 50 in
0 6 12 lbs. mass
Scale of Weights
Y = 19.4 in
I = H.H'.Y
= 100 x 50 x 19.4
= 97.000 lb. mass. x in²

0 400 800 lbs.
Scale of Forces
0 30 60 in
Scale of Lengths
693 lbs
479 "
256 "
250 "
320 "
55 "
82 "
98 "
110 "
108 "
200 "
205 "
1000 "
1000 lbs
Center of Gravity
(a)
SPACE DIAGRAM
(b)
H=1000 lbs.
2 3 4a 4b 5 6 7 8 9 10 11 12 13
0 400 800 lbs.
Scale of Shears
SHEAR DIAGRAM
(c)
2 3 4a 4b 5 6 7 8 9 10 11 12 13
0 30000 60000 in/lbs
Scale of Moments.
MOMENT DIAGRAM

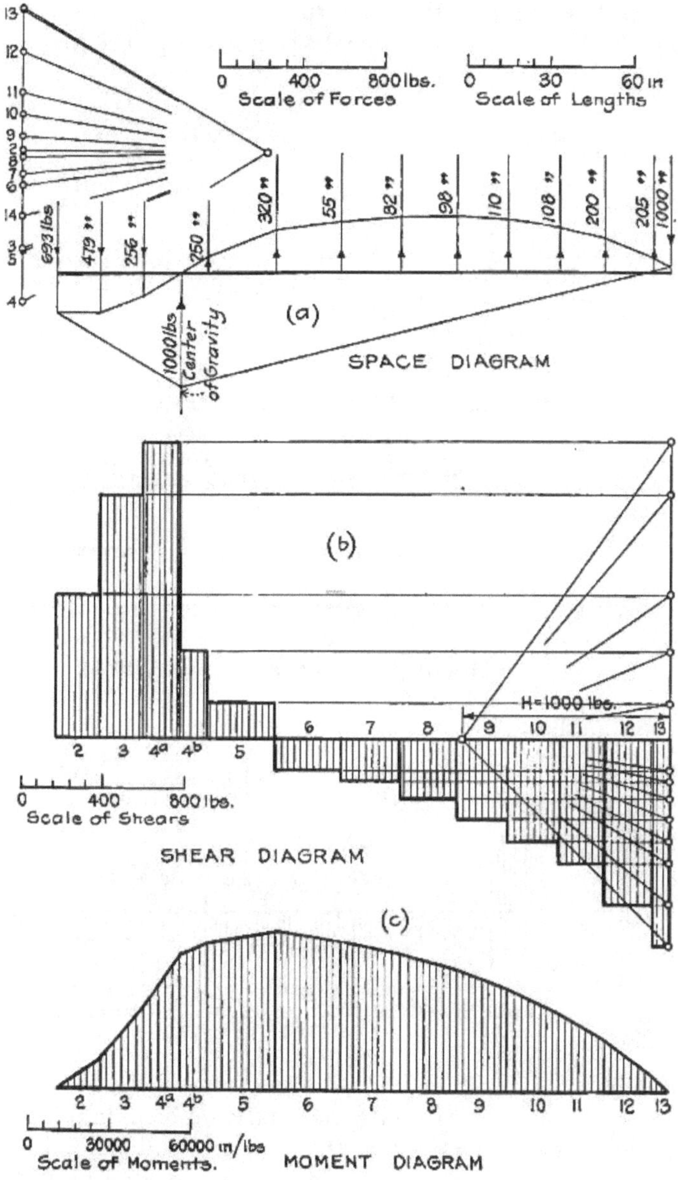

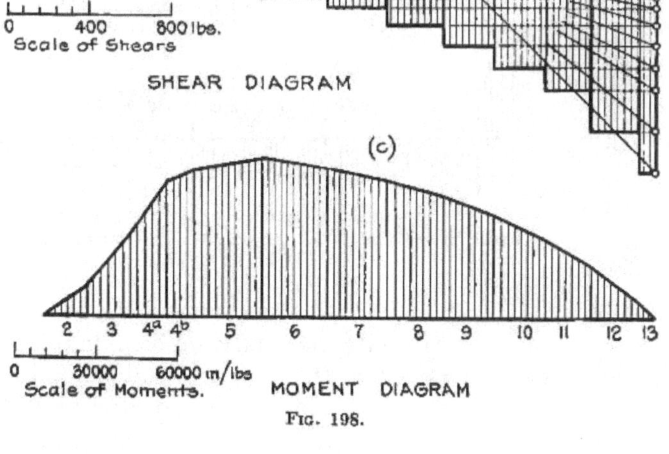

FIG. 198.

(*c*) For maneuvering the rudder the same applies as for the elevator. The same diagrams of Fig. 198 may also be used for this case.

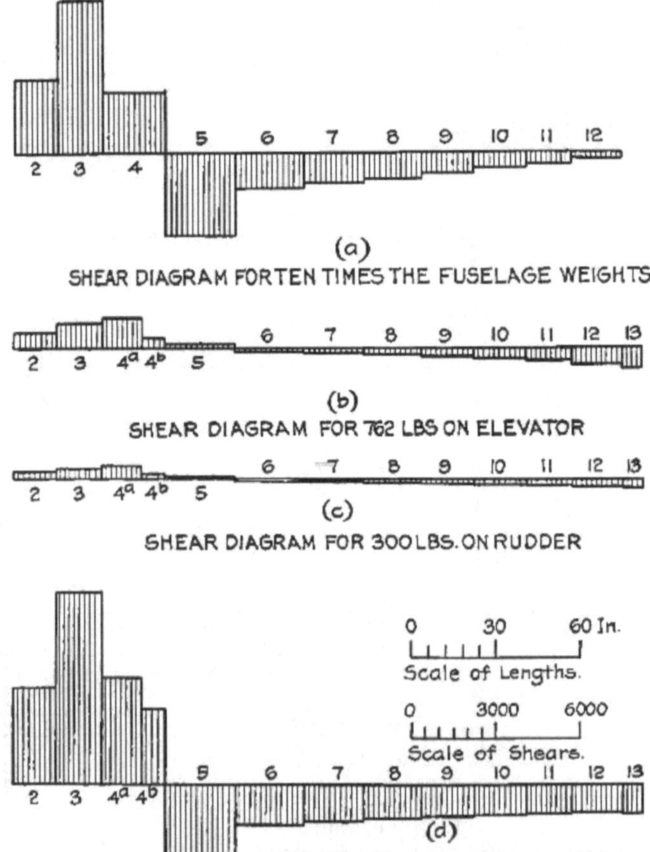

Fig. 199.

(*d*) In order to calculate the maximum breaking stresses in flight, let us suppose that the breaking load is applied at the same time upon the wings, the elevator, and the rudder. This is equivalent to make the following hypothesis:

1. to multiply the loads of the fuselage by 10,
2. to apply 762 lb. upon the elevator,
3. to apply 309 lb. upon the rudder.

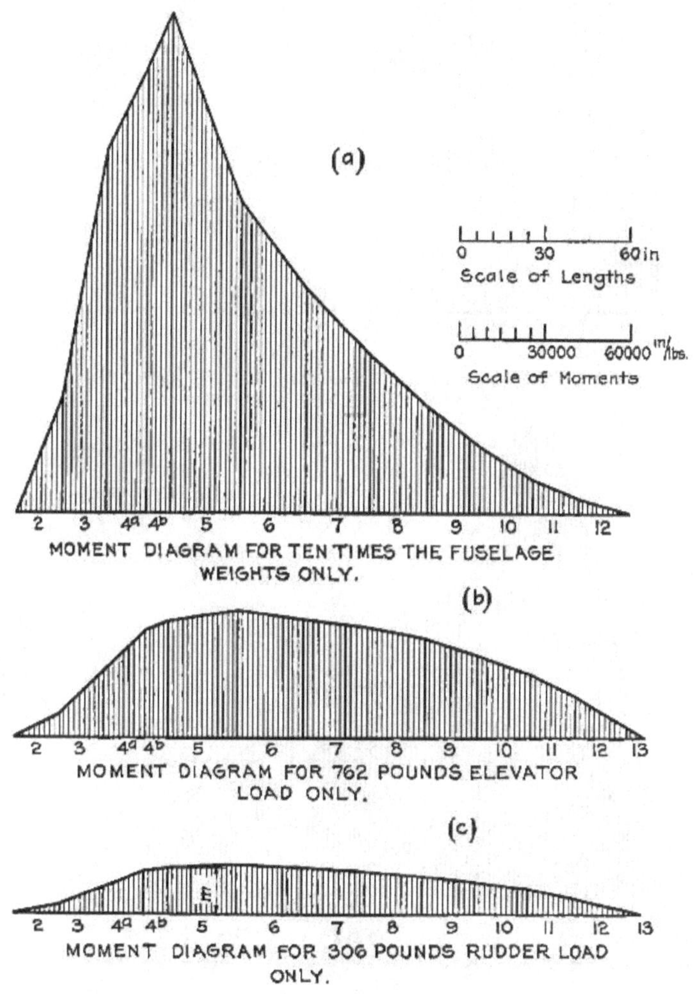

Fig. 200.

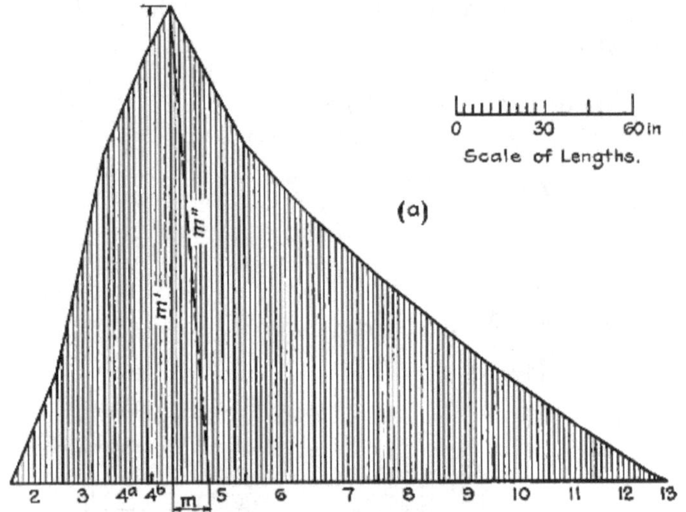

MOMENT DIAGRAM FOR ELEVATOR LOADS AND TEN TIMES THE FUSELAGE WEIGHTS

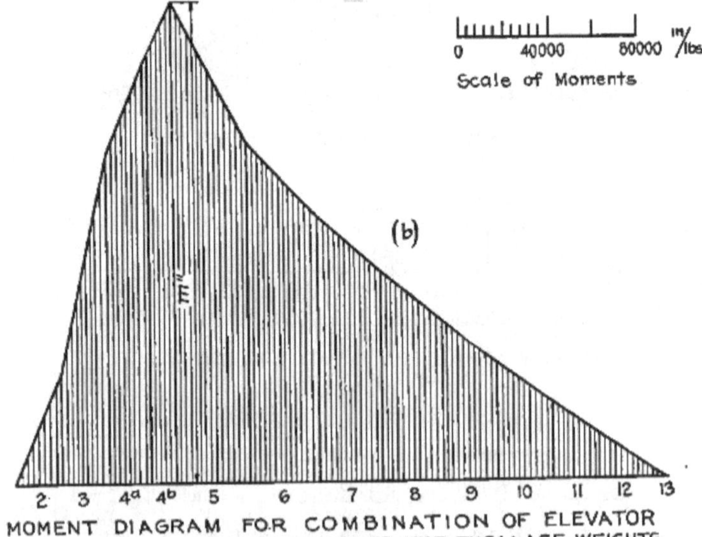

MOMENT DIAGRAM FOR COMBINATION OF ELEVATOR AND RUDDER LOADS AND TEN TIMES THE FUSALAGE WEIGHTS

Fig. 201.

It is then easy to draw the diagrams of the shearing stresses in this case (Fig. 199, *a*, *b*, *c*), and consequently, through their sum, the diagram of the total shearing stresses in flight (Fig. 199*d*).

In order to calculate the maximum bending moments, it is necessary to consider separately those produced by vertical forces (loads on the fuselage and on the elevator), and those produced by horizontal forces (loads on the rudder). In Fig. 200 *a*, *b*, *c*, the bending moments are shown due respectively to 10 times the loads on the fuselage, to the load of 762 lb. on the elevator, and to the load of 306 lb. on the rudder.

Fig. 201*a* shows a diagram obtained by the algebraic sum of the first two diagrams, Fig. 201*b* shows the total diagram whose ordinates m'' are equal to the hypotenuses of the right triangles having the sides corresponding to the ordinates m and n of diagrams 200*c* and 201*a*.

Having obtained in this manner, the diagrams of the maximum shearing stresses and maximum bending moments corresponding to the various sections, it is possible to proceed in the checking of the resistance of those sections.

In Fig. 202 the checking for section 4–5 has been effectuated. For simplicity it is customary to assume that the longerons resist to the bending and the veneer sides to the shearing stresses. The stress due to shearing is given immediately, dividing the maximum shearing stress by the sections of the veneer. As for the stresses in the longerons, it is necessary to determine their ellipse of inertia.

Let 1, 2, 3 and 4 be the four longerons constituting section 4–5. The maximum moment is equal to 216,600 lb. × inch, and its plane of stress makes an angle x with the vertical plain such that

$$\tan\alpha = \frac{\text{Horizontal moment}}{\text{Vertical moment}} = \frac{16{,}600}{215{,}300} = 0.076$$

Then a certain section is fixed for the longerons and with the usual methods of static graphics the moments of inertia of the four assembled longerons with respect to horizontal axis and to a vertical axis passing through the center of

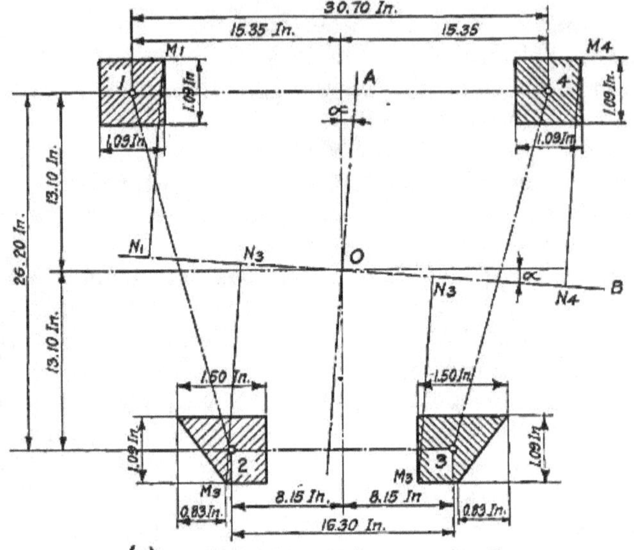

(a) TRANSVERSE SECTION AT 4-5

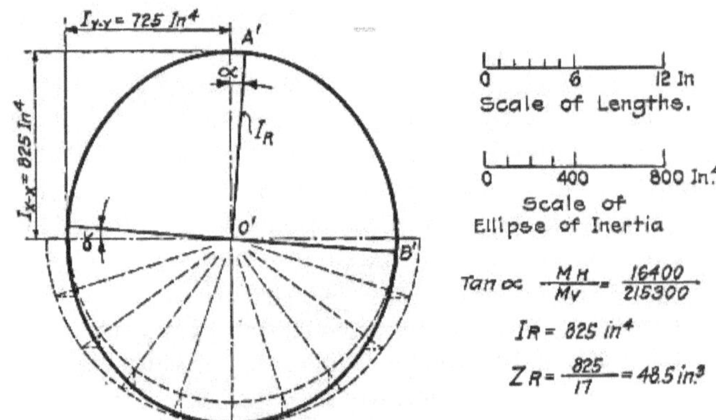

$$\tan\alpha \frac{M_H}{M_V} = \frac{16400}{215300}$$

$$I_R = 825 \text{ in}^4$$

$$Z_R = \frac{825}{17} = 48.5 \text{ in}^3$$

(b) ELLIPSE OF INERTIA AT SECTION 4-5

Maximum Moment at Section 216600 in lbs.

Maximum Extreme Fiber Stress $= \frac{216600}{48.5} = 4470$ *lbs/in²*

Modulus of Rupture for Spruce $= 9700$ *lbs/in²*

Factor of Safety $\frac{9700}{4470} \times 10 = 21.7$

FIG. 202.

gravity of the system are determined (Fig. 202*a*). Then the ellipse of inertia may be drawn (Fig. 202*b*). The vector radius OA' of such an ellipse which makes the angle α with the vertical gives the moments of inertia to be used in the calculations. In order to have the section modulus, it is necessary to draw $B'O'$ the conjugate diameter to $O'A'$. For the center of gravity O of the four longerons draw OB parallel to diameter $O'B'$; from the four points M_1, M_2, M_3, and M_4 draw the parallels to OA, to meet the straight line OB in N_1, N_2, N_3 and N_4. By dividing the moments of inertia measured by $O'A'$ by the largest of the 4 segments M_1N_1, M_2N_2, M_3N_3, M_4N_4 the section modulus Z_R is obtained. We can then compute the unit stresses and therefore the coefficient of safety.

(*e*) In landing, the fuselage is supported by the landing gear and by the tail skid. The system of acting forces, with coefficient 1, is then that shown in Fig. 203.

Fig. 204 shows the diagrams of the shearing stresses and bending moments corresponding to that case. Since, as it will be seen, the coefficient of resistance of the landing gear is usually taken between 5 and 6, it will suffice to multiply the preceding stresses by 6 and verify that the sections of the fuselage are sufficient. In our case these stresses result lower than the maximum considered in flight.

B. Analysis of Landing Gear.—Let us consider the following particular cases:

1. Normal landing with airplane in line of flight.
2. Landing with tail skid on the ground.
3. Landing on only one wheel; that is, with the machine laterally inclined by the maximum angle which can be allowed by the wings.
4. Landing with lateral wind.

Figs. 205, 206, 207 and 208 illustrate respectively the construction for those four cases, giving for each the tension on compression stresses, the diagrams of the bending moments, and the member subjected to bending (axle and spindle). In the fourth case it has been assumed that the maximum horizontal stress is not greater than 400 lb.

1
2
3
4
5
6
7
8
9
10
11
12
13
265 LBS.
284 "
324 "
451 "
162 "
20 "
18 "
17 "
16 "
14 "
23 "
21 "
15 14
1470 LBS.
14 3
145 LBS.
15
2
3
4
5
14
13
1470 LBS.
145 LBS.
0 30 60 in.
Scale of Lengths
0 200 400 Lbs.
Scale of Weights.

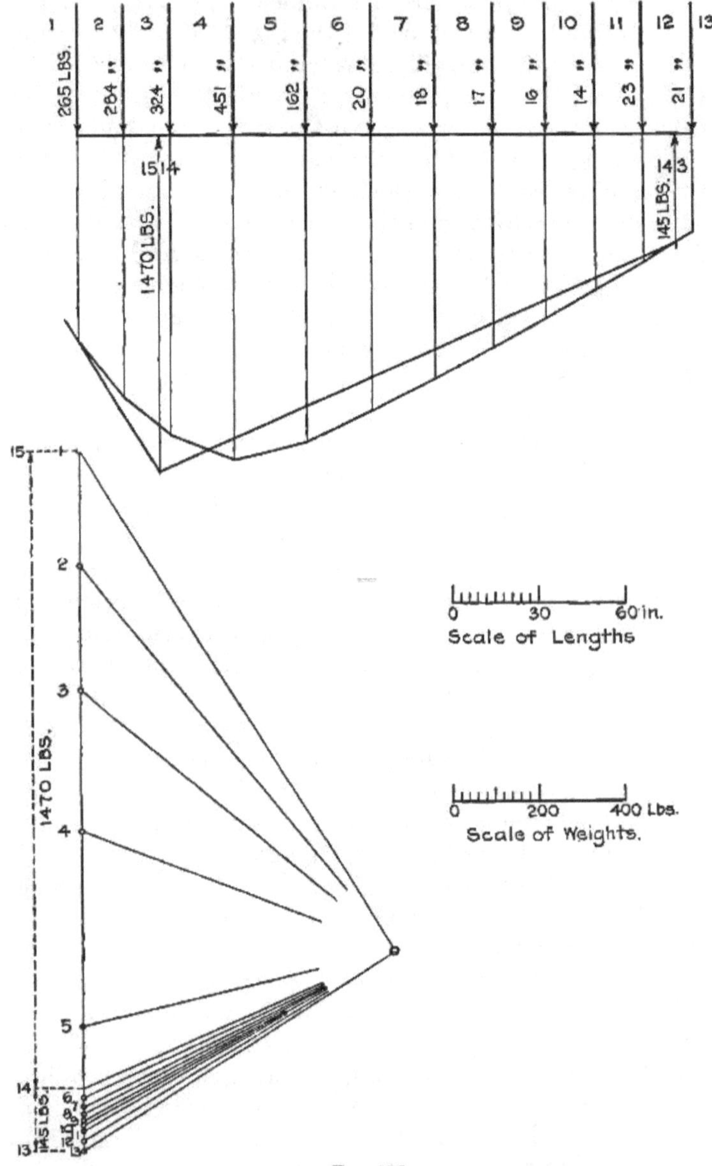

Fig. 203.

265 LBS.
284
324
451
162
20
18
17
16
14
23
21
1470 LBS.
145 LBS.
SPACE DIAGRAM
0 30 60 inches
Scale of Lengths
SHEAR DIAGRAM
250 Lbs
0 200 400 Lbs.
Scale of Shears
0 7500 15000 in. lbs
Scale of Moments
MOMENT DIAGRAM.

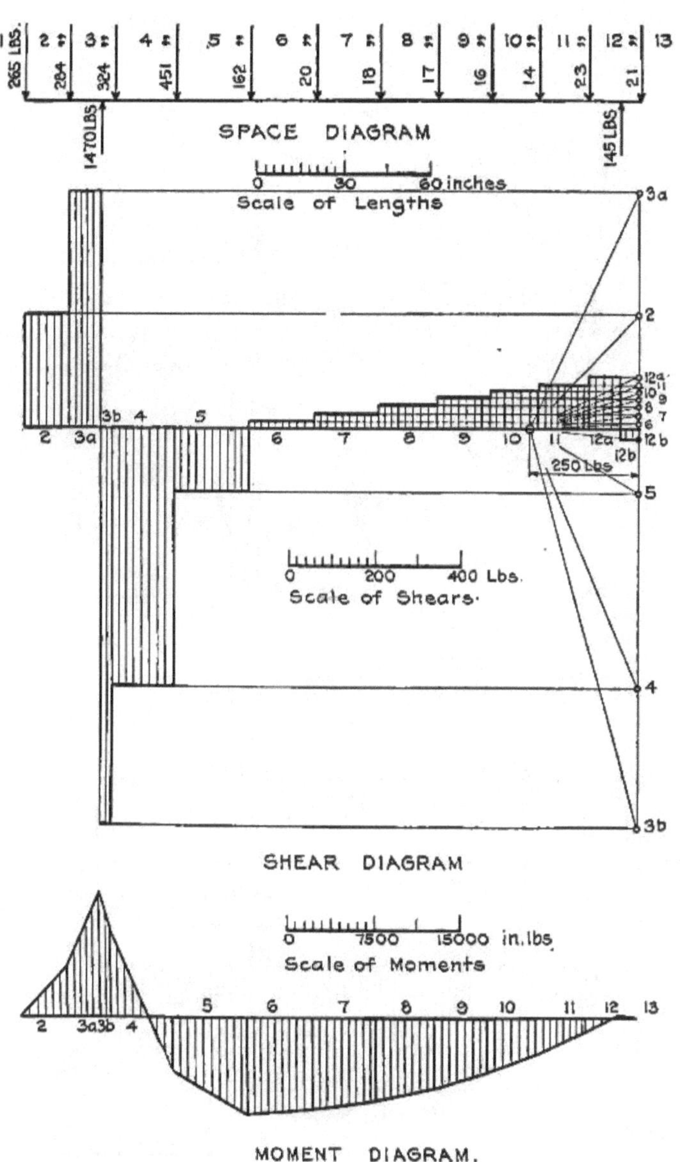

Fig. 204.

CASE. I
-1100 LBS.
-125 LBS.
1065 LBS.
SIDE ELEVATION
DIAGRAM
OF
LANDING GEAR
+500 LBS
C.L. of Fuselage
HALF FRONT ELEVATION
0 20 40 in
Scale of Lengths.
500 LBS.
(a)
1065 LBS.
1175 LBS
125 LBS.
(b)
1100 LBS
1175 LBS.
0 300 600 lbs.
Scale of Forces
FORCE POLYGONS
33.46 in.
33.46 in
5.9 in.
27.56 in.
27.56 in
5.9 in.
13.78 in.
13.78 in.
13.78 in.
13.78 in.
1520 LBS
455"
455"
1520"
1065 LBS.
1065 LBS
SPACE DIAGRAM
FORCES ACTING ON
SPINDLES
0 10 20 in.
Scale of Lengths.
455 LBS
455"
455 LBS
6275 in lbs.
6275 in.lbs.
455 lbs
AXLE MOMENT DIAGRAM
0 4000 8000 in/lbs.
Scale of Moments

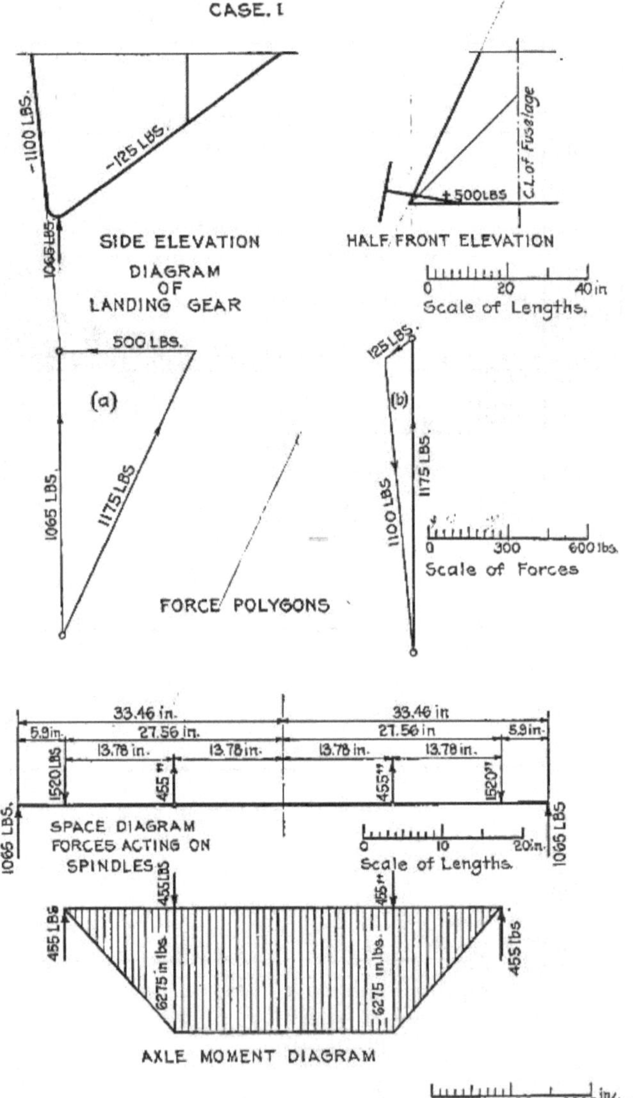

Fig. 205.

CASE 2.
-1200 LBS.
8°
-50 LBS.
8°
1065 LBS.
SIDE ELEVATION
DIAGRAM OF LANDING GEAR
C.L. of Fuselage
+ 500 LBS.
HALF
FRONT ELEVATION
0 20 40 in.
Scale of Lengths
500 LBS.
(a)
1065 LBS.
1175 LBS.
50 LBS.
(b)
1175 LBS.
1200 LBS.
FORCE POLYGONS
0 300 600
Scale of Forces
33.46 in.
33.46 in.
5.9 in.
27.56 in.
27.56 in.
5.9 in.
13.78 in.
13.78 in.
13.78 in.
13.78 in.
1520 LBS.
455 LBS.
455 LBS.
1520 LBS.
1065 LBS.
1065 LBS.
SPACE DIAGRAM FORCES ACTING
ON SPINDLES.
0 10 20 in.
Scale of Lengths
455 LBS.
455 LBS.
455 LBS
455 LBS
6275 IN. LBS.
6275 IN. LBS.
AXLE MOMENT DIAGRAM
0 4000 8000 in/lbs.
Scale of Moments

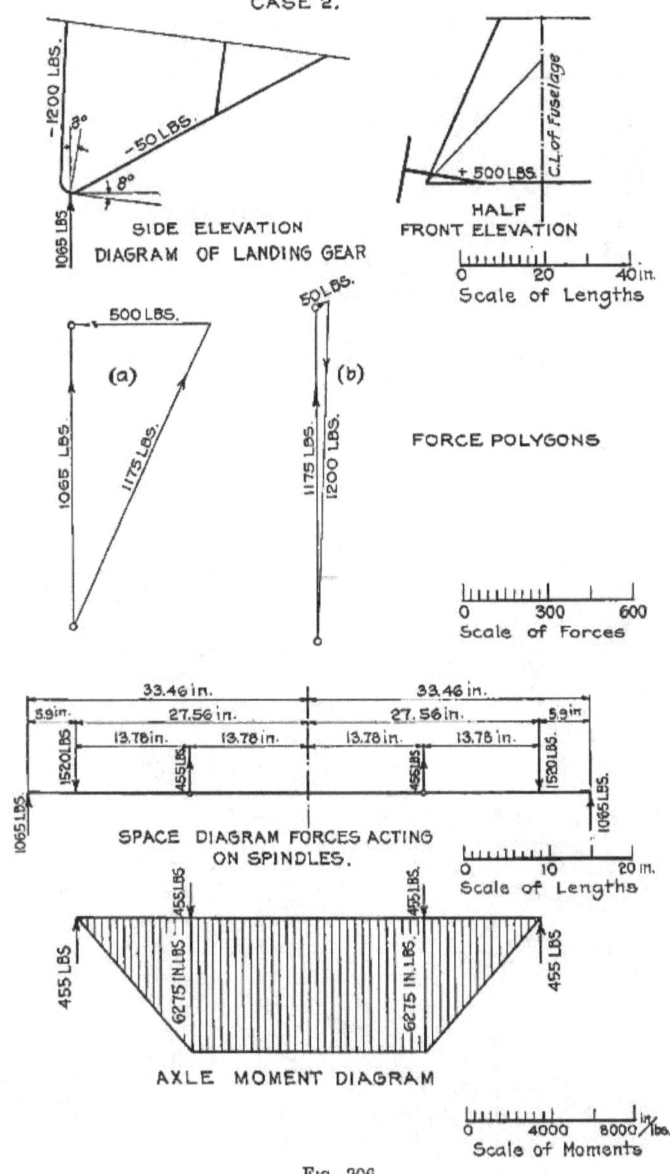

Fig. 206.

CASE 3

-2850 lbs.
-400 lbs.
2130 lbs.
SIDE ELEVATION
DIAGRAM OF LANDING GEAR
+1010 lbs.
C.L. of Fuselage
15°
HALF FRONT ELEVATION
0 20 40 In.
Scale of Lengths.
3060 lbs.
2130 lbs.
1010 lbs.
400 lbs.
2850 lbs
3060 lbs
0 600 1200 lbs.
Scale of Forces
FORCE POLYGONS
230 lbs
13.78 In.
13.78 In.
33.46 In.
13.78 In.
13.78 In
2810 lbs.
5.9 In.
15°
0 10 20 In.
Scale of Lengths
SPACE
DIAGRAM
FORCES ACTING ON SPINDLES
910 lbs
2130 lbs
5.7 In.
13.31 In.
13.31 In.
26.62 In
32.32 In.
910 lbs
230 lbs.
0 6000 12000 In.lbs.
Scale of Moments
9050 in/lbs
680 lbs.
AXLE MOMENT DIAGRAM

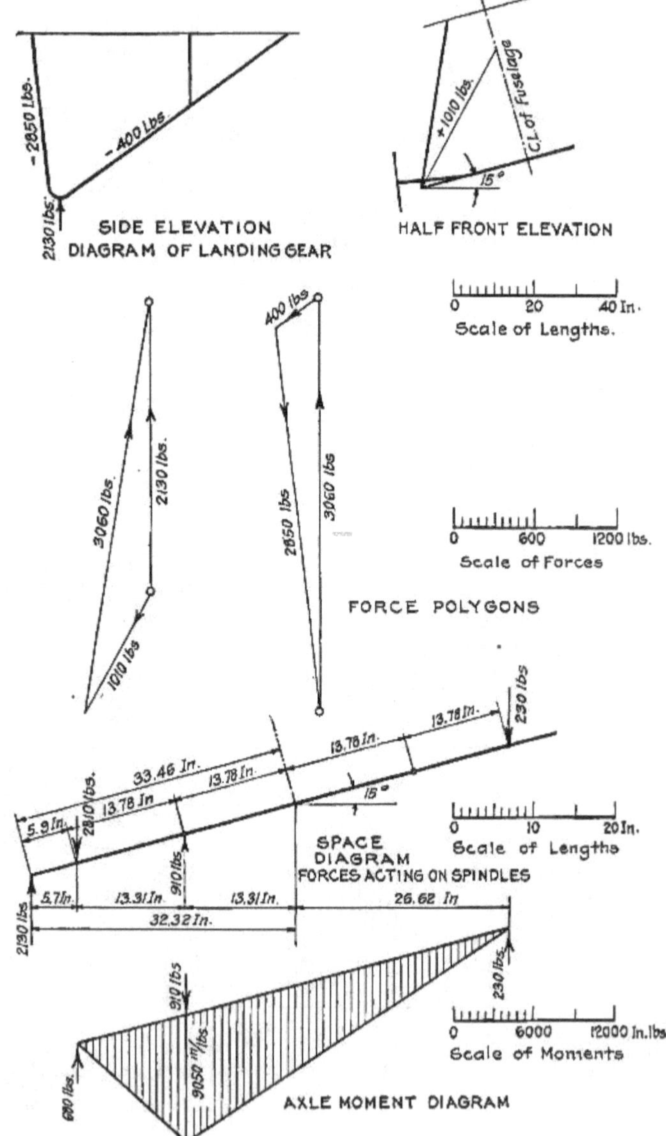

because with a great transversal load the wheel would break. In Fig. 209 the sections of the various members have been given, the results of the analysis having been

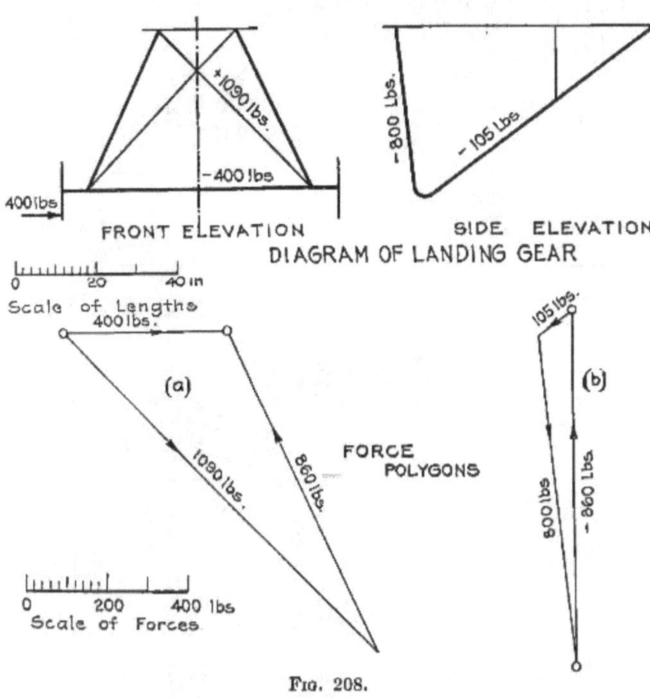

Fig. 208.

grouped in table 44. The table gives the following elements for each member:

P = compression or tension stress
M_f = Bending moment
I = Moment of inertia
Z = Section Modulus
A = Area of the section
F_c = unit load due to compression or tension
F_m = Unit load due to bending
F_t = Total unit load
Modulus of rupture
Coefficient of safety

TABLE 44

	Member	P, lb.	Mf, in lbs	I, in.⁴	Z, in.³	A, sq. in.	F_c, lb. per sq. in.	F_m, lb. per sq. in.	F_t, lb. per sq. in.	Mod. of rupture, lb. per sq. in.	Coef. of safety
Case 1	Spindle	0	6,275	0.326	0.326	0.736	0	19,250	19,250	110,000	5.7
	Axle	+ 500	6,275	0.660	0.610	0.680	740	10,300	11,040	60,000	5.4
	Front strut	−1100	.....	0.048	0.108	0.386	...	......	......	7,200	6.5
	Rear strut	− 125	.....	0.005	0.025	0.187	...	......	......	815	6.5
Case 2	Spinled	0	6,275	0.326	0.326	0.736	0	19,250	19,250	110,000	5.7
	Axle	+ 500	6,275	0.660	0.610	0.680	740	10,300	11,040	60,000	5.4
	Front strut	−1200	.....	0.048	0.108	0.386	...	......	......	7,200	6.0
	Rear strut	− 50	.....	0.005	0.025	0.187	...	......	......	815	16.8
Case 3	Spindle	0	9,050	0.326	0.326	0.736	0	27,700	27,700	110,000	4.0
	Axle	0	9,050	0.660	0.610	0.680	0	14,800	14,800	60,000	4.0
	Front strut	−2850	.....	0.048	0.108	0.386	...	......	......	7,200	2.5
	Rear strut	− 400	.....	0.005	0.025	0.187	...	......	......	815	2.0
	Wire	+1010		⅛ in. φ steel wire cable.			Breaking strength 2,000 lb.				2.0
Case 4	Spindle	− 400	0	0.326	0.326	0.736	540	0	540	110,000	20.0
	Axle	− 400	0	0.660	0.610	0.680	580	0	580	60,000	10.3
	Front strut	− 800	.....	0.048	0.108	0.386	...	......	......	7,200	9.0
	Rear strut	− 105	.....	0.005	0.025	0.187	...	......	......	815	7.8
	Wire	+1090		⅛ in. φ steel wire cable.			Breaking strength 2,000 lb.				1.8

As for the criterions to be followed in the selection and computation of the shock absorbers, reference is to be made to what has been said in Chapter XVI.

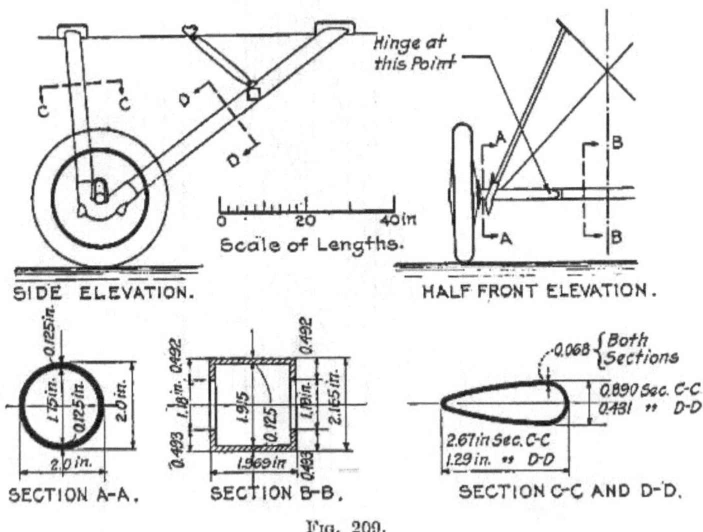

Fig. 209.

C. Analysis of the Propeller.—In the following chapter it will be seen that for the airplane of our example the adoption of a propeller having a diameter of 7.65 ft. and a pitch of 9 ft. is convenient. We shall then see the aerodynamic criterions which have suggested that choice. In this chapter we shall limit ourselves to static analysis of the propeller. This static analysis is usually undertaken as a checking; that is, by first drawing the propeller based upon data furnished by experience and afterward verifying the sections by a method which will be explained now.

Supposing a propeller is chosen having the profile shown in Figs. 210, 211, 212 and 213.

Fig. 210 gives the assembly of only one half the propeller blade the other half being perfectly symmetrical. Furthermore it gives six sections of the propeller which are reproduced on a larger scale in Figs. 211, 212 and 213. It

D = 7.65 ft. - p.v. = 9 ft. - 1 - 7.08 in.
P
1a Sez. = 8.825 ft.
2a " = 8.74 "
3a " = 8.64 "
4a " = 8.70 "
5a " = 8.85 "
6a " = 9.00 "
2.52"
7.23"
7.23"
7.23"
7.23"
7.23"
7.23"
r = 45.9 in.

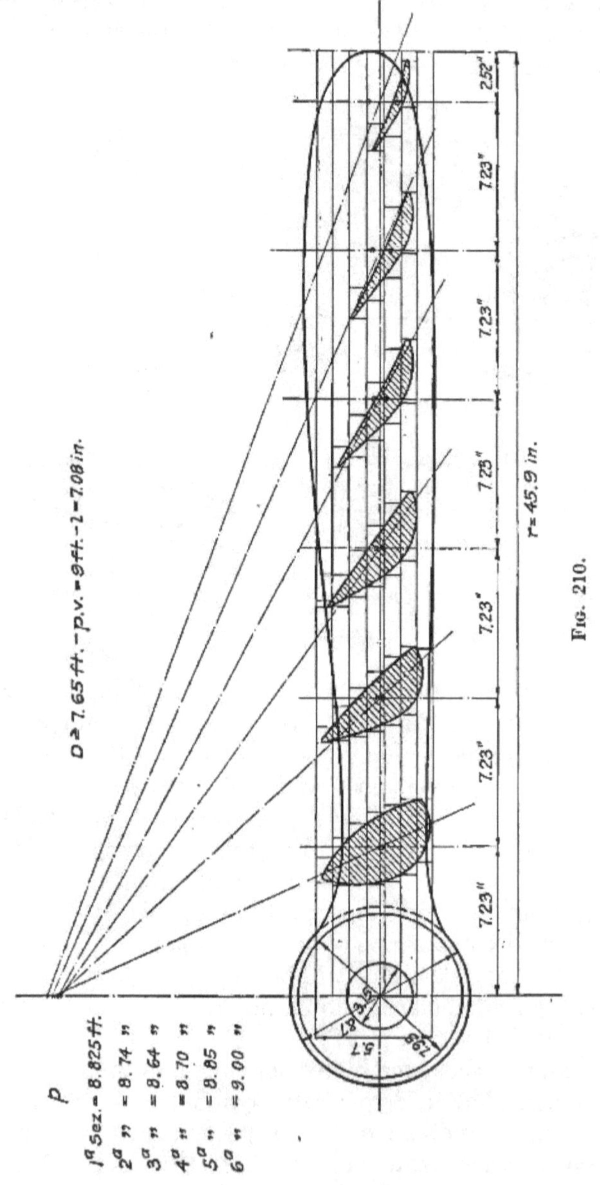

FIG. 210.

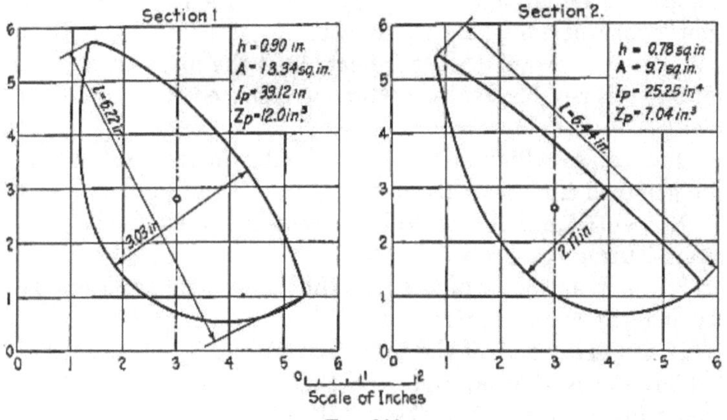

Fig. 211.

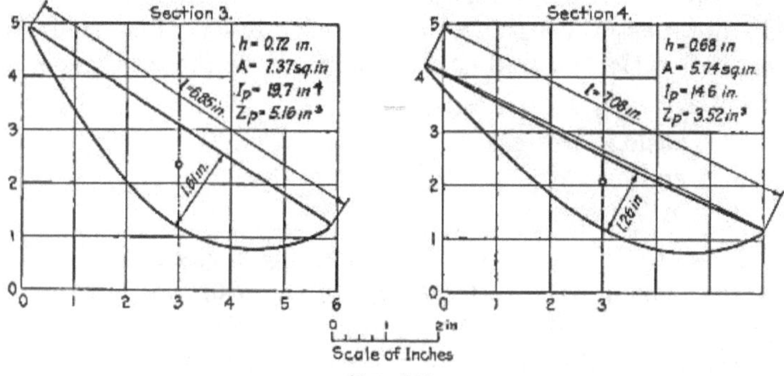

Fig. 212.

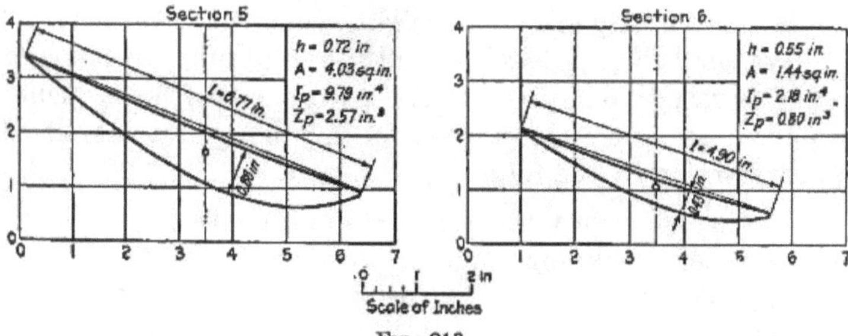

Fig. 213.

should be noted that in that type of propeller the pitch is not constant for the various sections, but increases from the center toward the periphery until the maximum value of 9 feet is reached which is the one assumed to characterize the propeller.

The forces which stress the propeller in its rotation can be grouped into two categories:

1. Centrifugal forces which stress the various elements constituting the propeller mass.
2. Air reactions which stress the various elements constituting the blade surface.

If any section A of the propeller is considered, the forces which stress that section are then the resultants of the centrifugal forces and the resultants of the air reactions pertaining to that portion of the propeller included between section A and the *periphery*. In general, these resultants do not pass through the center of gravity of section A, so their action on that section produces in the most general case:

1. Tension stresses.
2. Bending stresses.
3. Torsion stresses.

It is immediately seen that by giving a special curvature to the neutral axis or elastic axis of the propeller blade it is possible to equilibrate the bending moment in each section produced by the centrifugal force, with that produced by the air reaction.

The stresses will then be those of tension and torsion, resulting thereby in a greater lightness for the propeller.

We shall then proceed to find the total unit stresses and the curvature to be given to the neutral axis of the propeller blade. In order to proceed in the computations, it is necessary to fix the following elements:

N = number of revolutions of the propeller,

ω = corresponding angular velocity,

P_p = power absorbed by the propeller when turning at N revolutions,

Δ = density of the material out of which the propeller is to be made.

In our case, $N = 1800$, and therefore

$$\omega = \frac{2\pi N}{60} = 188 \ 1/\text{sec.}$$

Furthermore $P_p = 300$ H.P.

As for the material, the propellers can be made of walnut, mahogany, cherry, etc. Suppose that we choose walnut, for which $\Delta = 0.0252$ lb. per cu. in. Let us now find the expression for the centrifugal force $d\Phi$ which stresses an element of mass dM, and for the reaction of the air dR which stresses an element $l \cdot dS$ of the blade surface. The elementary centrifugal force $d\Phi$ has, as is known, the expression

$$d\Phi = dM \times \omega^2 \times r$$

since we can place

$$dM = \frac{\Delta}{g} \times A \times dr$$

where

g is the acceleration due to gravity = 386 in./sec.^{2},
A is any section whatever of the propeller, and
dr is an infinitesimal increment of the radius.

We shall then have

$$d\Phi = \frac{\Delta}{g} \times \omega^2 \times A \times r \times dr$$
$$= 2.3 \times A \times r \times dr$$

from which

$$\frac{d\Phi}{dr} = 2.3 \times A \times r \qquad (1)$$

Then by determining the areas of the various sections A, we shall be able to draw the diagram $A = f(r)$ of Fig. 214, which by means of formula (1) permits drawing the other one

$$\frac{d\Phi}{dr} = f(r)$$

whose integration gives the total centrifugal forces Φ which stress the various sections (Fig. 215).

The elementary air reaction dR has the following expression

$$dR = K \times dS \times U^2$$

where K is a coefficient which depends upon the profile of the blade element and upon the angle of incidence, dS is a

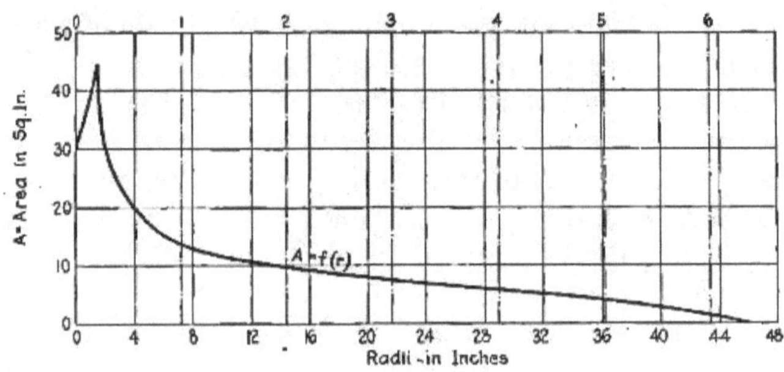

Fig. 214.

surface element of the blade, and U is the relative velocity of such a blade element with respect to the air.

Calling l the variable width of the propeller blade, we may make

$$dS = l \times dr$$

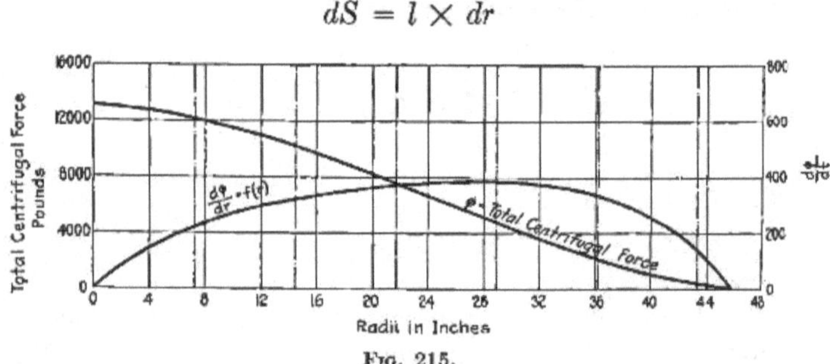

Fig. 215.

on the other hand, velocity U is the resultant of the velocity of rotation r and of velocity of translation V, of the airplane. The direction of these velocities being at right angles to each other we shall have

$$U^2 = \omega^2 \times r^2 + V^2$$

therefore

$$dR = K \times (\omega^2 \times r^2 + V^2) \times l \times dr$$

from which

$$\frac{dR}{dr} = K \times (\omega^2 \times r^2 + V^2) \times l$$

It is immediately seen that it would be very difficult to take into consideration the variation of coefficient K from one section to the other, and therefore with sufficient

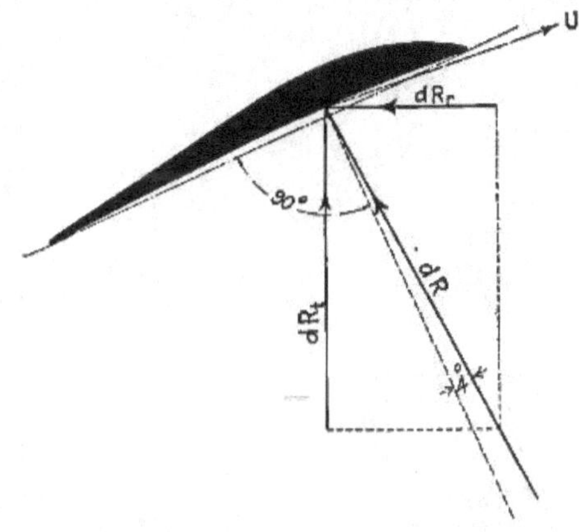

Fig. 216.

practical approximation K may be kept constant for the various sections and equal to an average value which will be determined.

We note that dR being inclined backward by about 4° with respect to the normal to the blade cord, changes direction from section to section; it will consequently be convenient to consider the two components of dR, component dR_t perpendicular to the plane of propeller rotation and component dR_r contained in that plane of rotation (Fig. 216).

The expression $\frac{dR}{dr}$ can also be put in the following form:

$$\frac{dR}{dr} = K \times \omega^2 \times (r^2 + \frac{V^2}{\omega^2}) \times l$$

Scale of Segment C-D
Scale of Segment A-B
Scale of Radii
Scale of Forces
Scale of Moments
Scale of Pressures
RADII

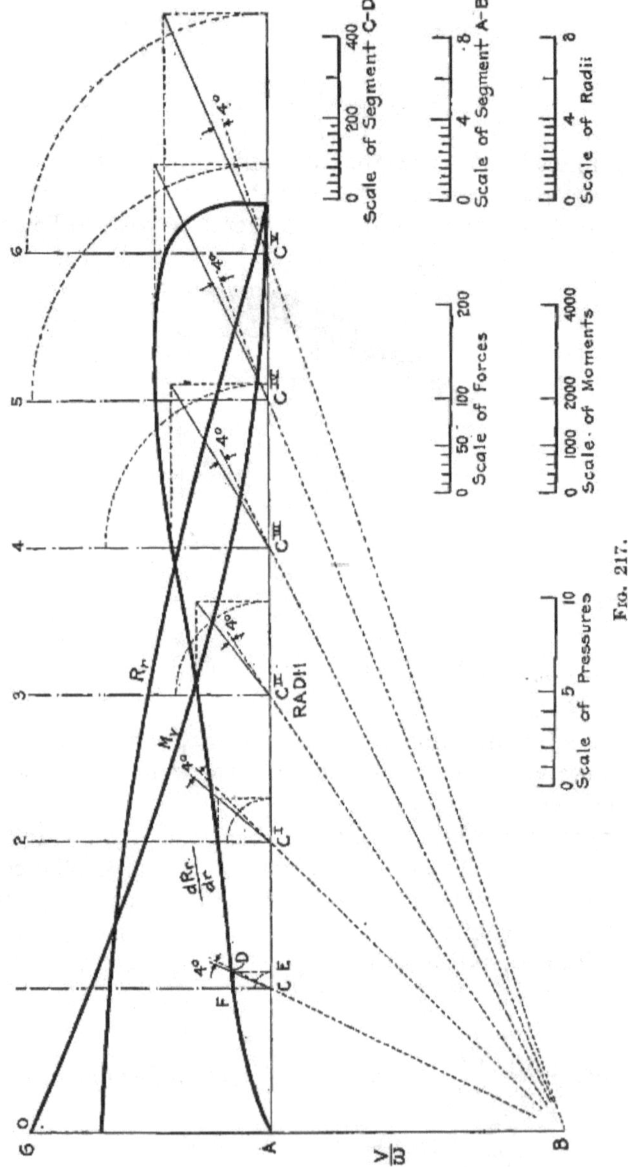

FIG. 217.

In our case $\omega = 188$ and $V = 156$ m.p.h. $= 2800$ in. per sec.

On an axis AX lay off the various radii (Fig. 217), make $AB = \frac{V}{\omega} = \frac{2800}{188} = 1.49$ perpendicular to AX, and from B draw segment BC. We shall evidently have

$$\overline{BC}^2 = \overline{AB}^2 + \overline{AC}^2$$

that is,

$$\overline{BC}^2 = \frac{V^2}{\omega^2} + r^2$$

Analogously by drawing BC', BC'', etc. the squares of these segments will give the terms $\frac{V^2}{\omega^2} + r^2$. In this manner $\frac{dR}{dt}$ may be calculated, except for the constant K.

Make $\frac{dR}{dt}$ equal to CD, so that CD makes an angle of 4° with the prolongation of BC. Projecting D in E and F, we shall have

$$DE = \frac{dR_r}{dr} \text{ and } DF = \frac{dR_t}{dr}$$

We may then draw the two diagrams

$$\frac{dR_r}{dr} = f(r) \text{ and } \frac{dR_t}{dr} = f(r)$$

whose integration gives the value of components R_r and R_t corresponding to the various sections; that is, gives the shearing stresses. For clarity, these diagrams have been plotted in two separate figures for components R_r and R_t, the former having been plotted in Fig. 217 and the latter in Fig. 218.

The shearing stresses R_r and R_t being known, by means of a new integration, the diagrams of the bending moments M_r and M_t can easily be obtained. It should be noted that the maximum value of M_r equals one-half of the motive couple. The power being 300 H.P. and the angular velocity $\omega = 188$, the motive couple will equal

$$\frac{300 \times 500}{188} = 800 \text{ lb.} \times \text{ft.} = 9600 \text{ lb.} \times \text{inch}$$

Scale of Segment C-D
0 200 400
Scale of Segment A-B
0 4 8
Scale of Radii
0 4 8 in.
Scale of Forces.
0 200 400
Scale of Moments
0 4000 8000
Scale of Pressures.
0 5 10
RADII
R_t
M_t
$\frac{dR_t}{dr}$
$\frac{V}{\omega}$
A
B
C
D
E
F
4°
0 1 2 3 4 5 6

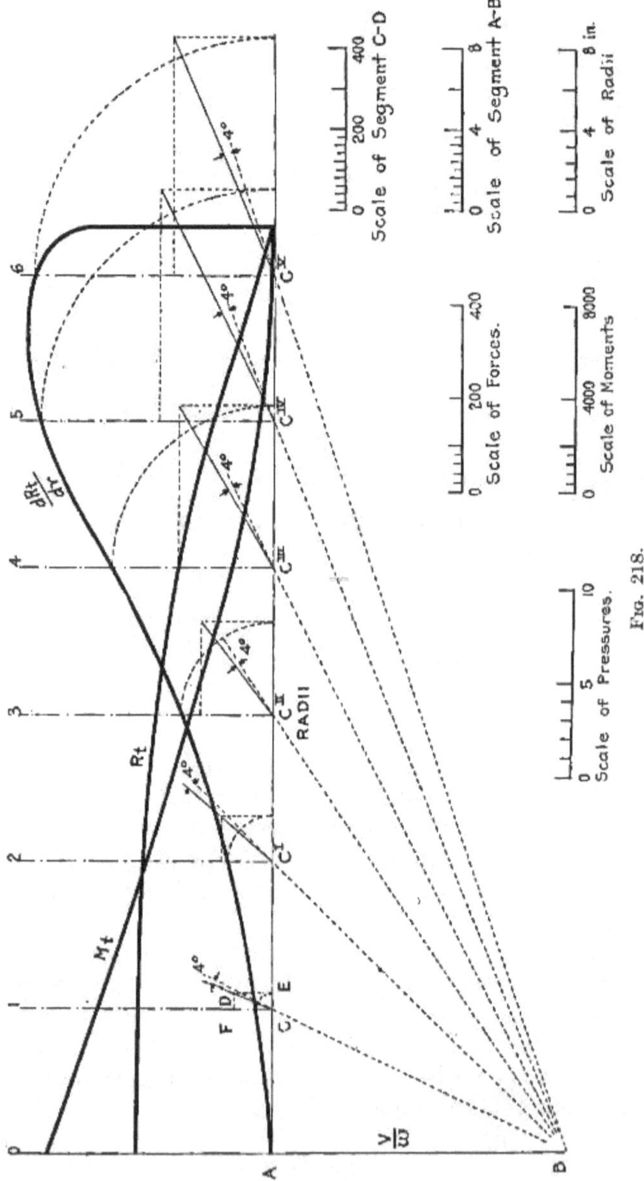

FIG. 218.

therefore

$$M_r = \frac{1}{2} \times 9600 \text{ lb.} \times \text{inch} = 4800 \text{ lb.} \times \text{inch}.$$

The scale of moments is fixed in this manner and consequently that of the shearing stresses; and thus the value of the coefficient K is also determined. Then, for each section, the resultant stress due to the centrifugal force, the shearing stresses R_r and R_t, and the moments M_r and M_t due to the air reaction, are known.

If the moment produced in any section whatever by the centrifugal force is somehow made to be in equilibrium with the moments M_r and M_t, the deflection stresses will be avoided.

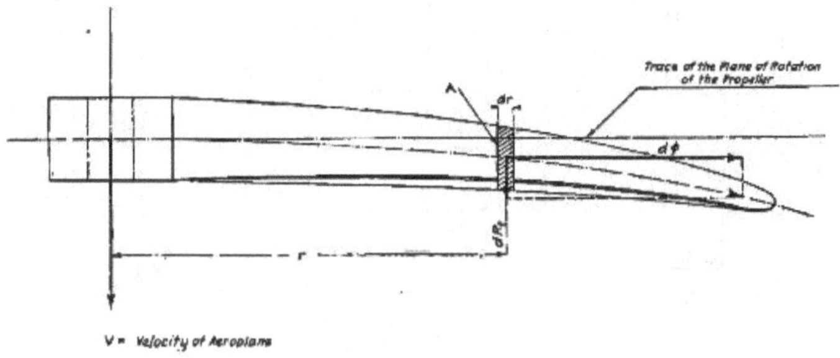

Fig. 219.

Let us first of all consider the moments M_t which are the greatest and consequently the most important, especially because they stress the blade in a direction in which the moment of inertia is smaller than that corresponding to the direction in which the blade is stressed by the bending moments M_r.

Let us call $\frac{dy}{dx}$ the inclination of any point whatever of the neutral axis curve of the propeller. We shall then consider any section A whatever of the propeller blade, and the elementary forces $d\Phi$ and dR applied to it. The elementary force $d\Phi$ follows a radial direction, while the elementary force dR_t follows a direction perpendicular to the plane of rotation of the propeller (Fig. 219); while

$d\Phi$ is applied to the center of gravity of the element $A \times dr$, the air reaction dR_t is not applied to the center of gravity, but falls at about 33 per cent. of the chord. However, from known principles of mechanics, this force can be replaced by an elementary force dR_t applied to the center of gravity, and by an elementary torsion couple dT_t. The effect of this couple will again be referred to, and for the moment we shall suppose dR_t applied to the center of gravity. Let us assume then the condition

$$\frac{d\Phi}{dR_t} = \frac{dy}{dr}$$

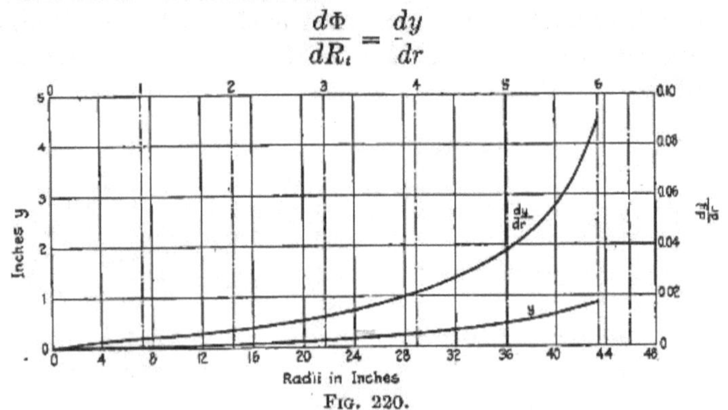

FIG. 220.

that is, that the resultant of $d\Phi$ and dR_t be tangent to the neutral curve of the propeller blade. Under these conditions, supposing that this be true for every element $A \times dr$ of the propeller, all the various sections will be stressed only to tension.

Since we may write

$$\frac{d\Phi}{dR_t} = \frac{d\Phi/dr}{dR_t/dr}$$

it is easy to draw the diagram

$$\frac{dy}{dr} = f\ (r)$$

and, by graphically integrating this diagram, obtain

$$y = f(r)$$

which gives the shape that the center of gravity axis of the propeller blade must have in elevation (Fig. 220).

With an analogous process, the shape in plan is found by considering the forces $d\Phi$ and dR_r; in Fig. 221 the relative diagrams have been drawn for $\frac{dy}{dr} = f(r)$ and $y = f(r)$. Thus the propeller may be designed. In Fig. 210 the neutral axis has been drawn following this criterion.

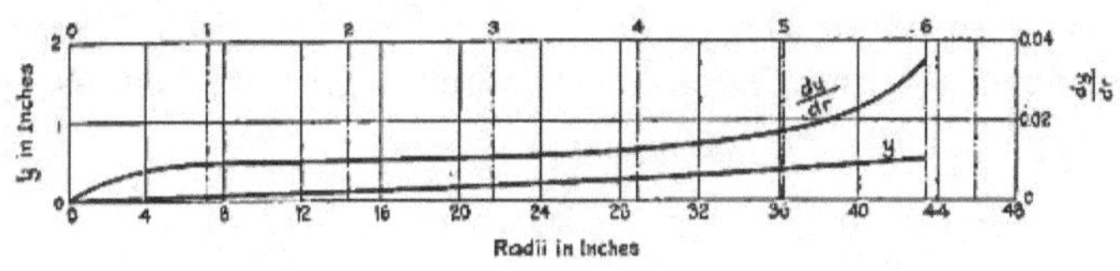

FIG. 221.

Let us now determine the unit stresses corresponding to the case of normal flight.

These stresses are of two types:

1. tension stresses,
2. torsion stresses.

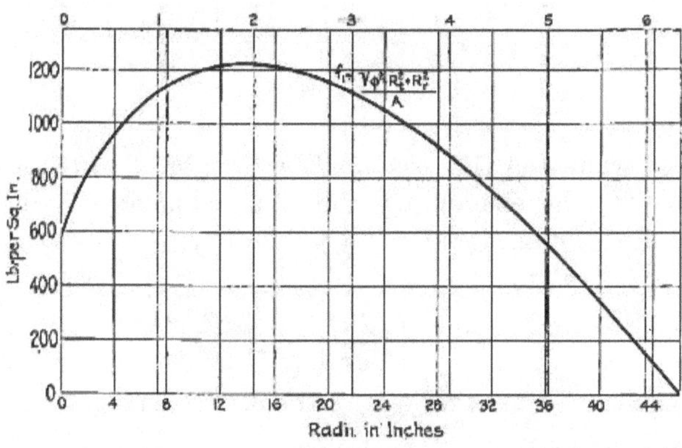

FIG. 222.

Tension stresses are easily calculated, in fact, for every section A they are equal to

$$f_1 = \frac{(\Phi^2 + R_t^2 + R_r^2)^{1/2}}{A}$$

In Fig. 222 the diagram of f_1 obtained by the preceding equation has been drawn.

As to the torsion stresses, they depend only upon the air reaction. Let us consider a section A and the air reaction dR which acts upon the blade element $l \cdot dr$ corresponding to this section. Evidently

$$dR = (dR_t^2 + dR_r^2)^{1/2}$$

The point of application of dR falls, as we have seen, at 0.33 of the width of the blade l; therefore dR will in general produce a torsion about the center of gravity; let us call

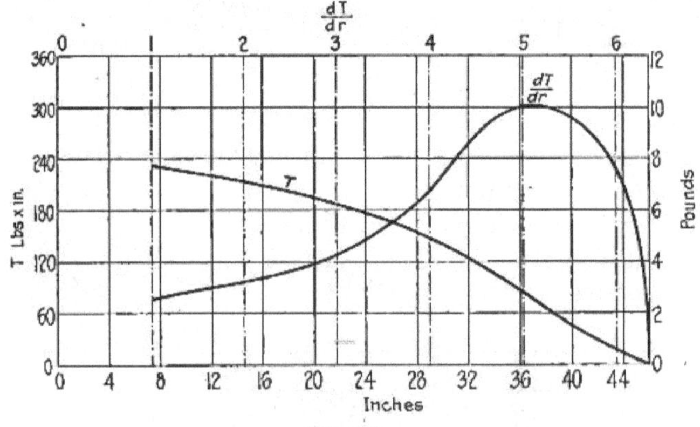

FIG. 223.

h the lever arm of the axis of dR with respect to the center of gravity; the elementary torsional moment will be

$$dT = h \times dR = h \times (dR_t^2 + dR_r^2)^{1/2}$$

and consequently

$$\frac{dT}{dr} = h \times \left[\left(\frac{dR_t}{dr}\right)^2 + \left(\frac{dR_r}{dr}\right)^2\right]^{1/2}$$

The values of h are marked on the sections (Figs. 211, 212 and 213); the values $\frac{dR_t}{dr}$ and $\frac{dR_r}{dr}$ are given by the diagrams of Figs. 217, 218; thus in Fig. 223 the diagram may be drawn of

$$\frac{dT}{dr} = f(r)$$

and by integrating, that of

$$T = f(r)$$

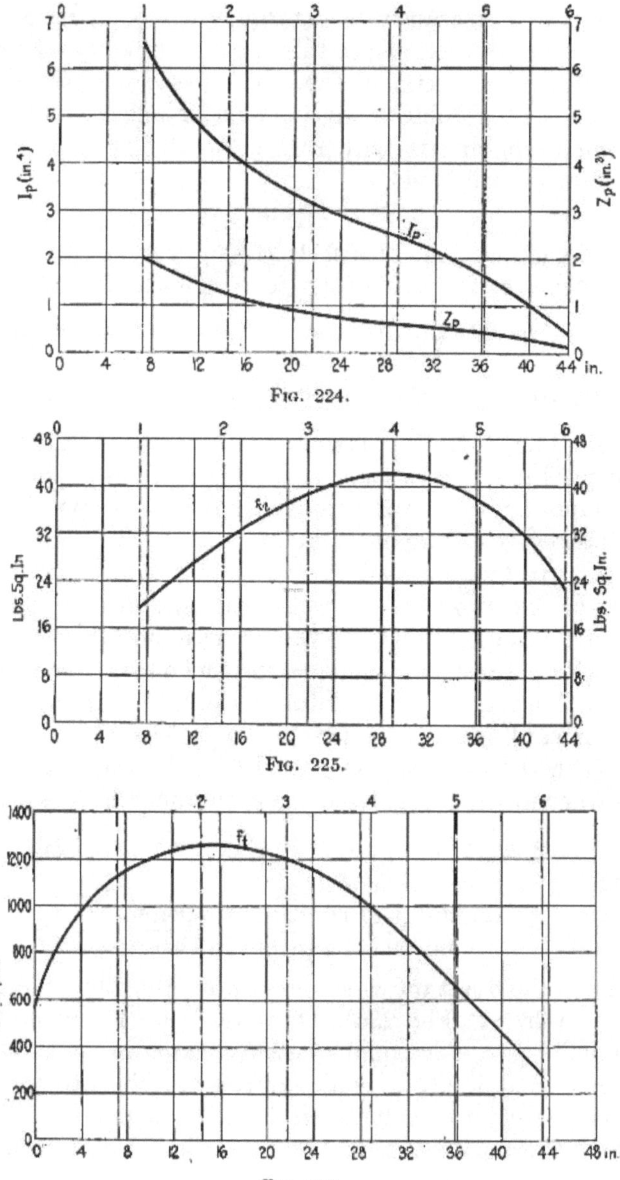

Fig. 224.

Fig. 225.

Fig. 226.

It is now necessary to determine the polar moments I_p of the various sections; to this effect it suffices to determine the ellipse of inertia of the various sections by the usual methods of graphic analysis; then calling I_x and I_y the moments of inertia with respect to the principal axis of inertia, we will have

$$I_p = (I_x^2 + I_y^2)^{1/2}$$

For each section (Figs. 211, 212 and 213), we have shown the values of the area, of the polar moment I_p and of $Z_p = \frac{I_p}{x}$. In Fig. 224 the diagram I_p for the various sections and the diagram $\frac{I_p}{x} = Z_p$ have been drawn.

Dividing, for each section, the corresponding values of the total moment of torsion T by the values of the section modulus for torsion Z, we shall have the values f_2 of the unit stresses to torsion (Fig. 225). It is immediately evident that this method is exact only when the neutral axis of the propeller is rectilinear and in the direction of the radius, which, however, does not correspond to practice. In effect though, as the torsion stresses represent a small fraction of the total stresses, the approximation which can be reached is practically sufficient.

When the unit stresses f_1 and f_2 to tension and torsion are known, the total stress f_t is determined by the formula

$$f_t = 0.35 \times f_1 + 0.65 \times (f_1^2 + 4 \times \alpha^2 \times f_2^2)^{1/2}$$

where

$$\alpha = \frac{\text{modulus of rupture in tension}}{1.3 \text{ modulus of rupture in shearing}} = \sim 7$$

Then the diagram which gives f_t for the various sections may be drawn (Fig. 226). It is seen that the value of the maximum stress is equal to 1280 pounds per square inch; that is, to about 1/6 the value of the modulus of rupture.

As a safety factor between 4 and 5 is practically sufficient for propellers, it may be concluded that the aforesaid sections are sufficient.

CHAPTER XX

DETERMINATION OF THE FLYING CHARACTERISTICS

Once the airplane is calculated and designed, it becomes possible to determine its flying characteristics. The best method for this determination would undoubtedly be that of building a scale model of the designed airplane and of testing it in an aerodynamic laboratory. This, however, is often impossible, and it is therefore necessary to resort to numeric computation.

Let us remember that the aerodynamical equations binding the variable parameters of an airplane are

$$W = 10^{-4}\,\lambda A V^2 \text{ and}$$
$$550P_1 = 147 \times 10^{-9}(\delta A + \sigma)V^3$$

where

W = weight in pounds,
A = surface in square feet,
V = speed in miles per hour,
P_1 = theoretical power in horsepower necessary for flight,
σ = coefficient of total head resistance, and
λ and δ = coefficient of sustentation and of resistance of the wing surface.

Let us assume, as in Chapter VIII, that

$$\Lambda = 10^{-4}\,\lambda A$$
$$\Delta = 10^{-4}(\delta A + \sigma)$$

The preceding equations can then be written

$$\frac{W}{V^2} = \Lambda$$
$$\frac{550P_1}{V^3} = 147\ \Delta$$

Since A and σ are constant and λ and δ are functions of the angle of incidence i, Λ and Δ will also be functions of i.

W
5000 lb.
4000
3000
2000
1000
900
800
Δ
0.50
0.40
0.30
0.20
0.10
0.09
0.08
0.01
0.02
0.03
0.04
0.05
0.06
0.07
0.08
0.09
0.10
1.47 Δ
0.20
Hp
20
40
80
100
150
200
300
400
V-100
P 66
A'
A''
B'
B''
V
9°
7°
5°
3°
2°
1°
70
80
90
110
120
130
140
150
160

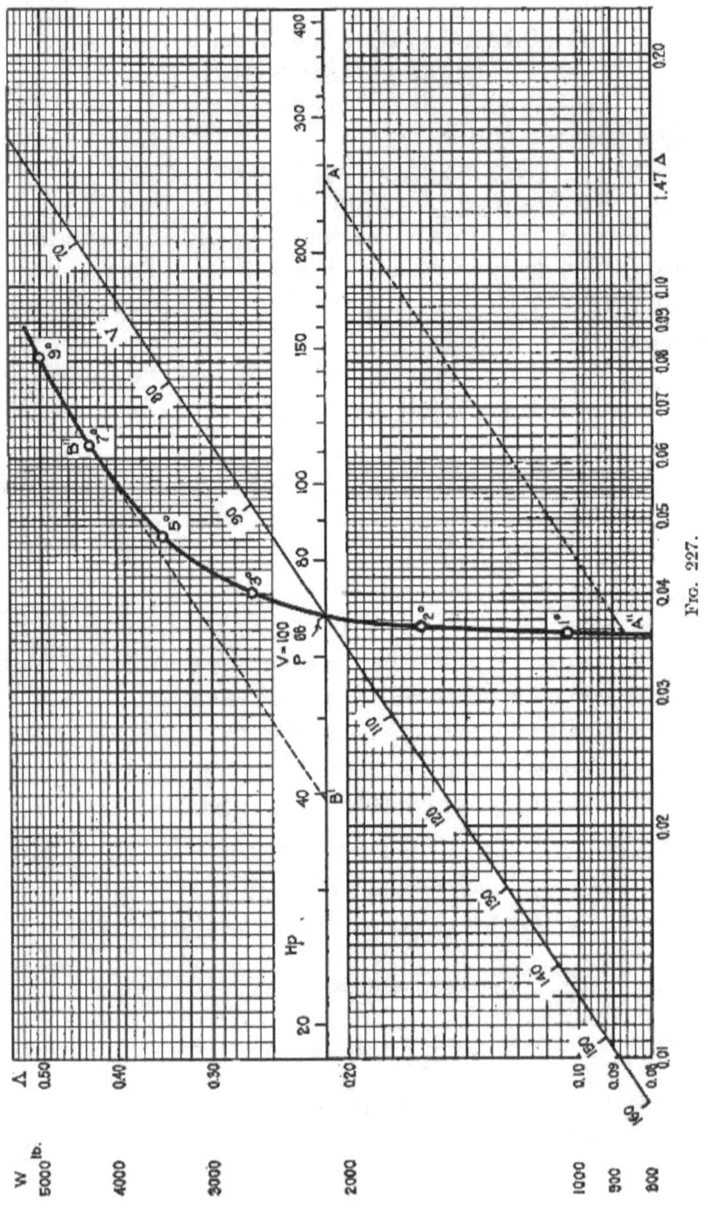

Fig. 227.

Then, λ, δ and σ being known, it is possible to obtain a pair of values of Λ and Δ corresponding to each value of i, and the logarithmic diagram of Λ as function of Δ can then be drawn.

Let us suppose that λ and δ are given by the diagram of Fig. 155 (Chapter XVII). The value of σ is calculated by remembering that

$$\sigma = \Sigma K \times A$$

that is, it is equal to the sum of the head resistances of the various parts of which the airplane is composed. This, however, does not always hold true, because of the fact that the head resistance offered by two or more bodies close to each other and moving in the air is not always equal to the sum of the head resistances the bodies encounter when moving each one separately, but it can be either greater or smaller. Thus, an exact value of the coefficient σ can be obtained only by testing a model of the airplane in a wind tunnel. However, if such experimental determination cannot be available, the value σ can be determined approximately by calculation as has been mentioned above. Table 45 shows the values of K, A and $K \times A$ for the various parts constituting the airplane in our example. This table gives $\sigma = 132.5$. It is then easy to compile Table 46 which gives the couples of values corresponding to Λ and Δ and consequently enables us to draw the logarithmic diagram of Λ as function of Δ (Fig. 227).

TABLE 45

Parts	A	K	KA
Fuselage	7.68	8.0	61.5
Cables	2.22	8.0	17.6
Struts	5.17	3.5	18.1
Landing gear	1.98	3.5	6.9
Wheels	1.68	14.0	23.5
Control surfaces	1.45	3.5	4.9

$\sigma = \Sigma KA = 132.5$

The scales of W, P_1 and V of this diagram are easily found with processes analogous to those used in Chapters VIII and IX.

The diagram then enables us to immediately find the pair of values V and P_1 corresponding to sea level; this makes possible the immediate determination of the maximum speed which can be reached. Thus it is necessary to know the power of the engine (which in our case is 300 H.P.) and the propeller efficiency; supposing, as it should always be, that the number of revolutions of the propeller may be selected, we can reach an efficiency of $\rho = 0.815$; then the maximum useful power is $0.815 \times 300 = 244$ H.P.; making $P_1 = 244$ we have $A'A''$ the segment which represents $V_{max.}$; laying this segment off on the scale of speeds we have $V_{max.} = 153$ m.p.h.

It is also seen that the minimum speed at which the airplane can sustain itself is given by the segment $B'B''$ which, read on the scale of speeds, gives $V_{min.} = 72$ m.p.h.; that is, it is lower than the value 75 m.p.h. imposed as a condition. Then our airplane can fly at speeds between 72 and 153 m.p.h. If we wish to study its climbing speed it is necessary to draw the diagram which gives ρP_2 as function of the various speeds. Thus it is necessary to know the characteristics of the engine and propeller.

TABLE 46

i	−1°	0°	1°	2°	3°	4°	5°	6°	7°	8°	9°
λ	2.3	4.0	6.0	8.0	10.0	11.4	13.0	14.5	16.0	17.8	19.2
δ	0.41	0.42	0.43	0.47	0.53	0.60	0.73	0.88	1.08	1.32	1.57
Λ	0.061	0.106	0.159	0.212	0.265	0.302	0.344	0.384	0.423	0.472	0.509
Δ	241×10^{-4}	244×10^{-4}	246×10^{-4}	257×10^{-4}	274×10^{-4}	291×10^{-4}	326×10^{-4}	365×10^{-4}	428×10^{-4}	483×10^{-4}	550×10^{-4}

$A = 265$ sq. ft. $\quad \sigma = 132.5$

$\Lambda = 10^{-4}\lambda A \quad \Delta = 10^{-4}(\delta A + \sigma)$

Let us suppose that the characteristics of the engine be the same as those given in Fig. 228. We see that the maximum power of 300 H.P. is developed at 1800 revolutions per

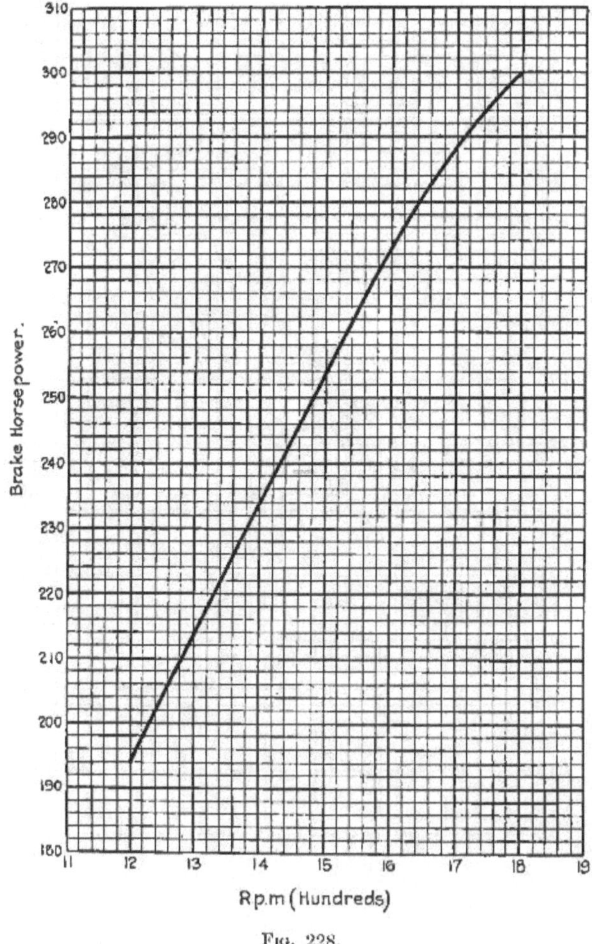

FIG. 228.

minute; on the other hand, if we wish to reach the maximum efficiency of $\rho = 0.815$, it is necessary to satisfy a certain ratio between the translatory velocity of the airplane and the peripheric velocity of the propeller. In Fig. 71 (Chapter VI), which is repeated in Fig. 229 are shown the values of the maximum obtainable efficiencies with

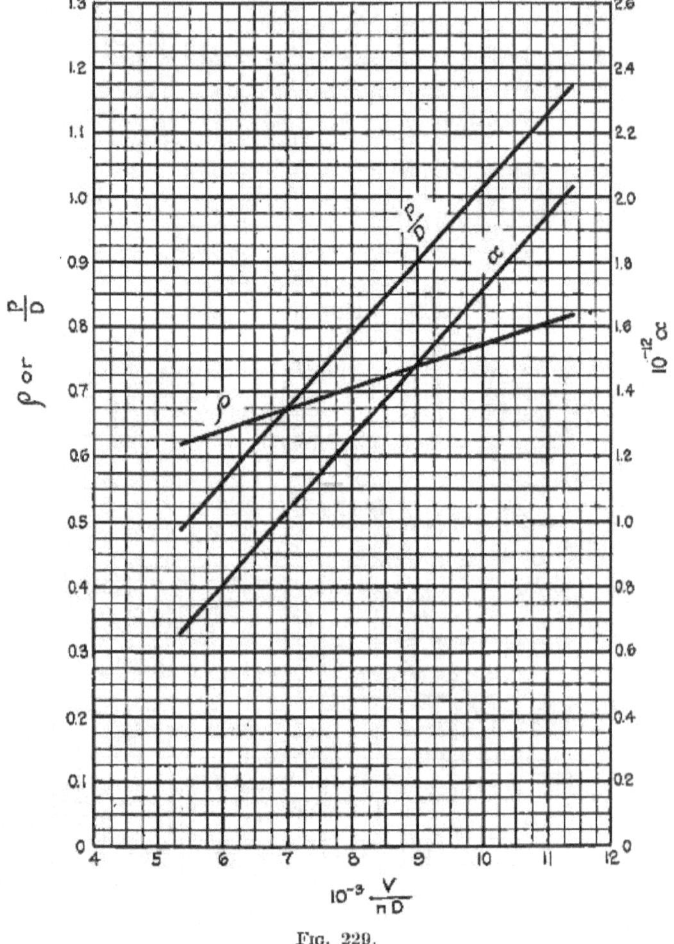

Fig. 229.

propellers of the best known type to-day, with the indication of the values $\frac{V}{nD}$, $\alpha = \frac{P}{n^3D^5}$ and $\frac{p}{D}$ corresponding to the value of maximum efficiency, adopting as units, however, m.p.h. for V, r.p.m. for n, feet for p and D, and H.P. for P.

Since we want $\rho = 0.815$, and consequently we have seen that $V_{max.} = 153$ m.p.h., the diagrams of Fig. 229 allow us to obtain the number of revolutions and the diameter of the propeller. In fact for $\rho = 0.815$ we find

$$\frac{V}{nD} = 11.4 \times 10^{-3}$$

$$\frac{p}{D} = 1.18$$

$$\frac{P}{n^3D^5} = 2.02 \times 10^{-12}$$

Knowing that $V = 153$ and $P = 300$ H.P. we have as unknowns n, D and p, whose values are defined by the preceding equations. Solving these equations we obtain:

$n = 1690$ revolutions per minute,
$D = 7.92$ feet, and
$p = 9.35$ feet.

Since the number of revolutions found is very near to the average R.p.m. of the engine, it will be convenient in our case to connect the propeller directly with the crank-shaft.

Having obtained the propeller, it is necessary to know the characteristic curve of the propeller family to which it belongs. It should be remembered that all propellers having the same blade profile and the same ratio between pitch and diameter, have the same characteristics (see Chapter IX).

Let the characteristics of a family to which our propeller belongs be those given in the logarithmic diagram of Fig. 230. Then with the same criterions which have been explained in Chapters, IX, XIII, and XIV, it is possible to draw the diagram of ρP_2 as a function of V for any altitude; for instance, the altitudes 0, 16,000, 24,000, and 28,000 ft. For this purpose the diagrams have been drawn in Fig. 230, which give the values $\frac{P_p}{n^3D^5}$ corresponding to these altitudes and in Fig. 231 the diagrams of P_2 of the same heights.

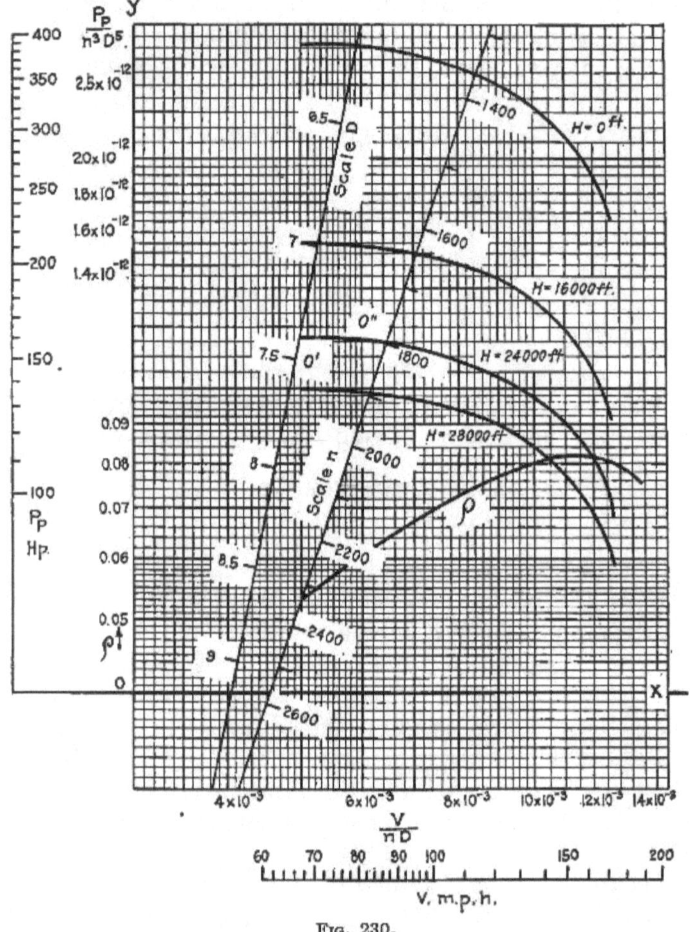

FIG. 230.

By using these diagrams those of Fig. 232 have been drawn from which it is seen that the maximum velocity at sea level is only 150 m.p.h. with a corresponding useful power of 225 H.P. This depends upon the fact that a propeller has been directly connected which should have been used with a reduction gear having a ratio of $\frac{1690}{1800}$. We will

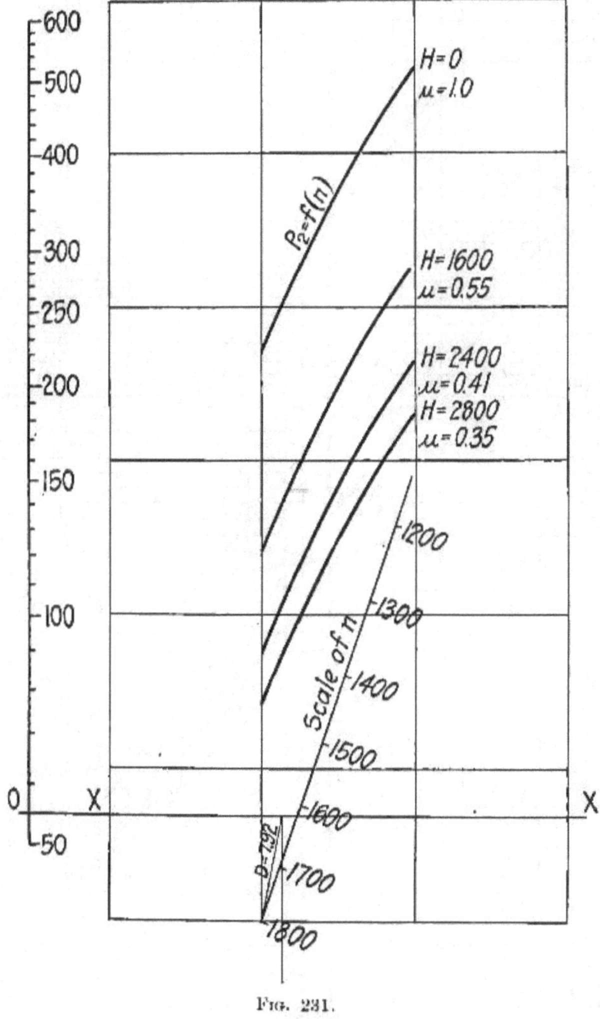

Fig. 231.

immediately see that if we wish to adopt a direct connection it is more convenient to choose a propeller which, although

W lb.
Λ
5000
4000
3000
2000
1000
800
0.50
0.40
0.30
0.20
0.08
0.01
0.02
0.03
0.04
0.05
0.06
0.07
0.08
0.09
0.10
0.20
1.47 Λ
20
40
100
200
300
400
Hp
Scale of μ
μ = .55

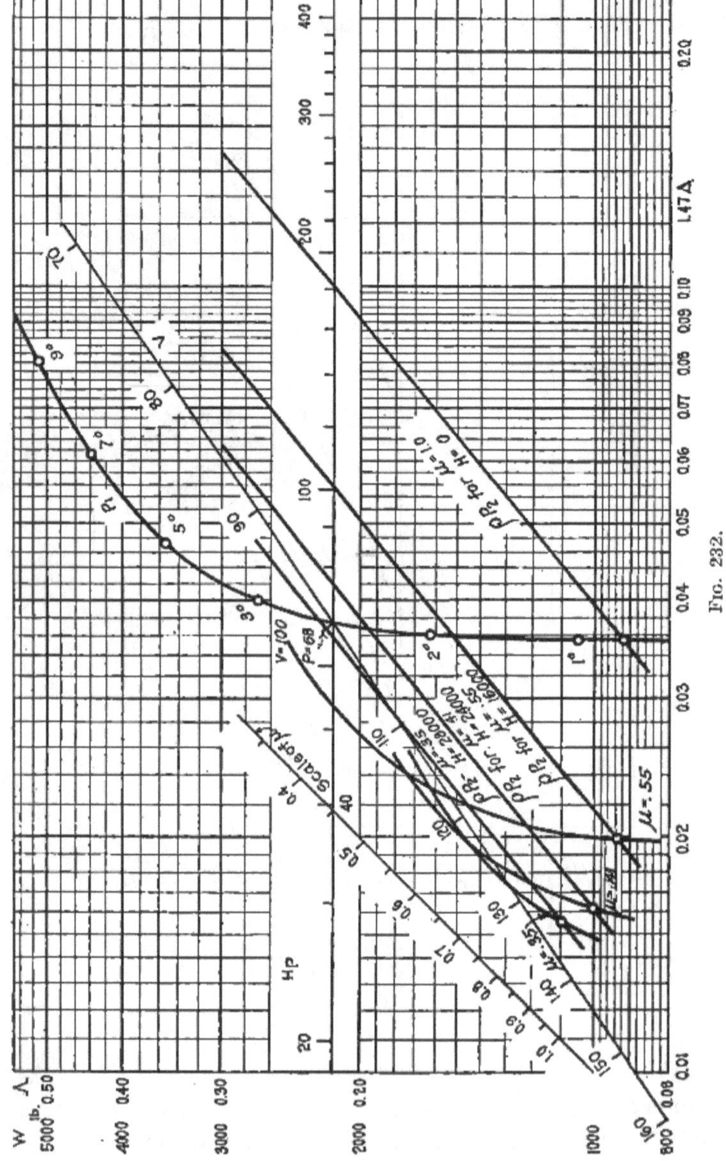

FIG. 232.

belonging to the same family, is of smaller dimensions so as to permit the engine to reach the most advantageous number of revolutions and therefore to develop all the power of which it is capable. It is interesting, however, to first study the behavior of the propeller having a diameter of 7.92 ft. in order to compare it to that of a smaller diameter.

The diagrams of Fig. 232 show that the maximum horizontal velocities at the various altitudes with the propeller of 7.92 ft. of diameter are

at 0 ft., 150 m.p.h.
at 16,000 ft., 148 m.p.h.
at 24,000 ft., 144 m.p.h.
at 28,000 ft., 138 m.p.h.

These diagrams allow us to obtain the differences ${}_\rho P_2 - P_1$ and therefore to compute the values of the maximum climbing velocities v at the various heights. These velocities are plotted in Fig. 233; on the ground the ascending velocity is equal to 29.5 ft. per second. At 28,000 ft. it is equal to 1.7 per second; that is, equal to a little more than 100 ft. per minute; the height of 28,000 ft. must then be considered as the ceiling of our airplane if equipped with the above propeller.

From the diagram of $v = f(H)$ it is easy to obtain that of $\frac{1}{v} = f(H)$ (Fig. 234*a*), and therefore by its integration, we obtain that of $t = f(H)$, which gives the time of climbing (Fig. 234*b*). It can be seen that with this particular propeller, the airplane can reach a height of 28,000 ft. in 3000 seconds; that is, in 50 minutes.

Let us now suppose that a propeller is adopted of such diameter as to permit the engine to reach its maximum number of revolutions. By using the diagram of Figs. 227 and 230 we find with easy trials and by successive approximation that the most suitable propeller will have a diameter of 7.65 ft. and therefore as $\frac{p}{D} = 1.18$, a pitch of about

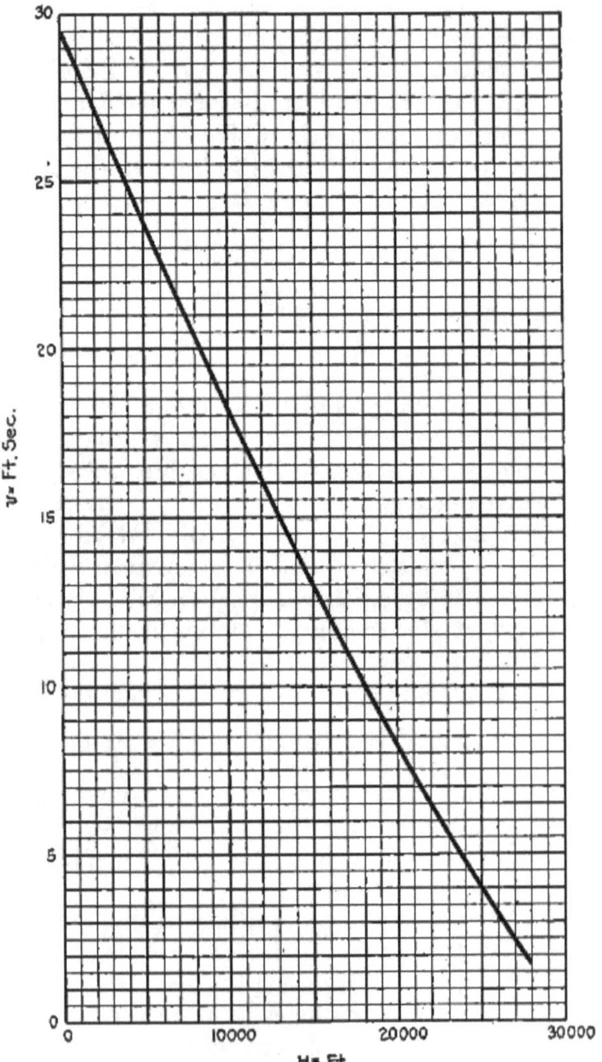

FIG. 233.

0.6
0.5
0.4
0.3
0.2
0.1
0
$\frac{1}{v}$ (v = Ft. Sec.)
T_0 = 59°F
0
10000
20000
30000
H = Ft.
(a)
3200
2400
1600
800
0
t (Sec.)
T_0 = 59°F
0
10000
20000
30000
H = Ft
(b)

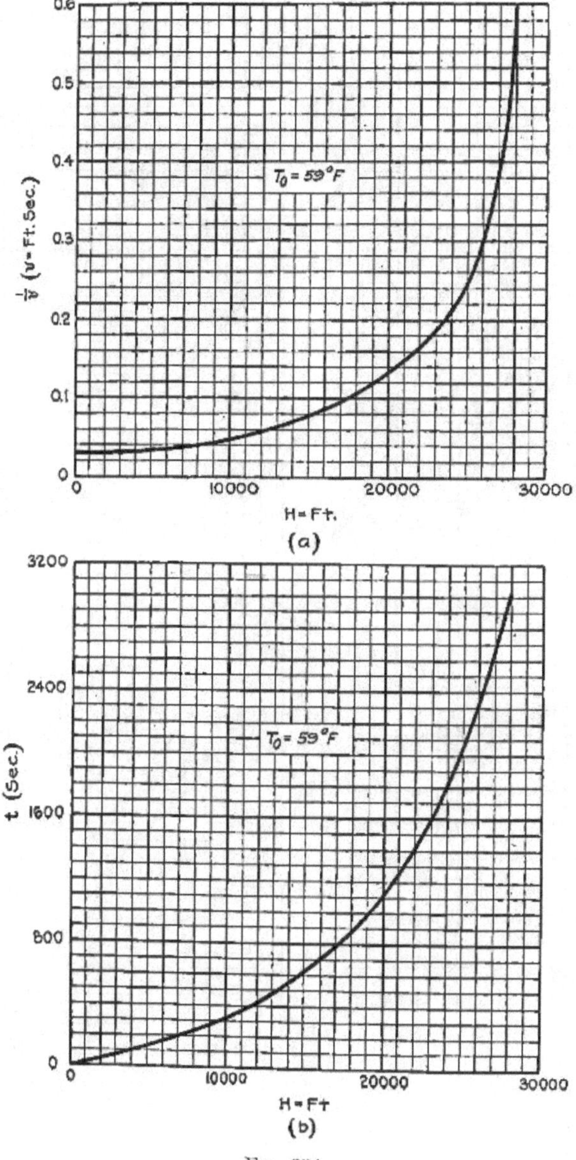

Fig. 234.

W
5000 lb.
4000
3000
2000
1000
800
Λ
0.50
0.40
0.30
0.20
0.10
0.08
0.01
0.02
0.03
0.04
0.05
0.06
0.07
0.08
0.09
0.10
1.47 Δ
0.15
0.20
Hp.
20
40
60
80
100
150
200
300
400
V=100
P=68
μ=0.55
μ=0.41
μ=0.235

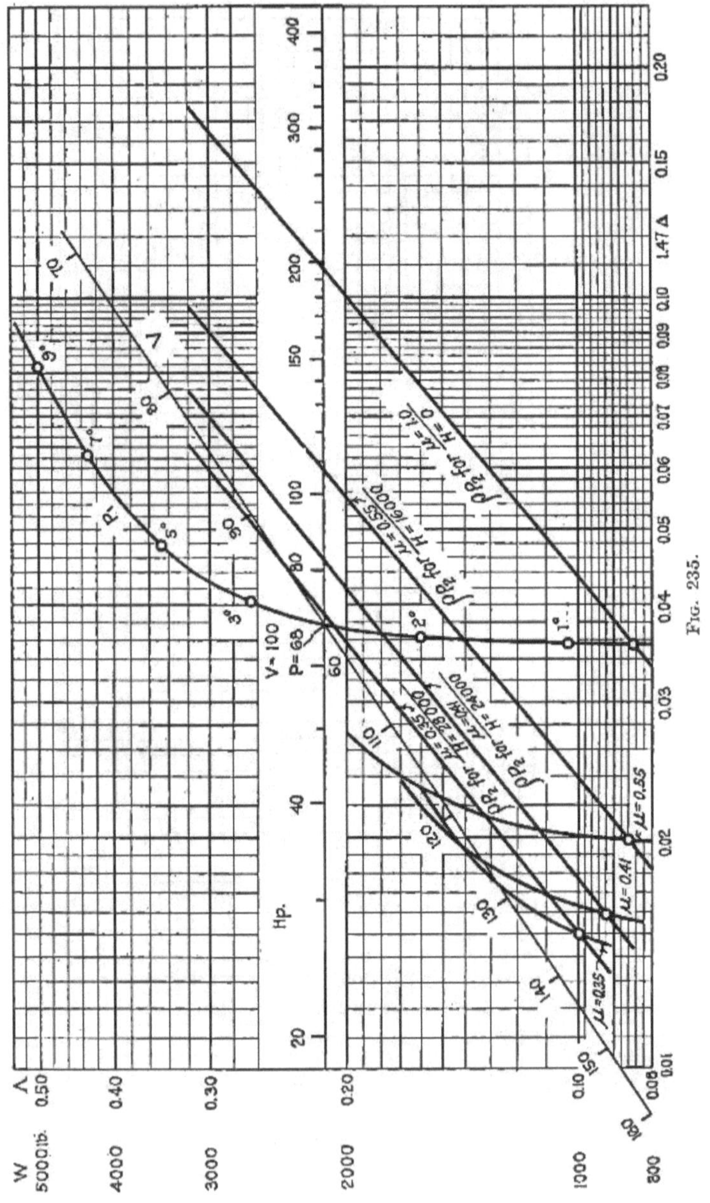

Fig. 235.

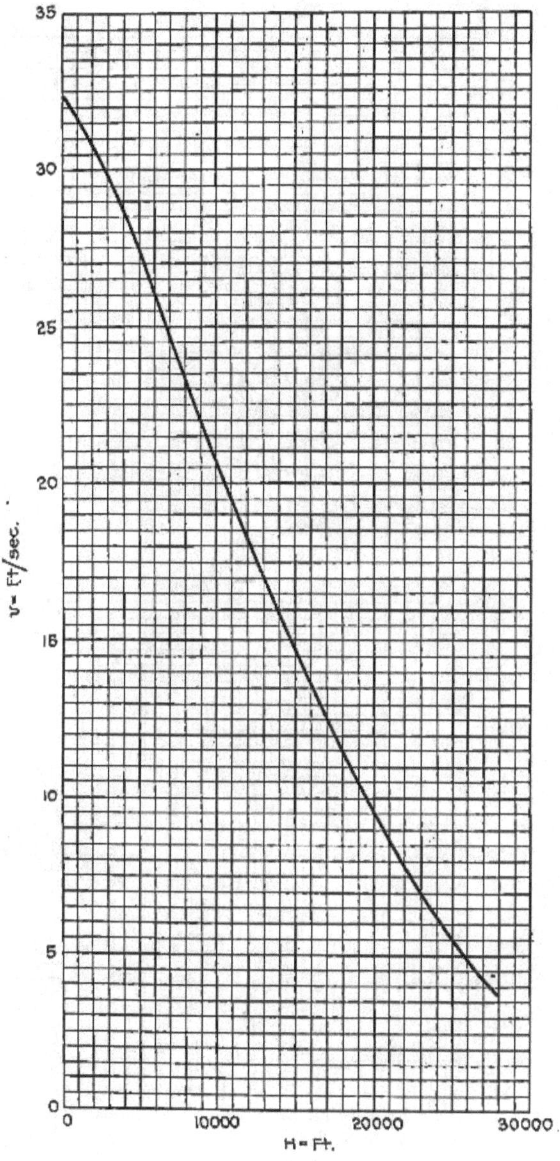

FIG. 236.

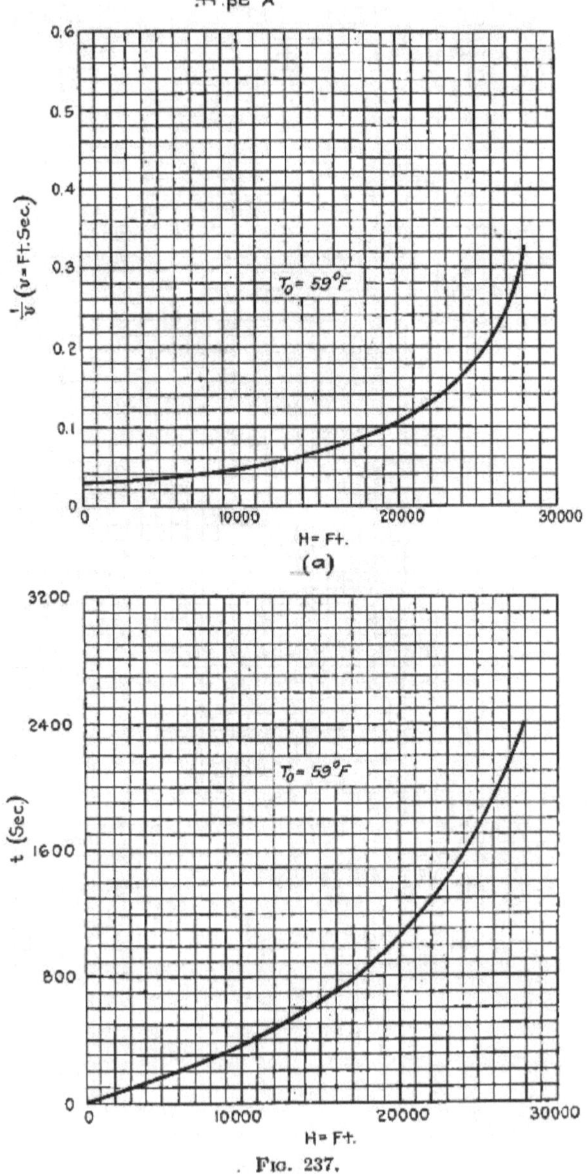

FIG. 237.

9 ft. This propeller is the one for which the static analysis was given in the preceding chapter. For such a propeller the logarithmic diagrams of $_{\rho}P_2$, the diagram $v = f(H)$ and those of $\frac{l}{v} = f(H)$ and $t = f(H)$ have been plotted in figures 235, 236 and 237*ab* respectively.

The diagrams of Fig. 235 show that the new maximum velocities are

at 0 ft., 156 m.p.h.
at 16,000 ft., 155 m.p.h.
at 24,000 ft., 150 m.p.h.
at 28,000 ft., 144 m.p.h.

The diagram of Fig. 236 shows that at an altitude of 28,000 ft., $v = 3.7$ ft. per second = 222 ft. per minute; that is, the ceiling has become greater than 28,000 ft.

The diagram of Fig. 237 finally shows how the height of 28,000 ft. is reached in 2400 seconds; that is, in only 40 minutes.

The second propeller, therefore, is decidedly better than the first one.

The question now arises: What is the maximum load that can be lifted with our airplane? It is therefore necessary to suppose the efficiency of the propeller to be known. Supposing $\rho = 0.815$, then the maximum useful available power will be 244 H.P.

Let us again examine the diagram $\Lambda = f(1.47\ \Delta)$ (Fig. 238) for our airplane at the point corresponding to 244 H.P. on the scale of powers, draw a perpendicular to meet tangent t in B drawn from the diagram parallel to the scale of velocities. From B draw the parallel BC to the scale of powers. Point C gives the maximum theoretical load which the airplane could lift, and which in our case would be about 7300 lb. The corresponding velocity is measured by segment BD which, read on the scales of velocity, gives $V = 132$ m.p.h.

Practically, however, the airplane cannot lift itself in this condition as it is necessary to have a certain excess of power in order to leave the ground.

W
7000 lbs.
6000
5000
4000
3000
2000
1800
1600
1400
1200
Λ
C
C''
0.70
0.60
0.50
0.40
0.30
0.20
0.18
0.16
0.14
0.12
Hp.
400
300
200
150
100
80
60
40
20
A
A' = 167.5
B
B'
C
C'
D
R1
V=100
Scale of P
Scale of H
1.47 Δ : 0.15
0.10
0.09
0.08
0.07
0.06
0.05
0.04
0.03
0.02
μ=0.685

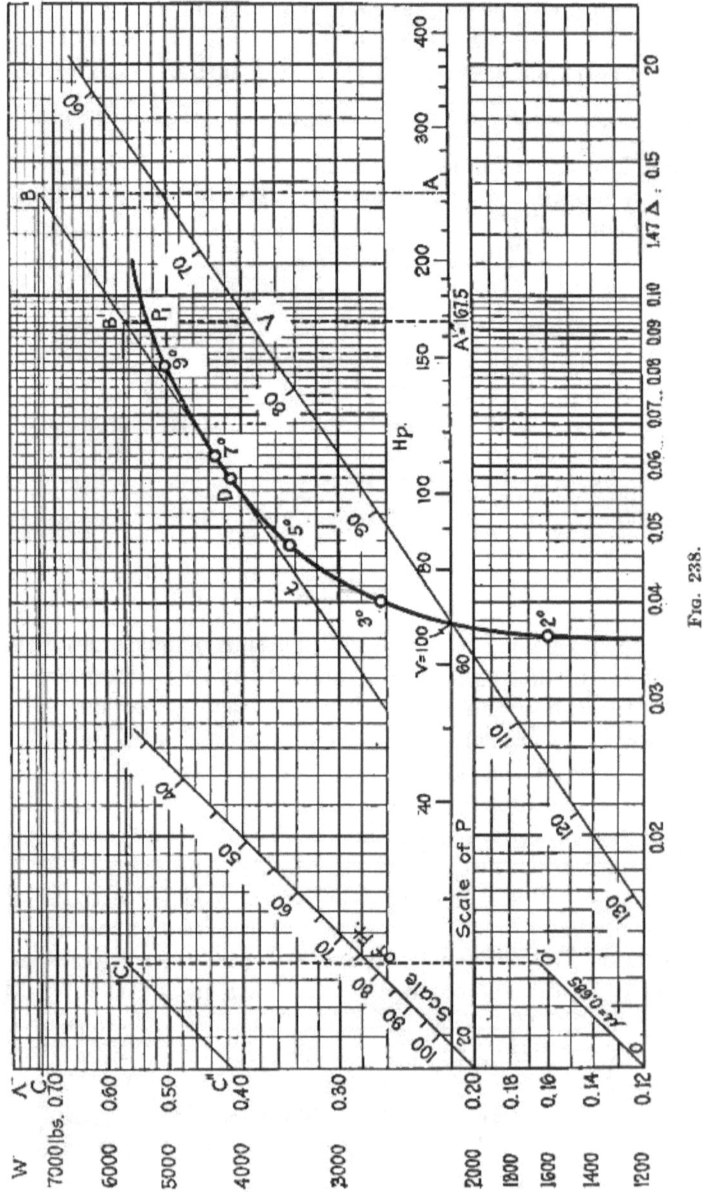

Fig. 238.

Supposing then we fix the condition that the airplane should be able to sustain itself at a height of 10,000 ft. As

$$H = 60{,}720 \log \frac{1}{\mu}$$

for $H = 10{,}000$ we will have $\mu = 0.685$, therefore in this case the useful power becomes $0.815 \times 0.685 \times 300 = 167.5$ H.P. Let us then draw a perpendicular from A' corresponding to 167.5 H.P. to meet tangent t in B'. From B' draw the parallel to the scale of power. From origin O of the diagram draw a segment OO' parallel to the scale of μ and which measures $\mu = 0.685$; from O' raise the perpendicular until it meets the horizontal line in C' drawn from BB'; from C' draw the parallel to OO' up to C''; this point defines the value of the maximum load which our airplane could lift up to 10,000 ft. and which in our case is about 4100 lb. The corresponding velocity is measured by $B'D$ and is equal to 116 m.p.h.

Let us now study what the effect would be of a diminution of the lifting surface. Until now we had supposed that $A = 265$ sq. ft.; that is, we had a load of 8 lb. per sq. ft. Now supposing this load is increased up to 10, 12, 14, and 16 lb. per sq. ft. respectively; that is, the lifting surface is reduced from 265 sq. ft. to 214, 178, 153 and 134 sq. ft. successively. For each of such hypotheses it will be necessary to calculate the new values of Λ and Δ; the results of these calculations are grouped in Table 47. By means of this table the diagrams of Fig. 239 have been drawn; let us then suppose that in each case a propeller having the maximum efficiency of 0.815 has been adopted. The useful power will be 244 H.P.; drawing from A, the point which corresponds to this power, the parallel p to the scale of velocity, on the intersection with this line and the diagram we shall have the point which defines the maximum velocities; drawing the tangent t parallel to the scale of V from each of the various curves the points of tangency which determine the minimum velocities will be obtained.

W
5000lb.
4000
3000
2000
1000
800
Λ
0.50
0.40
0.30
0.20
0.10
0.08
Vmin ●
Vmax o
A=265
A=214
A=178
A=153
A=134
V=100
P=68
Hp
A
20
40
60
80
100
150
200
300
400
70
80
90
110
120
130
140
150
160
0.01
0.02
0.03
0.04
0.05
0.06
0.07
0.08
0.09
0.10
1.47Δ
0.15
0.20

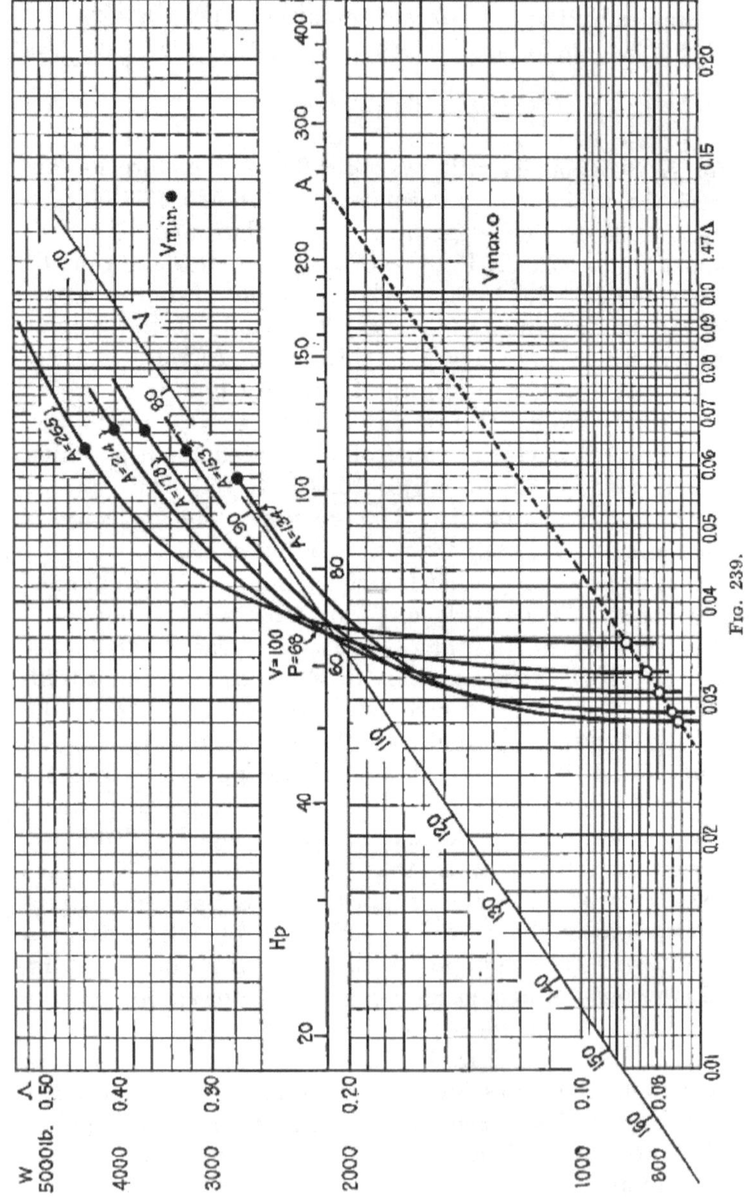

Fig. 239.

TABLE 47

$\Lambda = 10^{-4}\lambda A$ $\sigma = 132.5$
$\Delta = 10^{-4}(\delta A + \sigma)$

A, sq. ft.	i	−1°	0°	1°	2°	3°	4°
	λ	2.3	4.0	6.0	8.0	10.0	11.4
	δ	0.41	0.42	0.43	0.47	0.53	0.60
265	Λ	0.061	0.106	0.159	0.212	0.265	0.302
	Δ	241.0×10^{-4}	244.0×10^{-4}	246.0×10^{-4}	257.0×10^{-4}	274.0×10^{-4}	291.0×10^{-4}
214	Λ	0.049	0.086	0.128	0.171	0.214	0.244
	Δ	220.5×10^{-4}	222.5×10^{-4}	225.0×10^{-4}	233.5×10^{-4}	246.5×10^{-4}	261.5×10^{-4}
178	Λ	0.041	0.071	0.107	0.143	0.178	0.203
	Δ	205.5×10^{-4}	207.5×10^{-4}	209.5×10^{-4}	216.5×10^{-4}	227.5×10^{-4}	239.5×10^{-4}
153	Λ	0.035	0.061	0.092	0.122	0.153	0.174
	Δ	195.0×10^{-4}	197.0×10^{-4}	198.0×10^{-4}	204.5×10^{-4}	213.5×10^{-4}	224.5×10^{-4}
134	Λ	0.031	0.054	0.081	0.107	0.134	0.153
	Δ	187.5×10^{-4}	189.0×10^{-4}	190.5×10^{-4}	195.5×10^{-4}	204.0×10^{-4}	213.0×10^{-4}

A, sq. ft.	i	5°	6°	7°	8°	9°
	λ	13.0	14.5	16.0	17.8	19.2
	δ	0.73	0.88	1.08	1.32	1.57
265......	Λ	0.344	0.384	0.423	0.472	0.509
	Δ	326.0×10^{-4}	365.0×10^{-4}	428.0×10^{-4}	483.0×10^{-4}	550.0×10^{-4}
214......	Λ	0.278	0.310	0.342	0.381	0.411
	Δ	288.5×10^{-4}	321.5×10^{-4}	364.5×10^{-4}	415.5×10^{-4}	469.5×10^{-4}
178......	Λ	0.232	0.258	0.285	0.318	0.342
	Δ	262.5×10^{-4}	289.5×10^{-4}	324.5×10^{-4}	367.5×10^{-4}	412.5×10^{-4}
153......	Λ	0.199	0.222	0.245	0.272	0.293
	Δ	244.5×10^{-4}	267.5×10^{-4}	297.5×10^{-4}	334.5×10^{-4}	372.5×10^{-4}
134......	Λ	0.174	0.195	0.215	0.238	0.257
	Δ	230.5×10^{-4}	250.5×10^{-4}	279.5×10^{-4}	309.5×10^{-4}	343.5×10^{-4}

Table 48 gives the values of the maximum and minimum velocities corresponding to the various wing surfaces. This table sustains the point that while a reduction of surface increases the maximum velocities, it also increases the values of the minimum velocities. Figure 239 also clearly shows that a diminution of surface requires an increase in the minimum power necessary for flying, and therefore a diminution in the climbing velocity and in the ceiling.

TABLE 48

A	265	214	178	153	134
V(max.)	156	158	162	164	166
V(min.)	72	74	77	82	88

CHAPTER XXI

SAND TESTS—WEIGHING—FLIGHT TESTS

I

The ultimate check on static computations giving the resistance to the various parts of the airplane, is made either by tests to destruction of the various elements of the structure or by static tests upon the machine as a whole.

In general it is customary to make separate tests (*A*) on the wing truss, (*B*) on the fuselage (*C*) on the landing gear and (*D*) on the control system.

A. Sand Tests on the Wing Truss.—Two sets of tests are usually made on a wing truss to determine its strength; one assuming the machine loaded as in normal flight, the other loaded as in inverted flight.

In the first assumption, the inverted machine is loaded with sand bags, so that the weight of the sand exerts the same action on the wings as the air reaction does in flight; in the second assumption the machine is loaded with sand bags in the normal flying position. In both cases the machine is placed so as to have an inclination of 25 per cent. (Fig. 240), so that weight W, with its component L stresses the vertical trusses, and with its component D stresses the horizontal trusses.

During the test, the fuselage is supported by special trestles, constructed so as not to interfere with the deformation of the wing truss. The distribution of the load upon the wings must be made in such a manner that the reactions on the spars will be in the same ratio as those assumed in the computation. For the example of the preceding chapters it is well to remember that these reactions were due to the following loading:

Upper front spar.............1.98 lb. per linear inch.
Upper rear spar ....1.82 lb. per linear inch.
Lower front spar.............1.75 lb. per linear inch.
Lower rear spar..............1.62 lb. per linear inch.

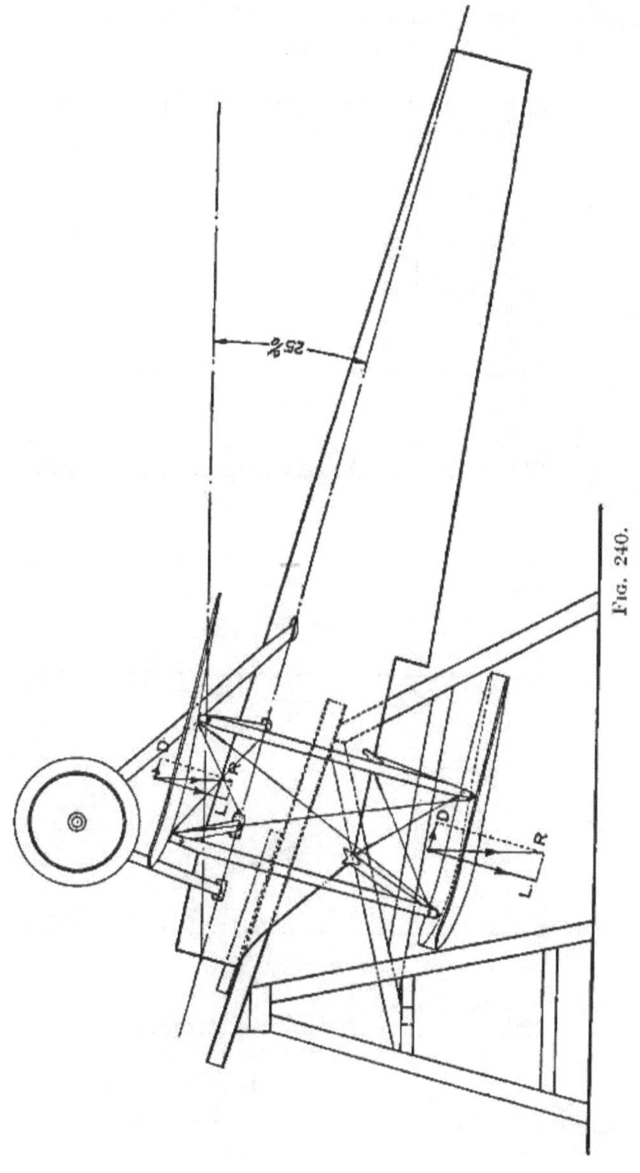

Fig. 240.

The sand is usually contained in bags of various dimensions, not exceeding a weight of 25 lb. in order to facilitate

UPPER RIB

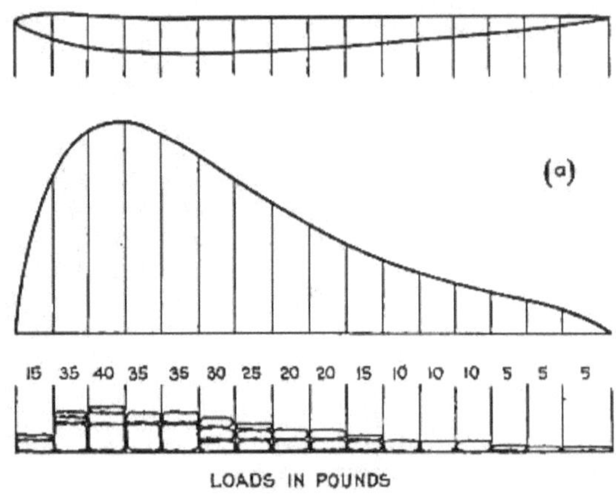

LOWER RIB.

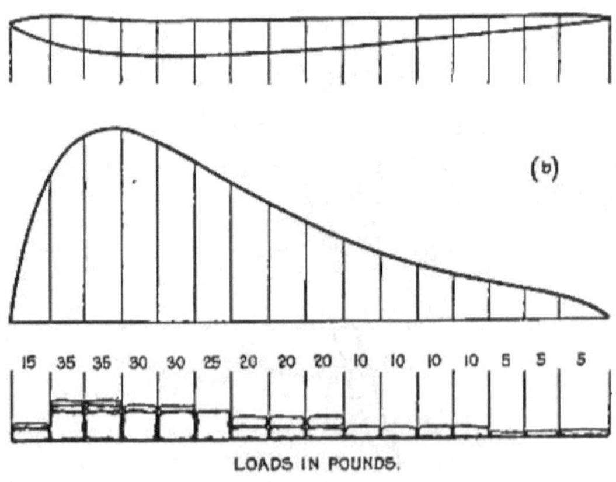

Fig. 241.

handling. These sand bags must be so placed that beside satisfying the preceding conditions, they give a loading diagram for the upper and lower rib analogous to those shown in Fig. 241 *a, b*.

In these figures, below the theoretical diagrams, the practical loading, using sand bags of 5, 10 and 25 lb. has been sketched. In the test corresponding to normal flight, the machine being inverted, it is necessary to consider the weight of the wing truss, which gravitates upon the vertical trusses and therefore must be added to the weight of the sand, while in actual flight it has an opposite direction to the air reaction.

These weights must be taken into consideration in determining the sand load corresponding to a coefficient of 1.

Before starting a static test it is customary to prepare a diagram of each wing with a table showing the loads corresponding to the various coefficients. For the airplane of our example, these diagrams are shown in Figs. 242 and 243, and tables 49 and 50.

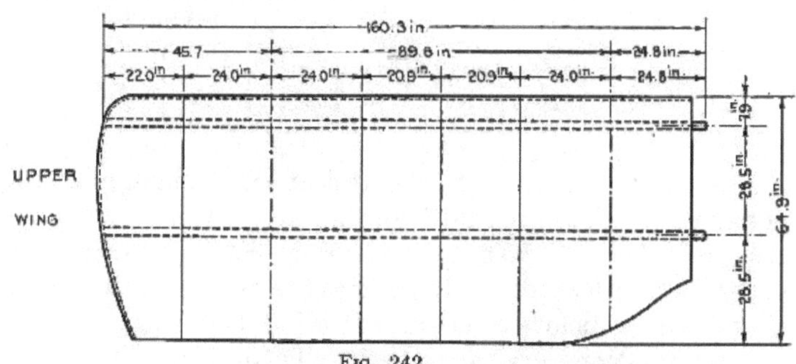

Fig. 242.

Table 49

Factor safety	Table of loads for sand test						
3	235	255	255	220	220	255	270
4	305	345	345	300	300	345	360
5	390	435	435	375	375	435	455
6	475	525	525	455	455	525	540
7	560	610	610	520	520	610	635
8	640	700	700	615	615	700	725
9	720	790	790	690	690	790	820
10	800	875	875	760	760	875	905

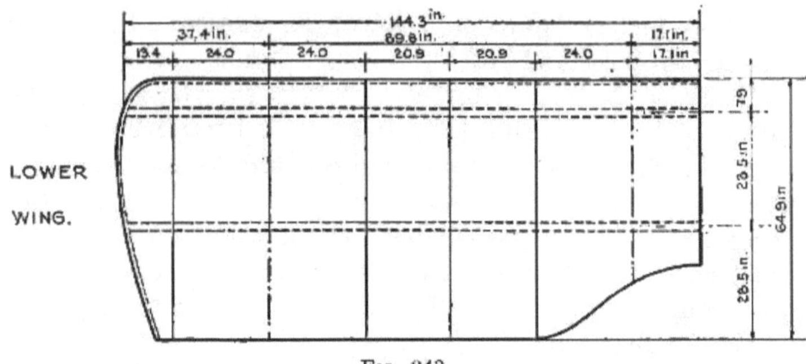

Fig. 243.

TABLE 50

Factor safety	Table of loads for sand test						
3	130	225	225	195	195	225	165
4	170	305	305	270	270	305	215
5	210	385	385	335	335	385	275
6	260	460	460	405	405	460	330
7	300	540	540	470	470	540	390
8	345	620	620	540	540	620	440
9	390	700	700	615	615	700	500
10	440	780	780	690	690	780	560

During the progress of the test it is of maximum importance to measure the deformation to which the spars are subjected in order to determine their elastic curves under various loadings. In general the determination of an elastic curve below a coefficient of 3 is disregarded, as the deformations are very small. To measure the deformations small graduated rulers are usually attached to the spars in front of which a stretched copper wire is kept as a reference line. Naturally, before applying the load, it is necessary to take a preliminary reading of the intersections of the graduated rulers with the copper wire, so as to compute the effective deformation. Then proceed as follows:

1. Start loading the sand bags on the wings, following the preceding instructions for a total load corresponding to a coefficient of 3, minus the weight of the wing truss.

2. When this entire load has been placed on the wings, take a reading of all the rulers.

3. Unload the wing truss gradually and completely.

4. Take a new reading with the machine unloaded.

5. Load the machine again so as to reach a total load equal to four times that corresponding to a coefficient of 1 minus the weight of the wing truss.

6. Take another reading.

7. Unload the machine completely.

8. Take another reading with the machine unloaded.

And so on for coefficients of 5, 6, 7, etc.

As the maximum coefficient for which the machine has been computed, and that corresponding to which the machine will brake, is approached, it is not safe to take further readings as the falling of the load which follows the braking may endanger the observer. The various readings of the deformations with the load and those after unloading, are usually put in tabular forms and serve as a basis for plotting the elastic curves. Furthermore the deformations with the load, allow the computation of deformations sustained both by struts and diagonals. Consequently all the elements are had by means of which the unit stresses in the various parts of the wing truss under different loadings can be computed.

B. Sand Test of the Fuselage.—In computing the fuselage, it was seen that the principal stresses are those produced in flight. Therefore the fuselage sand test is usually made by suspending it by the four fittings of the main diagonals of the wings, and subsequently loading it with sand bags and lead weights so as to produce loads equal to 3, 4, 5, etc., times the weight of the various masses contained in the fuselage. For the determination of the coefficient of safety the sum of the weights of these masses is taken as a basis. At the same time a load equal to the breaking load of the elevator itself is placed corresponding to the point at which the elevator is fixed; to equilibrate the moment due to this load the usual procedure is to anchor the forward portion of the fuselage. Fig. 244 clearly shows how the test is prepared.

Controls
Load on Stabilizer and Elevator.
1 Propeller
2 Radiator
3 Motor
4 Cowl
5 Oil
6 Gasoline
7 Armament
8 Gasoline
9 Fuselage & Controls
10 Pilot
11 Controls.

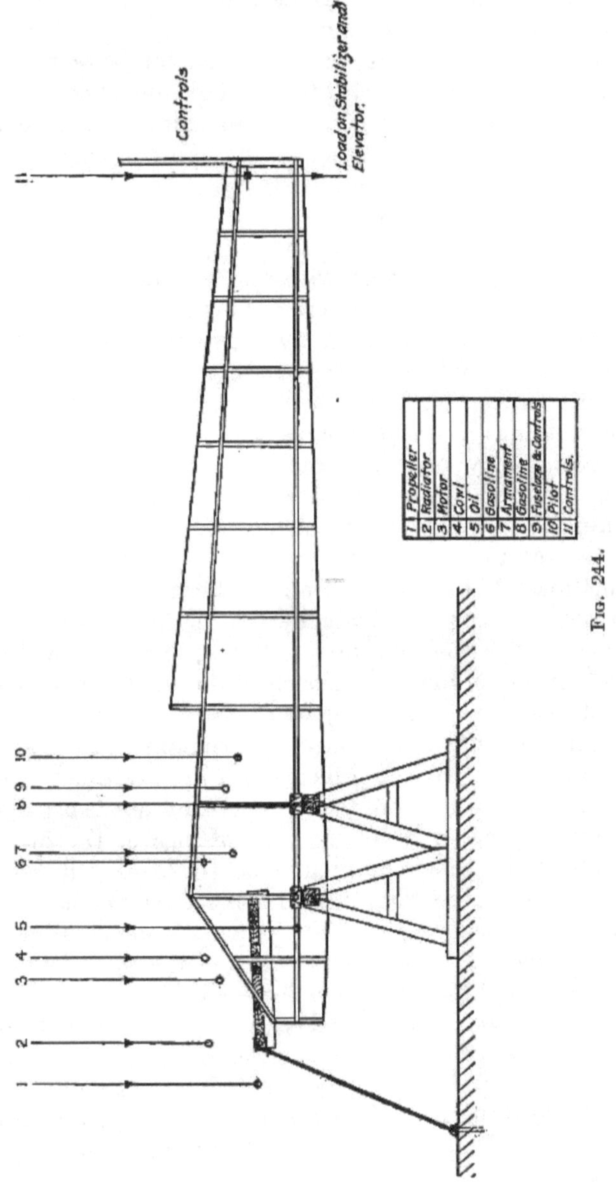

Fig. 244.

C. Sand Test of the Landing Gear.—This is done with the landing gear in a position corresponding to the line of flight and by loading it with lead weights.

The load assumed as a basis for the determination of the coefficient is taken equal to the total weight of the airplane with full load. If, corresponding to each value of load W, the corresponding vertical deformation f is determined, it is possible to plot the diagram of W as a function of f, whose area $\int W\,df$ gives the total work the shock absorbing system is capable of absorbing.

D. Sand Test of Control Surfaces.—This test is made with the control surfaces mounted on the fuselage, and loaded with the criterion explained in Chapter XVIII.

II

Weighing the Airplane.—The weighing of the airplane is necessary not only to determine whether the effective weights correspond to the assumed ones, but also to determine the position of the center of gravity both with full load and with the various hypothesis of loading which may happen in flight.

The center of gravity is contained in the plane of symmetry of the airplane. To determine this it suffices to determine two vertical lines which contain it, and for this it is only necessary to weigh the aeroplane twice, the first time with the tail on the ground (Fig. 245), and the second time with the nose of the machine on the ground (Fig. 246). Three scales are necessary for each weighing, two under the wheels, and one under the tail skid for the case of Fig. 245, and under the propeller hub for the case of Fig. 246.

Using W' and W'' to denote the weights read on the scales under the wheels and W''' for that read on the scale supporting the tail skid, the total weight will be

$$W = W' + W'' + W'''$$

The vertical axis v' passing through the center of gravity divides the distance l between the axis of the wheels and

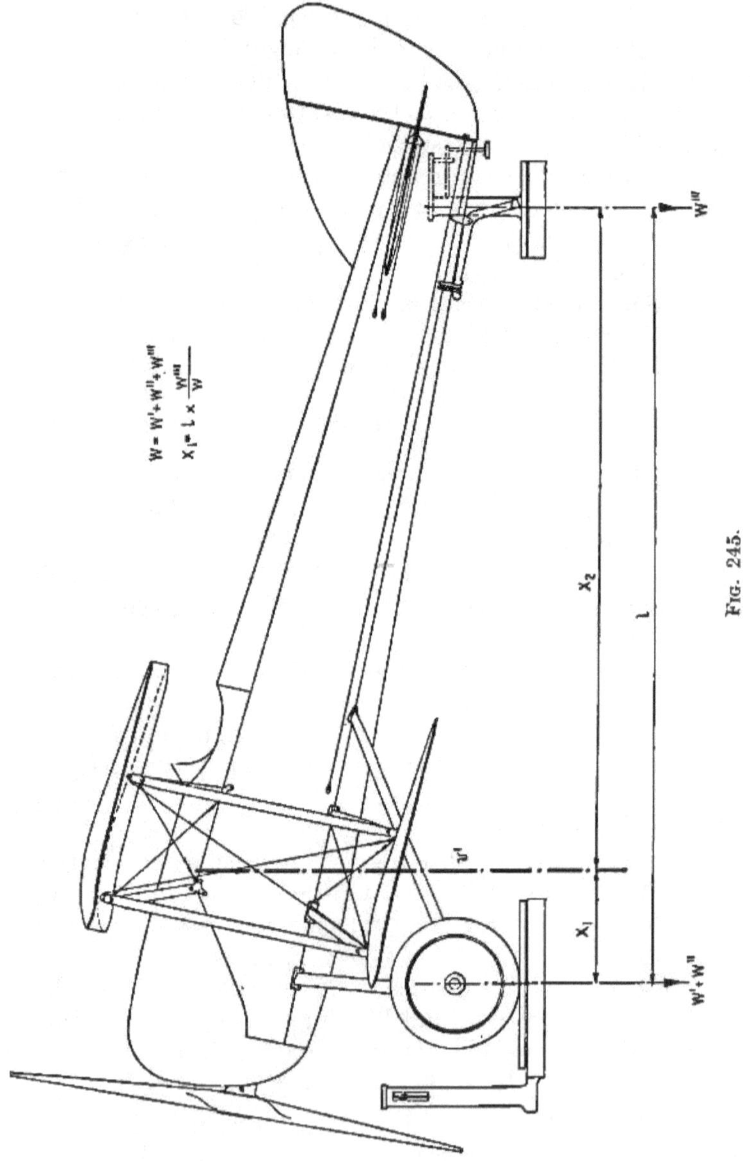

FIG. 245.

the point of support of the tail skid into two parts x_1 and x_2 so that

$$\frac{x_1}{x_2} = \frac{W'''}{W' + W''}$$

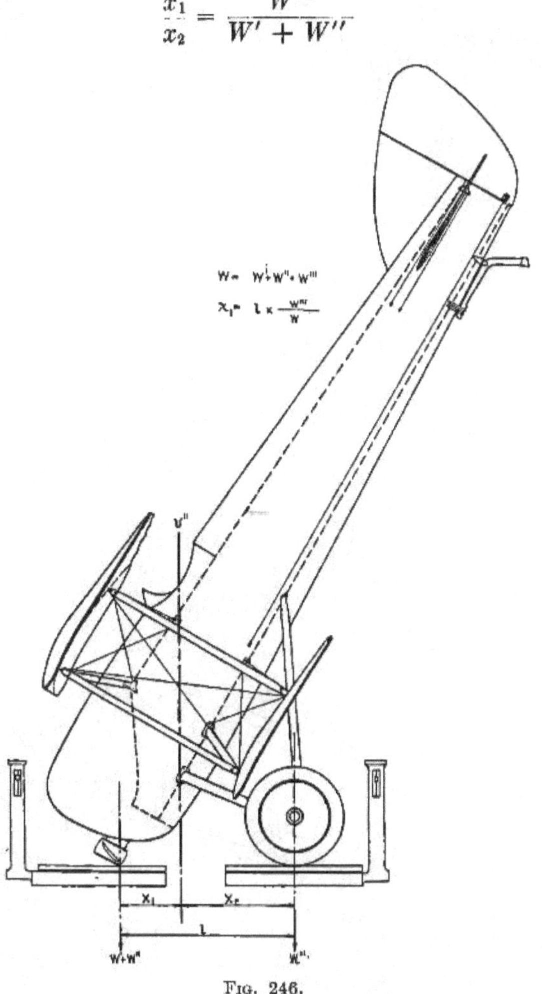

FIG. 246.

for which

$$\frac{x_1}{x_1 + x_2} = \frac{W'''}{W' + W'' + W'''}$$

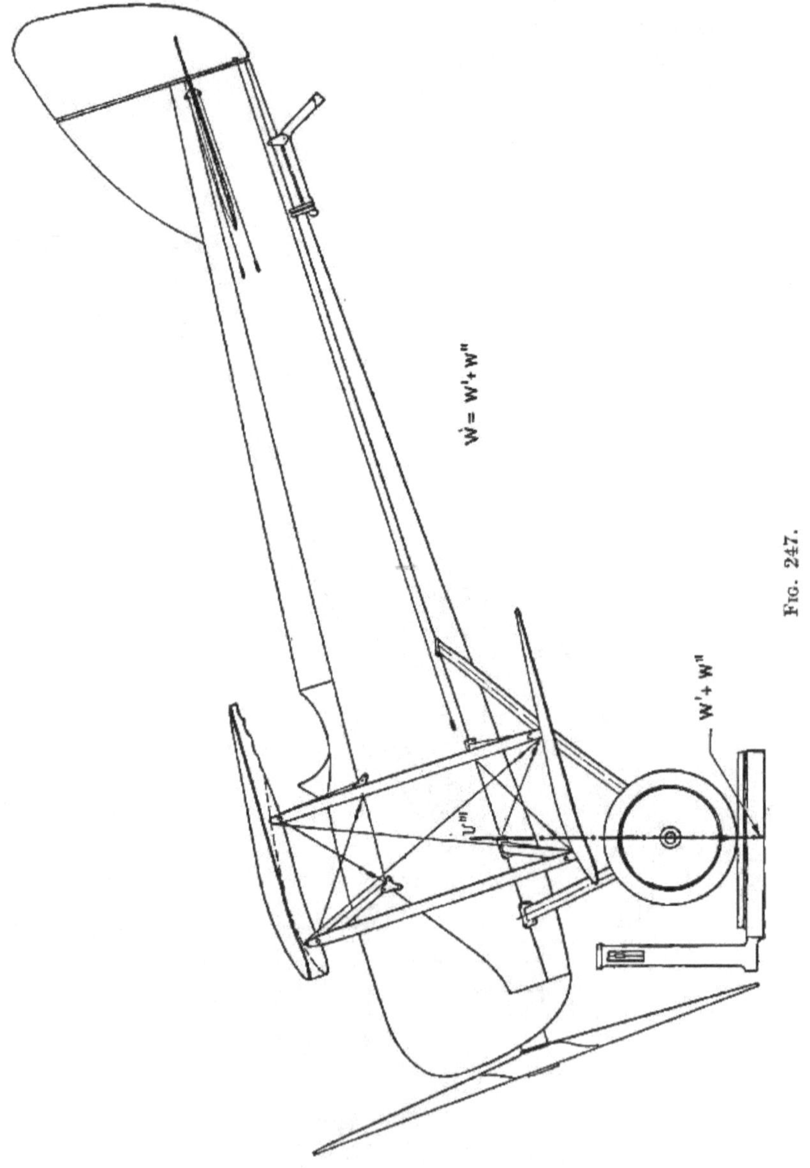

Fig. 247.

and since

$$x_1 + x_2 = l \text{ and } W' + W'' + W''' = W$$

we shall have

$$x_1 = l \times \frac{W'''}{W}$$

Let us proceed analogously for the case of Fig. 246. In this manner two lines v' and v'' are obtained whose intersection defines the center of gravity.

To eliminate eventual errors and to obtain a check on the work it is convenient to determine the third line v''', by

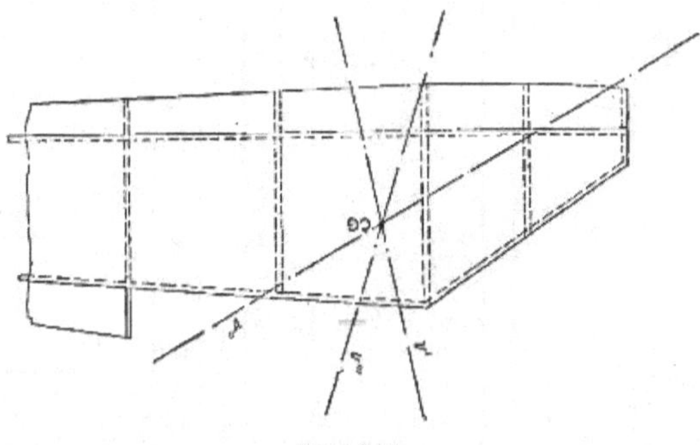

FIG. 248.

balancing the machine on the wheels; v''' will then be the vertical which passes through the axis of the wheels (Fig. 247). The three lines v', v'' and v''' must meet in a point (Fig. 248).

III

The flight tests include two categories of tests, that is;

A. Stability and maneuverability tests.

B. Efficiency test.

A. The purpose of the stability tests is to verify the balance of aeroplane when (*a*) flying with engine going, and when volplaning, (*b*) in normal flight and during maneuvers.

Chapter XI has stated the necessary requisites for a well-balanced airplane, therefore a repetition need not be given.

TABLE 51

AEROPLANE TEST DATA

Aeroplane: type..............No............

Motor: Type.............No.........H.P...... At. r.p.m............

Propeller No.........Diam.........P.........Max. width........No. of blades..........Location of test...........................

Weight { Empty......................
Useful load..................
Total........................

PROPELLER TEST ON GROUND

Date	Hour	Air temp.	Air press.	Air humidity, per cent.	R.p.m.	H.P. motor*	Density gasoline	Duration of test	Unusual conditions	Remarks

* See the power curve.

TEST IN FLIGHT (HORIZONTAL)

Date	Air temp.	Air press.	Weather conditions	Starting space in ft.	Altitude	Hour	Average speed	Motor r.p.m.	Motor water temp.	Speedom. reading	Landing space in ft.	Duration of flight	Remarks

TEST IN FLIGHT (CLIMBING)

Altitude in feet																																				Duration of flight	Remarks
Sea level				5,000				10,000				14,000				16,000				18,000				20,000				22,000				24,000					
Time	R.p.m.	Speedom. reading	Water temp.	Time	R.p.m.	Speedom. reading	Water temp.	Time	R.p.m.	Speedom. reading	Water temp.	Time	R.p.m.	Speedom. reading	Water temp.	Time	R.p.m.	Speedom. reading	Water temp.	Time	R.p.m.	Speedom. reading	Water temp.	Time	R.p.m.	Speedom. reading	Water temp.	Time	R.p.m.	Speedom. reading	Water temp.	Time	R.p.m.	Speedom. reading	Water temp.		

The same may be said of maneuverability tests, whose scope is to verify the good and rapid maneuverability of the airplane without an excessive effort by the pilot.

B. The scope of the efficiency tests is to determine the flying characteristics of the airplane, that is, the ascensional and horizontal velocities corresponding to various loads and eypes of propellers which might eventually be wanted for txperiments.

Table 51 gives examples of tables that show which factors of the efficiency tests are the most important to determine.

APPENDIX

The following tables are given for the convenience of the designer: Tables 52, 53, 54, 55 and 56 giving the squares and cubes of velocities. Table 57 giving the cubes of revolutions per minute and per second. Table 58 giving the 5th powers of the diameters in feet.

TABLE 52.—TABLE OF SQUARES AND CUBES OF VELOCITIES

V		V^2		V^3	
Miles per hr.	Ft. per sec.	Miles per hr.	Ft. per sec.	Miles per hr.	Ft. per sec.
50	73.33	2,500	5,377.7	125,000	394,430
51	74.80	2,601	5,595.0	132,651	418,510
52	76.27	2,704	5,817.1	140,608	443,670
53	77.73	2,809	6,042.0	148,877	469,640
54	79.20	2,916	6,272.6	157,464	496,790
55	80.67	3,025	6,507.6	166,375	524,970
56	82.13	3,136	6,745.3	175,616	553,990
57	83.60	3,249	6,988.9	185,193	584,280
58	85.07	3,364	7,237.0	195,112	614,270
59	86.53	3,481	7,487.5	205,379	647,890
60	88.00	3,600	7,744.0	216,000	681,470
61	89.47	3,721	8,004.9	226,981	716,200
62	90.93	3,844	8,268.2	238,328	751,830
63	92.40	3,969	8,537.8	250,047	788,800
64	93.87	4,096	8,811.8	262,144	827,140
65	95.33	4,225	9,087.8	274,625	866,340
66	96.80	4,356	9,370.2	287,496	907,040
67	98.27	4,489	9,657.0	300,763	948,990
68	99.73	4,624	9,946.0	314,432	991,920
69	101.02	4,761	10,205.0	328,509	1,030,920
70	102.67	4,900	10,541.0	343,000	1,082,260
71	104.13	5,041	10,843.0	357,911	1,129,090
72	105.60	5,184	11,152.0	373,248	1,177,580
73	107.07	5,329	11,464.0	389,017	1,227,450
74	108.53	5,476	11,779.0	405,224	1,278,350
75	110.00	5,625	12,100.0	421,875	1,331,000
76	111.47	5,776	12,426.0	438,976	1,385,080
77	112.93	5,929	12,753.0	456,533	1,440,220
78	114.40	6,084	13,088.0	474,552	1,497,200
79	115.87	6,241	13,426.0	493,039	1,555,654
80	117.33	6,400	13,766.0	512,000	1,615,203

TABLE 53.—TABLE OF SQUARES AND CUBES OF VELOCITIES

V		V^2		V^3	
Miles per hr.	Ft. per sec.	Miles per hr.	Ft. per sec.	Miles per hr.	Ft. per sec.
81	118.80	6,561	14,113	531,441	1,676,680
82	120.27	6,724	14,465	551,368	1,739,690
83	121.73	6,889	14,818	571,787	1,803,820
84	123.20	7,056	15,178	592,704	1,869,960
85	124.67	7,225	15,543	614,125	1,937,700
86	126.13	7,396	15,909	636,056	2,006,570
87	127.60	7,569	16,282	658,503	2,077,550
88	129.07	7,744	16,659	681,472	2,150,190
89	130.53	7,921	17,038	704,969	2,224,000
90	132.00	8,100	17,424	729,000	2,299,970
91	133.47	8,281	17,814	753,571	2,377,670
92	134.93	8,464	18,206	778,688	2,456,550
93	136.40	8,649	18,605	804,357	2,537,720
94	137.87	8,836	19,008	830,584	2,620,650
95	139.33	9,025	19,413	857,375	2,704,800
96	140.80	9,216	19,825	884,736	2,791,310
97	142.27	9,409	20,241	912,673	2,879,650
98	143.73	9,604	20,658	941,192	2,969,220
99	145.20	9,801	21,083	970,299	3,061,260
100	146.67	10,000	21,512	1,000,000	3,155,180
101	148.13	10,201	21,943	1,030,301	3,250,340
102	149.60	10,404	22,380	1,061,208	3,348,070
103	151.07	10,609	22,822	1,092,727	3,447,750
104	152.53	10,816	23,265	1,124,864	3,548,670
105	154.00	11,025	23,716	1,157,625	3,652,260
106	155.47	11,236	24,171	1,191,016	3,757,850
107	156.93	11,449	24,627	1,225,043	3,864,720
108	158.40	11,664	25,091	1,259,712	3,974,340
109	159.87	11,881	25,558	1,295,029	4,086,030
110	161.33	12,100	26,027	1,331,000	4,199,000

TABLE 54.—TABLE OF SQUARES AND CUBES OF VELOCITIES

V		V^2		V^3	
Miles per hr.	Ft. per sec.	Miles per hr.	Ft. per sec.	Miles per hr.	Ft. per sec.
111	162.80	12,321	26,504	1,367,631	4,314,820
112	164.27	12,544	26,985	1,404,928	4,432,770
113	165.73	12,769	27,466	1,442,897	4,552,010
114	167.20	12,996	27,956	1,481,544	4,674,220
115	168.67	13,225	28,450	1,520,875	4,798,580
116	170.13	13,456	28,944	1,560,896	4,924,790
117	171.60	13,689	29,447	1,601,613	5,053,080
118	173.07	13,924	29,953	1,643,032	5,184,000
119	174.53	14,161	30,461	1,685,159	5,316,310
120	176.00	14,400	30,976	1,728,000	5,451,780
121	177.47	14,641	31,496	1,771,561	5,589,520
122	178.93	14,884	32,016	1,815,848	5,728,620
123	180.40	15,129	32,544	1,860,867	5,870,960
124	181.87	15,376	33,077	1,906,624	6,015,660
125	183.33	15,625	33,610	1,953,125	6,161,700
126	184.80	15,876	34,151	2,000,376	6,311,120
127	186.27	16,129	34,697	2,048,383	6,462,920
128	187.73	16,384	35,243	2,097,152	6,616,080
129	189.20	16,641	35,797	2,146,689	6,772,720
130	190.67	16,900	36,355	2,197,000	6,931,820
131	192.13	17,161	36,914	2,248,091	7,092,280
132	193.60	17,424	37,481	2,299,968	7,256,320
133	195.07	17,689	38,052	2,352,637	7,422,860
134	196.53	17,956	38,624	2,406,104	7,590,790
135	198.00	18,225	39,204	2,460,375	7,762,390
136	199.47	18,496	39,788	2,515,456	7,936,570
137	200.93	18,769	40,373	2,571,353	8,112,120
138	202.40	19,044	40,966	2,628,072	8,291,470
139	203.87	19,321	41,563	2,685,619	8,473,440
140	205.33	19,600	42,160	2,744,000	8,656,800

TABLE 55.—TABLE OF SQUARES AND CUBES OF VELOCITIES

V		V^2		V^3	
Miles per hr.	Ft. per sec.	Miles per hr.	Ft. per sec.	Miles per hr.	Ft. per sec.
141	206.80	19,881	42,766	2,803,221	8,844,050
142	208.27	20,164	43,376	2,863,288	9,034,000
143	209.73	20,449	43,987	2,924,207	9,225,330
144	211.20	20,736	44,605	2,985,984	9,420,670
145	212.67	21,025	45,229	3,048,625	9,618,750
146	214.13	21,316	45,852	3,112,136	9,818,220
147	215.60	21,609	46,483	3,176,523	10,021,800
148	217.07	21,904	47,119	3,241,792	10,228,200
149	218.53	22,201	47,755	3,307,949	10,435,900
150	220.00	22,500	48,400	3,375,000	10,648,000
151	221.47	22,801	49,049	3,442,951	10,862,800
152	222.93	23,104	49,698	3,511,808	11,079,100
153	224.40	23,409	50,355	3,581,577	11,279,500
154	225.87	23,716	51,017	3,652,264	11,523,300
155	227.33	24,025	51,679	3,723,875	11,748,200
156	228.80	24,336	52,349	3,796,416	11,977,600
157	230.27	24,649	53,024	3,869,893	12,209,900
158	231.73	24,964	53,699	3,944,312	12,443,600
159	233.20	25,281	54,382	4,019,679	12,682,000
160	234.67	25,600	55,070	4,096,000	12,923,300
161	236.13	25,921	55,757	4,173,281	13,166,000
162	237.60	26,244	56,454	4,251,528	13,725,800
163	239.07	26,569	57,154	4,330,747	13,663,900
164	240.53	26,896	57,855	4,410,944	13,915,800
165	242.00	27,225	58,564	4,492,125	14,172,500
166	243.47	27,556	59,278	4,574,296	14,432,300
167	244.93	27,889	59,991	4,657,463	14,693,400
168	246.40	28,224	60,713	4,741,632	14,959,600
169	247.87	28,561	61,440	4,826,809	15,229,000
170	249.33	28,900	62,166	4,913,000	15,499,700

TABLE 56.—TABLE OF SQUARES AND CUBES OF VELOCITIES

V		V^2		V^3	
Miles per hr.	Ft. per sec.	Miles per hr.	Ft. per sec.	Miles per hr.	Ft. per sec.
171	250.80	29,241	62,901	5,000,211	15,775,800
172	252.27	29,584	63,640	5,088,448	16,054,500
173	253.73	29,929	64,379	5,177,717	16,334,850
174	255.20	30,276	65,127	5,268,024	16,620,500
175	256.67	30,625	65,880	5,359,375	16,908,500
176	258.13	30,976	66,631	5,451,776	17,199,500
177	259.60	31,329	67,392	5,545,233	17,495,000
178	261.07	31,684	68,158	5,639,752	17,794,000
179	262.53	32,041	68,922	5,735,339	18,094,300
180	264.00	32,400	69,696	5,832,000	18,399,800
181	265.47	32,761	70,474	5,929,741	18,709,800
182	266.93	33,124	71,252	6,028,568	19,029,750
183	268.40	33,489	72,038	6,128,487	19,335,400
184	269.87	33,856	72,830	6,229,504	19,655,500
185	271.33	34,225	73,620	6,331,625	19,975,500
186	272.80	34,596	74,420	6,434,856	20,301,900
187	274.27	34,969	75,224	6,539,203	20,631,500
188	275.73	35,344	76,027	6,644,672	20,962,900
189	277.20	35,721	76,840	6,751,269	21,300,000
190	278.67	36,100	77,657	6,859,000	21,640,750
191	280.13	36,481	78,473	6,967,871	21,982,500
192	281.60	36,864	79,299	7,077,888	22,330,500
193	283.07	37,249	80,129	7,189,057	22,682,000
194	284.53	37,636	80,957	7,301,384	23,034,750
195	286.00	38,025	81,796	7,414,875	23,393,500
196	287.47	38,416	82,639	7,529,536	23,756,000
197	288.93	38,809	83,481	7,645,373	24,120,000
198	290.40	39,204	84,332	7,762,392	24,490,850
199	291.86	39,601	85,182	7,880,599	24,861,500
200	293.33	40,000	86,043	8,000,000	25,239,000

TABLE 57.—TABLE OF CUBES OF R.P.M. AND R.P.S.

n		n^3	
Per min.	Per sec.	Per min.	Per sec.
500	8.33	125.0×10^6	578.7
550	9.17	166.4×10^6	768.5
600	10.00	216.0×10^6	1,000.0
650	10.83	274.6×10^6	1,271.4
700	11.67	343.0×10^6	1,588.0
750	12.50	421.9×10^6	1,953.3
800	13.33	512.0×10^6	2,370.4
850	14.17	614.1×10^6	2,843.2
900	15.00	729.0×10^6	3,375.0
950	15.83	857.4×10^6	3,969.4
1,000	16.67	$1,000.0\times10^6$	4,629.6
1,050	17.50	$1,157.6\times10^6$	5,359.1
1,100	18.33	$1,331.0\times10^6$	6,162.0
1,150	19.17	$1,520.9\times10^6$	7,025.0
1,200	20.00	$1,728.0\times10^6$	8,000.0
1,250	20.83	$1,953.1\times10^6$	9,042.1
1,300	21.67	$2,197.0\times10^6$	10,171.0
1,350	22.50	$2,460.4\times10^6$	11,364.0
1,400	23.33	$2,744.0\times10^6$	12,704.0
1,450	24.17	$3,048.6\times10^6$	14,114.0
1,500	25.00	$3,375.0\times10^6$	15,625.0
1,550	25.83	$3,723.9\times10^6$	17,241.0
1,600	26.67	$4,096.0\times10^6$	18,963.0
1,650	27.50	$4,492.1\times10^6$	20,797.0
1,700	28.33	$4,913.0\times10^6$	22,746.0
1,750	29.17	$5,359.4\times10^6$	24,812.0
1,800	30.00	$5,832.0\times10^6$	27,000.0
1,850	30.83	$6,331.6\times10^6$	29,313.0
1,900	31.67	$6,859.0\times10^6$	31,755.0
1,950	32.50	$7,414.9\times10^6$	34,329.0
2,000	33.33	$8,000.0\times10^6$	37,037.0
2,050	34.17	$8,615.1\times10^6$	39,885.0
2,100	35.00	$9,261.0\times10^6$	42,874.0
2,150	35.83	$9,938.4\times10^6$	46,011.0
2,200	36.67	$10,648.0\times10^6$	49,296.0
2,250	37.50	$11,390.6\times10^6$	52,736.0
2,300	38.33	$12,167.0\times10^6$	56,329.0
2,350	39.17	$12,977.9\times10^6$	60,083.0
2,400	40.00	$13,824.0\times10^6$	64,000.0
2,450	40.83	$14,706.1\times10^6$	68,084.0
2,500	41.67	$15,625.0\times10^6$	72,338.0

TABLE 58.—5TH POWERS OF DIAMETER IN FEET

	0	0.1	0.2	0.3	0.4	0.5	0.6	0.7	0.8	0.9
5	3,125	3,450	3,802	4,182	4,592	5,033	5,507	6,017	6,564	7,149
6	7,776	8,446	9,161	9,924	10,737	11,603	12,523	13,501	14,539	15,640
7	16,807	18,042	19,349	20,731	22,190	23,730	25,355	27,068	28,872	30,771
8	32,768	34,868	37,074	39,390	41,821	44,371	47,043	49,842	52,773	55,841
9	59,049	62,403	65,908	69,569	73,390	77,378	81,537	85,873	90,392	95,099
10	100,000	105,101	110,408	115,927	121,665	127,628	133,823	140,255	146,933	153,863
11	161,051	168,324	176,234	184,306	192,541	201,135	210,034	219,244	228,776	238,631
12	248,832	259,374	270,271	281,531	293,163	305,176	317,580	330,384	343,597	357,231
13	371,293	385,795	400,746	416,158	432,040	448,403	465,259	482,618	500,490	518,889
14	537,824	557,308	577,353	597,971	619,174	640,973	663,383	686,415	710,082	734,398
15	759,375	785,027	811,368	838,411	866,171	894,661	923,896	953,890	984,658	1,016,220

INDEX

Zeitfracht Medien GmbH
Ferdinand-Jühlke-Straße 7
99095 Erfurt, Deutschland
produktsicherheit@kolibri360.de